Contents

In memory of Liz
and
for Sandra, Jess and Tom

CAMBRIDGE UNIVERSITY PRESS
Cambridge, New York, Melbourne, Madrid, Cape Town, Singapore,
São Paulo, Delhi, Dubai, Tokyo

Cambridge University Press
The Edinburgh Building, Cambridge CB2 8RU, UK

Published in the United States of America by Cambridge University Press, New York

www.cambridge.org
Information on this title: www.cambridge.org/9780521129770

First published 1991
This digitally printed version 2009

A catalogue record for this publication is available from the British Library

Library of Congress Cataloguing in Publication data

Thomas, David, S. G.
The Kalahari environment / David S. G. Thomas and Paul A. Shaw.
p.   cm.
Includes bibliographical references and index.
ISBN 0-521-37080-9
1. Kalahari Desert.   I. Shaw, Paul A.   II. Title.
GB618.84.K2T47   1991
508.6883—dc20   90-40873 CIP

ISBN 978-0-521-37080-6 Hardback
ISBN 978-0-521-12977-0 Paperback

The Kalahari Environment

DAVID S. G. THOMAS

Department of Geography, University of Sheffield

and

PAUL A. SHAW

Department of Environmental Science, University of Botswana

The right of the
University of Cambridge
to print and sell
all manner of books
was granted by
Henry VIII in 1534.
The University has printed
and published continuously
since 1584.

CAMBRIDGE UNIVERSITY PRESS

Cambridge

New York   Port Chester   Melbourne   Sydney

THE KALAHARI ENVIRONMENT

The Kalahari is an unusual and elusive place, embracing an environment of great ecological and geomorphological diversity. Its complex climatic and geological history and its long association with human societies attempting to utilise its natural resources are aspects of increasing scientific interest.

This book has evolved from the authors' own research in the Kalahari and surrounding areas over the past decade, and provides an integrated, thorough and up-to-date review of the nature and development of the Kalahari environment. The authors attempt to provide explanations and answers to some of the many questions raised about the Kalahari, ranging from the commonly asked 'is it really a desert?', to more specific concerns about its evolution, dimensions and the impact of human activities. During recent years there has been an acceleration in the processes of change in the Kalahari, resulting from human activities set against a background of fluctuating climatic conditions. These changes have their origins in the distant past, but their impacts on the landscape are reaching unprecedented levels today as access to the Kalahari core improves and human activities intensify and diversify. Concern about these environmental changes and impacts have made apparent to the authors the need for an integrated scientific analysis of the Kalahari.

The interdisciplinary approach will make the book of interest to researchers, lecturers and advanced students in earth sciences, environmental studies, geomorphology and Quaternary science. The very extensive bibliography will also make it an important source of reference.

Preface

The Kalahari is an unusual and elusive place. In popular imagination it is a desert inhabited by 'bushmen' pursuing a life reminiscent of that lived in the Stone Age. Some, however, would query the classification of the Kalahari as a desert, pointing in evidence to its relatively well vegetated state and the diversity of human activities which it supports. In part such differences of opinion have arisen from the viewpoint and discipline of the observer and have been reflected in the subsequent application of the term Kalahari at a range of scales and to a variety of climatic, ecological and physiographic regions, a sedimentary basin and a group of extensive sediments. Each application employs its own geographical dimensions and environmental criteria.

What is clear is that the Kalahari embraces an environment of great ecological and geomorphological diversity, a complex climatic and geological history and a long association with human societies attempting to utilise its natural resources. These aspects are of increasing scientific interest, but have not been addressed *in toto* since the publication of Siegfried Passarge's *Die Kalahari* in 1904. A great deal has been learned since then. In this book we aim to provide an integrated, thorough and up-to-date review of the nature and development of The Kalahari Environment.

The origins of this book are twofold. First, it stems from a desire to provide explanations and answers to some of the many questions raised about the Kalahari, ranging from the commonly asked 'Is it really a desert?' to more specific and detailed concerns about its evolution, dimensions and the impact of human activities. None of these questions has a simple answer, but by carefully examining the available evidence we hope it will be possible to go at least part way to achieving explanations and, perhaps more importantly, identifying some of the gaps in understanding which require further research.

Second, the book has evolved from our own research interests in the Kalahari and surrounding areas, developed over much of the past decade. During this period there has been an acceleration in the processes of change in the Kalahari, resulting from human activities set against a background of fluctuating climatic conditions. These changes have their origins in the distant past, but their impacts on the landscape are reaching unprecedented levels today as access to the Kalahari core improves and human activities intensify and diversify. Concern about these environmental changes and impacts have made apparent to us the need for an integrated scientific analysis of the Kalahari.

The book is divided into three parts. Following the introduction, in which the Kalahari is defined and its context established, the chapters in Part One establish the physical background, the geologic and tectonic setting, the nature and distribution of Kalahari sediments and present-day patterns of climate, vegetation and soils. In Part Two we deal with the geomorphology and landforms of the Kalahari, factors which emphasise much of its unique character, and in the light of this information discuss the palaeoenvironmental history of the region. In the final part we assess the influence, variety and scale of human activities on the environment, from the Stone Age up to the last decades of the twentieth century.

One problem encountered in writing this book relates to the different spellings used for many place names. We have therefore attempted to standardise names within Botswana to the requirements of the 1984 Place Names Commission, with the exception of Sua and Nossop, which remain in common usage. We have attempted to employ similar standardisations for locations outside Botswana but it is likely that some do not correspond to official versions, either because they are still evolving or because some names, especially of rivers, change across political borders.

D.S.G.T., Sheffield
P.A.S., Gaborone

Acknowledgements

Many individuals and organisations have provided assistance to our Kalahari interests, both in carrying out research and in producing this book. Initial inspiration and individual introductions to the Kalahari were provided by Andrew Goudie, now Professor of Geography at Oxford University, and John Cooke, Professor of Environmental Science at the University of Botswana. Both have provided continued support and encouragement to our efforts. Other Kalahari enthusiasts and colleagues have provided help in a variety of ways, including valuable discussions on a myriad of topics, comments on drafts of chapters and through logistical support. In a far from complete list we would like to acknowledge Jaap Arntzen, Alec Campbell, Stewart Child, Jim Denbow, Co de Vries, Ambro Gieske, Sarah Metcalfe, Neil Parsons, Sue Ringrose, Alan Simpkins, Pete Smith, Sylvia Cooke of the Botswana Society and Eleanor Warr of the Kalahari Conservation Society. Richard Whitlow, Sue Childes and Dave Cumming are also thanked for their help in making research possible in western Zimbabwe in the early 1980s. We would also like to acknowledge the various bodies that have funded our research including NERC, The Royal Society and the Universities of Sheffield and Botswana. Our researches in the Kalahari are, however, far from complete and in this respect we would also like to mention and wish well to our two postgraduate students, Jem Perkins and Dave Nash, who are extending some of these research themes.

Above all, however, we would especially like to mention Liz and Sandra for their encouragement and patience, not only through our periods of absence in the field and the 18 months' gestation of this volume, but also on those trying occasions in Gaborone, Sheffield, Luton and other exotic locations when we have discussed endlessly everything from the past of the Kalahari to the future of English cricket.

With one exception (Figure 8.1) all the maps and diagrams either stem directly from our own research or have been developed and modified from other sources, which are acknowledged with the figures as appropriate. They have been expertly drawn by Paul Coles at Sheffield, with additional cartographic advice from Graham Allsopp. All the photographs were prepared for publication by Dave Maddison and John Owen. For permission to reproduce certain photographs we would like to thank J. du P. Bothma (Figures 4.7*f* and 6.2*b*), John Cooke (Figure 6.10*b*), Alec Campbell (Figures 8.4 and 8.5), Jeremy Perkins (Figure 9.5*a*, *b* and *c*) and the Government of Botswana for the aerial photographs reproduced as Figures 4.9, 5.6, 6.3*a* and *b* and 6.8*a* and *b*. All the other photographs were taken by the authors.

To all those above and to any whom we have inadvertently omitted, we extend our sincere thanks and best wishes.

1 An introduction to the Kalahari

The Kalahari is not an easy place to know.
It is still less easy to understand.

E.J. Wayland (1953)

THE TERM KALAHARI, applied to an ill-defined area of the interior of central and southern Africa, means different things to different people. The Kalahari has been described as a desert, a thirstland (e.g. Schwarz, 1920; Debenham, 1952) and a sandveld. It has been defined as an ecozone (e.g. Werger, 1978) and a physiographic region (e.g. Wellington, 1955), and has given its name to a group of geological sediments (e.g. Passarge, 1904) which cover a large portion of southern Africa. From the human viewpoint it has provided the basis for subsistence for small groups of people over a very long period of time, while others have regarded it as a valueless wasteland, or, at best, suitable for poor-quality grazing. Individuals of romantic bent have considered the Kalahari as a primaeval Utopia (e.g. Van Der Post, 1958); for many more it has been a tribulation, a major obstacle in their travels (e.g. Burchell, 1822) and to their expectations (e.g. Andersson, 1857).

Popular conceptions of the Kalahari are equally varied. Many perceive it as an endless and largely featureless sandy waste, uninhabited except for 'Bushmen'. Others, better aquainted with its geography, will know that it contains the world's largest inland delta, the Okavango, and lake beds, now dry, which cover an area equivalent to that of Belgium and the Netherlands combined, or the state of West Virginia in the USA. In the longer term some will know of the differences in the landscape between dry and wet seasons, and the changes brought about by cycles of drought and abundant rainfall. The Kalahari environment is therefore one of considerable contrasts. The delicate balances between geology, soils, vegetation, fauna and climate determine its unique and changing character.

The fragility of this environment is now becoming apparent; the Kalahari is changing in response to human usage and exploitation, and at present the demands of society on it, in terms of land and water resources for agriculture, ranching, mineral exploitation, wildlife conservation, tourist utilisation and a host of other uses, are greater than ever before.

One facet of this process is that as interest has increased, so too has our understanding, in scientific terms, of the Kalahari environment. This understanding tends to be selective; scientific advances tend to be made in certain disciplines and in certain geographical areas while other fields of knowledge remain largely untrodden. This is very much a response to the demands of society and science as a whole. An example is in the field of mineral exploration, where prospecting has led to a greater understanding of the sub-Kalahari geology of the Karoo and Precambrian rocks, while the geology of the Kalahari sediments themselves remains largely unmapped.

Despite the increase in scientific literature over recent years, there has been no book dedicated to the environment of the Kalahari since Siegfried Passarge's monumental work *Die Kalahari*, published in 1904. The present literature tends to be dispersed in a range of mono-disciplinary journals and government reports, many of which are not readily available to the interested reader. The aim of this book, therefore, is to draw together and summarise the advances made in the study of the Kalahari environment and of human responses to it. It would not be possible, in a book of this length, to be comprehensive; the sources are too vast for this. Rather, we aim to summarise, in subsequent chapters, the emergent picture in geological, geomorphologicial, climatological and ecological studies, in palaeoenvironmental studies, and archaeology and history, concluding with an examination of current issues and dilemmas encountered in the use of the Kalahari. These chapters are organised into three parts: the first presents the physical background, the second describes the evolution of the landscape and its component landforms, and the third examines the impact of human society upon the environment.

The first step, in this introductory chapter, is to place the Kalahari into a physical and human framework, looking at general characteristics, definitions and changing scientific perceptions.

1.1 Problems of definition and delimitation

There has been little consistency or precision in the use of the name Kalahari to delimit an area of the African land surface. In the Setswana language, the tongue of the Batswana, the largest population group in Botswana, the word 'Kgalagadi', of which Kalahari is a corruption, means 'always dry', and it is in the context of aridity that the word has been most widely used to describe an area of desert or semi-desert. 'The Kalahari' has, however, also been defined and located in physiographical, geological and ecological contexts, while sometimes being treated as synonymous with the territory of the Republic of Botswana, thereby providing a political delimitation.

Various geographers and earth scientists have attempted to define the location of the Kalahari. The German geologist Passarge, who was the first earth scientist to attempt a comprehensive stratigraphical and geomorphological account of a large portion of the Kalahari (Passarge, 1904), divided the region into northern, middle and southern portions. According to this scheme (Figure 1.1), the Southern Kalahari includes the 'Bakalahari Schwelle', the watershed between the Orange River system and channels draining to the north, and continues northwards to the Makgadikgadi depression; the Northern Kalahari is a poorly defined area to the north and west of the Okavango Delta. The Middle Kalahari is therefore the area between these two sub-regions, which includes all of the Makgadikgadi, the Okavango Delta, the Chobe swamps, and a large area of sand-covered territory, being bounded approximately by latitudes 17° 30' and 22° S and longitudes 20° and 27° E.

Debenham (1952: p. 12) noted that, territorially,

. . . the Kalahari of today . . . straddles over the somewhat uncertain political boundaries of South-West Africa, the western province of the Union of South Africa, and the protectorate of Bechuanaland [now Botswana], with its greater part in the last territory.

Grove (1969) attempted to clarify this area by defining its northern and southern boundaries, the northern running from the 'swamp zone' (the Okavango Delta and Chobe and Zambezi swamps) to the Etosha Pan, the southern as the Orange River in South Africa. Though this effectively excludes Passarge's Northern Kalahari, this delimitation was reiterated by Heine (1982), and accords with many other views that the major portion of the Kalahari is in Botswana (e.g. Baillieul, 1975), representing about 80 per cent or 450 000 km² of

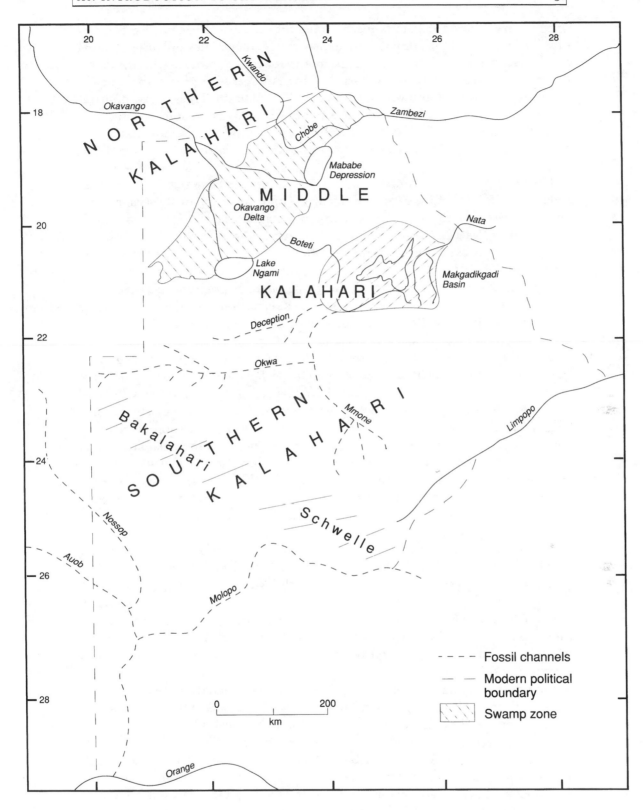

Figure 1.1. Passarge's Kalahari.

that country (Jones, 1982). However, within Botswana itself the name Kgalagadi is officially applied to the administrative district which occupies only the extreme southwestern part of the country. This, in turn, does not coincide with the use of the term central Kalahari, which is applied to a large game reserve in the sparsely occupied area to the southwest of the Makgadikgadi Basin. This lies, administratively, within the Ghanzi District. The central Kalahari has also been used in a geological sense to differentiate the sedimentary basin between the Bakalahari Schwelle and the Zambezi Trough from the Southern Kalahari (Molopo) Basin to the south.

The picture is further confused as other investigators have regarded the 'true Kalahari' (De Vos, 1975) as extending far beyond the limits of Botswana. This expansionist use of the name is based on geological criteria which have identified the extent of 'Kalahari sediments' and the structural basin which they occupy, covering an area extending from north of the equator to the Orange River in the south (Cooke, 1957). Likewise, biogeographical definitions (for summary see Werger, 1978) show considerable variety; the phytogeographical classifications of Bolus (1905) and Marioth (1908) extend the Kalahari south of the Orange River into Bushmanland, while others, such as White (1965), do not concede the existence of a separate Kalahari domain.

Barker (1983a, b) has addressed the question of the biogeographical integrity of the Kalahari, and concludes that, although the flora tends to be transitional towards the Zambezian Region, and the fauna has greater affinities with arid southwest Africa, there is sufficient evidence to support the existence of a Kalaharian Zone, with limited species diversity a result of the homogeneity of landscape components. However, the northern boundary of the Kalaharian Zone is placed at the transition to the Mopane (Colophospermum mopane) woodland of the Zambezian Region along the Okavango–Makgadikgadi line. Other boundaries include the limits of the tree Acacia haematoxylon in the southwest, the grass Schmidtia pappophoroides in the east, and, to the south, the limits of penetration of the common South African grass Themeda triandra.

Archaeological investigations have tended towards a further set of definitions, in which Botswana east of the Kalahari Sand boundary has been defined as Kalahari, although this area lies in the 'hardveld', where mature soil profiles are developed directly from underlying rock (e.g. Denbow, 1984). As the communities which have existed in this peripheral zone have had considerable impact in the region, the definition is acceptable in the archaeological context. The problem then lies in defining the cultural boundaries of these communities to the east, beyond the Limpopo River.

Obviously, the range of applications of the name Kalahari gives rise to great potential for confusion, yet each of the usages mentioned above has its own justifications. Although the definitions are not in themselves important, they are significant when they are used to ascribe characteristics to a particular area and to provide explanations for such attributes. It is therefore important to clarify some of the major issues pertaining to the various applications of the name before it is possible to describe and explain the nature and significance of the Kalahari environment itself. Overall, it is probably physical and environmental factors which are most useful in delimitations and definitions of the Kalahari, especially as it is these which actually contribute to the nature and major characteristics of this area, and also influence human occupation and usage.

1.2 Definitions and characteristics clarified

The application of the name Kalahari to areas of the African land surface of different extents is not, in fact, contradictory when considered in the context of environmental characteristics. The geologist Alex Du Toit, who made many valuable and perceptive contributions to the study of southern African environments during the first half of the twentieth century, noted that while the Kalahari was not well delimited, its dominant distinguishing features were its flatness, its waterlessness and the mantle of sand which blankets most of the solid geology of the area (Du Toit, 1927). These features usefully contribute to explanations for the application of the name Kalahari in both the 'confined' manner employed by those such as Grove (1969) and the more 'extensive' usage of De Vos (1975) and others.

1.2.1 The Kalahari Desert or Thirstland

The area delimited by Debenham (1952), Grove (1969) and others has often been termed the Kalahari Desert (Figure 1.2). The first policeman of the Bechuanaland Protectorate, Lieutenant A.W. Hodson, described this area as

> not quite an ordinary desert, for it has many varieties of country. Some parts it is covered with thick bush, in others it consists of open plains upon which good grasses flourish, whilst the remainder is made up of a sea of sand-hills. This is known as the sand dune country, and constitutes by far the most dreary and depressing part of the desert. (Hodson, 1912: p. 21).

The environmental variations and relatively well-developed vegetation cover which Hodson noted have subsequently led a number of authorities to regard the classification 'desert' as inappropriate (e.g. Debenham, 1952; Grove, 1969; Jones, 1982), preferring 'thirstland'. Although the term 'desert' does not have a clear scientific or popular meaning (see Heathcote, 1983) and is applied to a wide range of environments on earth, 'thirstland' is neither less ambiguous nor more enlightening, so that there is no real logic in rejecting the label 'desert'.

Deserts are perhaps most frequently regarded as synonymous with aridity. Arid environments, according to the widely used criteria of Meigs (1953: Table 1.1) cover a large part of southern Africa (Figure 1.3), embracing the hyper-arid Namib Desert, the Karoo, the Kalahari Desert (as delimited by Debenham, 1952 and Grove, 1969), and the intervening areas of the Great Escarpment and southern African Highveld. Although it is true to say that much of the Kalahari Desert delimited in this way is semi-arid rather than arid, it can be noted, following Shantz (1956) that the only safe assumption about a semi-arid environment is that any year could be extremely dry.

The Kalahari Desert is mantled by unconsolidated Kalahari Sand, being part of the larger Kalahari physiographic region (see below). The northern boundary of the desert, as already noted in Grove's (1969) definition, is the Etosha–Okavango–Zambezi swamp zone and its southern one the Orange River. In eastern Botswana its limit occurs at the transition from the 'sandveld' to the 'hardveld', roughly coincident with both the Kalahari–Limpopo watershed and the alignment of the Kalahari–Zimbabwe Axis of uplift, though this occurs far from abruptly and consequently is not clearly defined. Sand outliers lie beyond this; for example, some authors (e.g. Helgren, 1979) have considered the terrestrial sands found in association with the Vaal River sedimentary sequences to be generally related to the Kalahari Sand.

In northeastern Botswana its boundary is even less obvious. Though the Kalahari sand extends east into western Zimbabwe, to end abruptly where erosion has cut into the rim of the sand mantled basin to create a marked scarp and a series of upstanding outliers (Du Toit, 1927; Thomas and Shaw, 1988), this area receives between 550 and 660 mm mean annual precipitation and much of it is vegetated by dry deciduous forest (e.g. Boughey, 1963; Weare and Yalala, 1971). This can hardly be considered desert country, although rainfall is highly seasonal and surface water virtually absent during the dry season. A more

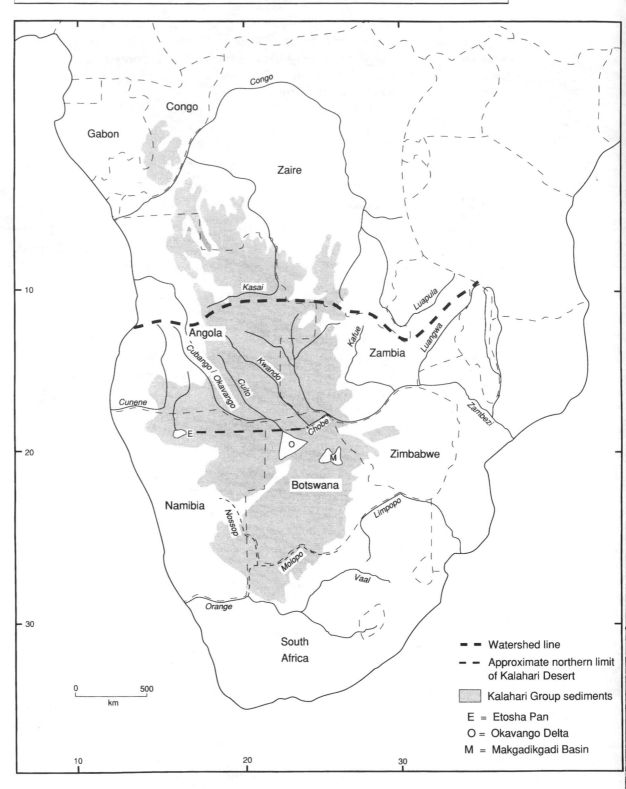

Figure 1.2. Kalahari regions. The shaded area, defined on geological and structural grounds, is the Mega Kalahari; the Kalahari Desert corresponds to the part of the Mega Kalahari between the Orange River in the south and the Etosha–Okavango–Zambezi line in the north. Major drainage lines are shown.

Table 1.1. *Classification of arid environments*

	Im^1	Mean annual precipitation (mm)[2]
Semi-arid	-20 to -40	200–500
Arid	< -40 to -56	25–200
Hyper-arid	< 56	12 consecutive months recorded without rainfall. No seasonal regime

After Meigs (1953).

[1] Im = Thornthwaite's (1948) Index of moisture availability.

$$Im = (100S - 60D)/PE$$

where PE is potential evapotranspiration; S and D are respectively the moisture surplus and moisture deficit, aggregated on an annual basis from monthly data, taking stored soil moisture into account.

[2] Values suggested by Grove (1977).

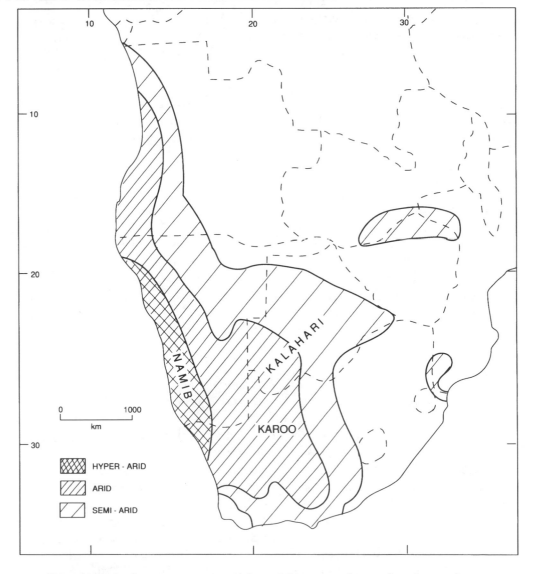

Figure 1.3. *Arid environments in southern Africa, according to the scheme of Meigs (1953).*

appropriate boundary to the desert, if somewhat transitional, could be a line from the eastern end of the distal margin of the Okavango Delta to the northernmost point of the Makgadikgadi basin, following the Maun–Nata road at 20° S.

In the drier west much of the desert's margin is marked by higher country which ascends to the Great Escarpment, but along the stretch between the lower Molopo River and Mariental in Namibia the Kalahari Sand ends at the brink of an erosion scarp (e.g. Du Toit, 1927), comparable to that in western Zimbabwe.

1.2.2 The Kalahari physiographic basin or Mega Kalahari

The Kalahari Desert defined above is part of a more extensive area which is physiographically and sedimentologically unified, and to which the name Kalahari is also applied. This is the 'true Kalahari' of De Vos (1975), but is perhaps better termed the 'Mega Kalahari' (Thomas, 1984b) to avoid confusion and to emphasise its great size. The structural development of this physiographic region is discussed in detail in Chapter 2.

This region is a downwarped basin, or series of contiguous sub-basins (Thomas, 1988a), into which continental sediments have been deposited since the Jurassic. The most common surface unit of the Kalahari Group (SACS, 1980) of sediments is the Kalahari Sand, which extends from latitude 1° N to 29° S (Cooke, 1957). This represents an area of over 2.5 million km² of central southern Africa (Figure 1.2), and makes it the largest continuous sand sea, or *erg*, on earth (Baillieul, 1975).

The Mega Kalahari, which encompasses the Kalahari Desert, impinges on the territories of nine countries: from north to south these are Gabon, Congo, Zaïre, Angola, Zambia, Namibia, Botswana, Zimbabwe and South Africa. The largest portion occurs within the boundaries of Angola, of which about 850 000 km², or 68 per cent, is blanketed by *Sables Ochres*, the name attached to the Kalahari Sand in Angola by Cahen and Lepersonne (1952).

Not suprisingly, a region as extensive as the Mega Kalahari and extending through 30 degrees of latitude embraces a wide range of climates and vegetation communities. The northern extremities in Gabon, Congo and western Zaïre have typical humid tropical regimes and tropical forest vegetation, with rainfall almost all year round. Moving southwards the climate becomes more seasonal; although the Mega Kalahari in central Angola experiences only limited monthly changes in temperature regimes and receives in excess of 1000 mm mean annual precipitation, this occurs in a rainy season from October to May. In the Mega Kalahari, south of the humid tropical zone, the rainy season is always during the southern hemisphere summer months, but obviously both the length of the season and the precipitation amounts decrease, and interannual variability increases, in the direction of the arid zone in the southwestern part of the Kalahari Desert. This is discussed in detail in Chapter 4.

Although the Kalahari Desert is notable for its lack of modern surface drainage, the Mega Kalahari is drained in its northern and central parts by components of the mighty Congo and Zambezi river systems. The watershed line between these two systems and their smaller counterparts, sometimes termed the Southern Equatorial Divide, crosses the Mega Kalahari from east to west in central Angola (Figure 1.2). The 700–800 km stretch of the watershed on the Kalahari Sand represents its lowest part, reaching a minimum altitude of 1082 m above sea level (asl) (Wellington, 1955). Following the observations of Dixey (1943), Wellington (1955) proposed that it be called the 'Kalahari Col' because of its position between higher topography of Karoo granites in the west and

Precambrian Katanga Group rocks in the east. However, despite the Kalahari tract being the low point of the east–west watershed line, it represents one of the highest areas of the north–south trending Mega Kalahari, which has a mean altitude of about 1000 m asl (Jones, 1982).

1.2.3 Core and periphery components of the Kalahari

These brief accounts suggest that there are major physical contrasts within the Kalahari physiographic region or Mega Kalahari. The core area is undoubtedly the predominantly semi-arid to arid Kalahari Desert, as previously defined, centred upon Botswana but extending into neighbouring territories. To the north of the Kalahari Desert, two further major physical sub-regions can be suggested: the first is the humid tropical area north of the east–west watershed line, dominated by part of the Congo (Zaïre) River drainage system and mantled by Kalahari sediments (Claeys, 1947; Cahen and Lepersonne, 1952), forming part of the Congo basin of Wellington (1955); the second is the region south of that watershed and north of the Kalahari Desert, with its southern limits marked approximately by a line running from the Etosha Pan to the Okavango Delta and the confluence of the Chobe and Zambezi rivers (Figure 1.2). This coincides approximately with the Northern Kalahari of Passarge (1904).

The core and periphery can also be defined in human terms. In human geography core areas are usually concentrations of human settlement and activity. The reverse is the case in the Kalahari, where the core is essentially a wilderness area of low population density (Figure 1.4), in which specialised strategies are required for survival. Again the core coincides with the area defined as the Kalahari Desert, focusing in particular in central Botswana.

A larger core, also centred on Botswana, can be discerned within scientific disciplines themselves, largely as a result of external events. Over the past 30 years intra-regional conflict, political introspection and economic difficulty in the subcontinent have impeded research activity. Such impedance takes many forms: access to the field becomes limited, basic documents such as topographic maps become unavailable, research itself becomes a luxury in the face of more immediate concerns. This problem has been particularly acute to the north and west of Botswana's borders, and appears in many forms and degrees of severity. For example, geomorphological research in Etosha Pan, northern Namibia, is restricted to the area within the National Park (Rust, 1985), while in Angola even basic hydrological data for the major rivers has not been available since Independence in 1975. Again, most research has been carried out in the core area of the Kalahari Desert and its southern periphery.

1.2.4 Scope and definitions in the present volume

The preceding sections show the considerable variation in definitions of the Kalahari, reflecting the diversity in conditions between the Kalahari core and the peripheral sub-regions to the north. As it is not feasible to cover all parts of the Mega Kalahari in the same detail, the strategy has thus been to focus on the Kalahari Desert and areas adjacent to it, while adopting a wider coverage where appropriate, as in discussions of structure, geology and climate.

Consequently, a number of the definitions cited are used in the text within their original contexts. These terms include Mega Kalahari, Kalahari Desert, and Passarge's (1904) Northern, Middle and Southern Kalahari. The term Central Kalahari is used when discussing the area now occupied by the game reserve of that name (Figure 1.4), while central Kalahari (with a small 'c') is used as sparingly

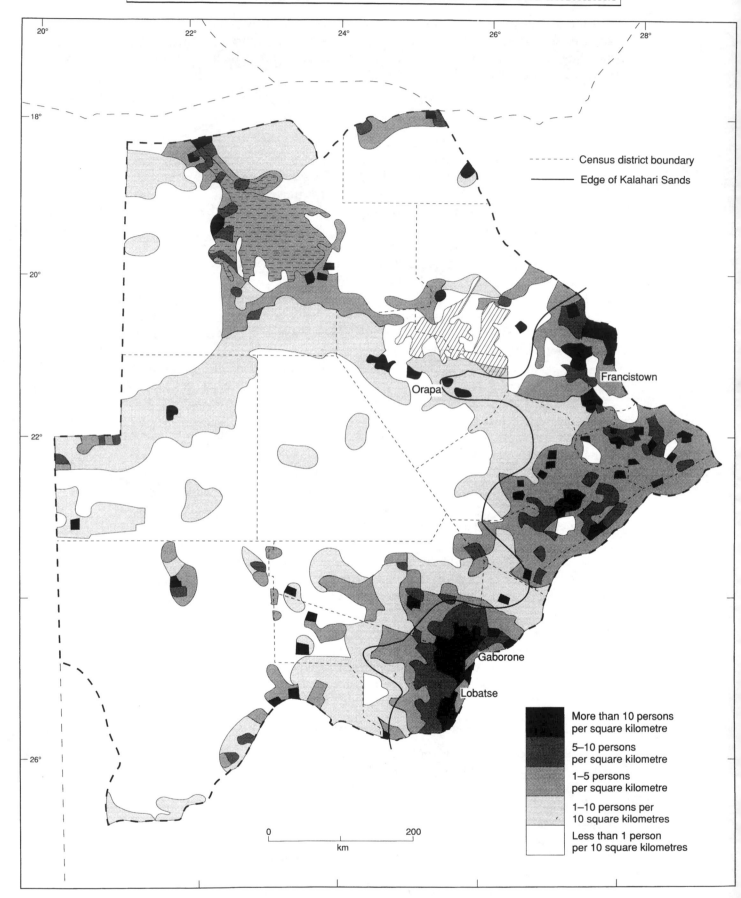

Figure 1.4. Human population densities in Botswana, showing the sparse occupation of the Kalahari.

as possible to encompass the area between the Bakalahari Schwelle and the Okavango–Makgadikgadi line.

The term 'sandveld' has also been retained to distinguish areas of the Kalahari Desert which are mantled with Kalahari Sand. It does not include the wetlands or lake basins, and can be distinguished readily from the 'hardveld' of eastern Botswana.

1.3 Characteristics of the Kalahari Desert environment

Although all aspects will be considered in detail in subsequent chapters, it is useful at this stage to examine the general characteristics of the Kalahari Desert environment. As Lieutenant Hodson noted, the Kalahari does not meet the layman's expectations of a desert: it is not a sea of shifting sand dunes, though extensive systems of dunes are present, and it generally possesses a significant natural vegetation cover. The Kalahari Desert is, however, notable for its lack of permanent, and even seasonal, water courses. Although the Okavango River contains water all year round, with a mean annual inflow to Botswana of $10\,500 \times 10^6$ m^3 (UNDP/FAO, 1977), it only touches upon the desert's northern boundary before 96 per cent of its waters are dissipated in its large terminal delta (Shaw, 1984). Likewise, the Zambezi and its major tributary the Chobe, with a mean annual flow of $44\,000 \times 10^6$ m^3 at Victoria Falls, merely traverse the northern perimeter of the Middle Kalahari.

There are also a number of channel systems which focus on the Makgadikgadi basin in the northeastern part of the desert. These systems, such as the Nata, Boteti, Mmone and Okwa, are, however, ephemeral or even fossil features which contribute little to the hydrological budget of the Kalahari, especially as their sources lie within the desert, or in the case of the Nata, just beyond its eastern margin. Interestingly, as all these fluvial features drain internally, the Kalahari fulfils the desert criterion of De Martonne and Aufrére (1928) as an area of internal drainage.

In the southwestern desert area, the Auob, Kuruman and Nossop drainage lines are components of the Molopo system which is linked to the Orange River and hence to the Atlantic Ocean. Although the Molopo and Kuruman have perennial spring sources, none of these rivers carries water into the Kalahari except in years of prolonged rainfall, while a link with the Orange River has not occurred during historical times, nor for at least 1000 years, according to Lewis (1936). Instead, the main repositories of surface water, mostly ephemeral, are confined basins known as pans, ranging in size from a few square metres to several square kilometres.

The climate (Chapter 4) is characterised by hot summers and winters with warm days and cold nights. Precipitation occurs mainly between October and April, the summer months, with mean annual rainfall increasing from about 150 mm in the driest southwestern areas to about 650 mm in the northeastern corner of Botswana. The high rates of evaporation which prevail, well in excess of 2000 mm yr^{-1} in most parts, result in a moisture deficit in all but the wettest months. Despite the mean deficit, and the absence of surface water, the Kalahari is a well-vegetated desert. This is partly due to the nature of the Kalahari Sand (Chapter 3) which mantles the surface, attaining thicknesses in excess of 400 m in some places.

Although sand (over 90 per cent quartz, with particles of between 2 mm and 0.063 mm in diameter) is not usually regarded as a good medium for plant growth, in arid and semi-arid environments it has the potential to trap and maintain moisture from bouts of rainfall, and therefore to support vegetation, though the low nutrient status of the Kalahari Sand (Sims, 1981) does not favour attempts to cultivate it. Studies of the ground water potential and recharge of the Kalahari Sand have yielded conflicting results (section 3.7), but there is a general consensus on the availability of soil moisture within the top 6 m, while deep-rooting species are capable of reaching the water table at considerable depth.

The natural vegetation of the Kalahari largely comprises a range of savanna types (e.g. Weare and Yalala, 1971; Chapter 4). In the drier southwest this is grass- and shrub-dominated, with the importance of trees generally increasing in a northeasterly direction. Superimposed on this pattern are local and regional variations induced by sedimentary and geomorphological factors, with the Okavango Delta possessing important swamp grassland and aquatic plant communities.

Du Toit (1927) noted that one of the major characteristics of the Kalahari was its great flatness. Locally, isolated hills or inselbergs occur, with highest concentrations in the northwest near the Okavango Delta. Additionally, a long, low (but notable in the context of the relief of the Kalahari) finger of protruding Precambrian rocks, the Ghanzi Ridge, extends northeastwards from Namibia to the southwestern edge of the Delta. Negative relative relief is provided by the fault-controlled basins within the Okavango rift zone, and the Etosha and Makgadikgadi basins, also believed to be structurally determined (e.g. Mallick, Habgood & Skinner, 1981), but the gradients of the land leading to these features are commonly so low that changes in relief are not always readily apparent.

Beyond these protrusions and basins, local relief is provided exclusively by landforms which have been developed by processes operating upon the surface units of the Kalahari Group of sediments themselves. These landforms and their significance as indicators of changing climatic and environmental conditions in the Kalahari are discussed in detail in Chapters 5, 6 and 7, and include the development of ancient lake shorelines, valley networks, pan depressions and fringing dune features, and extensive fields of partially or extensively vegetated linear dune systems. It is, however, uncommon for the relief consequent upon these landforms to exceed 20 m.

1.3.1 Early settlement and land use in the Kalahari Desert

Despite the limitations of the Kalahari environment from the human viewpoint, man has had a long association with the region, going back, perhaps, as far as the early hominids, of which *Australopithecus africanus*, first discovered at Taung on the southern fringes of the Kalahari, is an example (Dart, 1926; Butzer, 1974). Until recent decades the Desert has been occupied by scattered human groups practising lifestyles well adapted to the harsh conditions, and coexisting with both peripheral societies and the considerable variety of indigenous wildlife.

Archaeological evidence (Chapter 8) indicates human occupation from the Early Stone Age to the present day (e.g. Helgren and Brooks, 1983). Early and Middle Stone Age (ESA and MSA) sites are prolific and widespread; E.J. Wayland, Director of the Geological Survey of Bechuanaland in the 1940s, assembled a collection of over 6000 artefacts (Wayland, undated). Unfortunately, the sites tend to be surface deposits without stratigraphic context, though they provide sufficient evidence to suggest, not suprisingly, that population numbers waxed and waned in conjunction with climatic changes and fluctuations in water availability (e.g. Beaumont, 1986).

The Late Stone Age (LSA) has, until recently, been viewed as a culture personified by the 'bushmen' (baSarwa or Khoisan), maintaining a forager life-style in isolation from Bantu-speaking peoples until as recently as 200 years ago (e.g. Lee and De Vore, 1976; Lee, 1979; Silberbauer, 1981). Recent research (Denbow, 1984; Denbow and Wilmsen, 1986) has indicated that contact between Khoisan and Early Iron Age (EIA) cultures took place at a much earlier time, with the introduction of cattle and ceramics into the Kalahari about 2000 years ago, and the development of metallurgy, grain cultivation and transcontinental trade by 800 AD. This development was focused on riverine sites within the Kalahari, and in the kingdoms which sprang up in the eastern hardveld, centred on Toutswe. The Khoisan were by no means aloof from this cultural intrusion, but took an active part in trade and pastoralism, a relationship which persisted into the seventeenth and eighteenth centuries as successive waves of Tswana groups moved into the Kalahari periphery, or, in the case of the baTawana, to Ngamiland (Tlou, 1985). However, these traditional relationships declined in the eighteenth century as the Batswana imposed a system of clientship on the baSarwa and other subsidiary tribes. The decline was accelerated in the nineteenth century as a result of European intervention, drought cycles and the rinderpest epizootic, which decimated cattle and game alike.

In recent decades, the development of the Kalahari as a resource in a national and even global setting has brought, and is continuing to bring, changes in the patterns of occupation by animals and humans. The growing integration of the Kalahari into wider economic spheres has been a major factor in the demise of the traditional hunter–gatherer existence of the Khoisan. Recent estimates suggest that of the total 55 000–60 000 San living in the Kalahari and its perimeters (Tobias, 1964; Campbell in Vossen, 1984), only about 2000 now live fully as hunter-gatherers (Main, 1987).

1.3.2 European activity and scientific perception in the nineteenth century

The past 200 years have seen environmental changes on an unprecedented scale. These are, in part, the result of contact between the Kalahari region and its peoples and the industrialising world, a process that has been greatly influenced by the activities and perceptions of Europeans and the interests which they have represented.

Europeans first approached the Kalahari from the south and east, arriving in the region in the early nineteenth century, at the time of the period of tribal conflict and migration known as the *Difaqane*, and the succeeding unrest induced by Boer expansion. The historical milieu of this period is of great interest in itself, and certainly influenced the pattern of European penetration of the Kalahari, particularly by imposing difficulties over and above the normal physical hazards of travel. It must, however, be referred elsewhere (e.g. Parsons, 1982; Tlou and Campbell, 1984).

The initial exploratory phase was dominated by the activities of missionaries, hunters and traders, a trinity of purpose exemplified by the group of Europeans who first visited Lake Ngami on 1 August 1849: David Livingstone and his companions Murray and Oswell, sportsmen both, and Wilson, a trader. The firm establishment of the London Missionary Society (LMS) mission at Kuruman under the aegis of Robert Moffat in 1820 led to the rapid dissemination of Christianity and trade throughout the Molopo area, and Kuruman became the terminus of the 'Missionaries' Road', extending up the eastern fringe of the Kalahari via Kolobeng, Shoshong and hence to the Ndebele kingdom beyond the Shashe River. Livingstone, stationed at Kolobeng from 1846 onwards as missionary to the baKwena, made three journeys northwards into the middle Kalahari, before a fourth expedition in 1852 took him beyond the Zambezi and on to his great *traversa*. The first two journeys followed the same route to Shoshong, then northwest to Ntwetwe Pan, the Boteti River and baTawana territory around Lake Ngami and the Thamalakane River, while on the third he set out directly northwards from Ntwetwe Pan to the Chobe River, experiencing considerable hardship en route (Figure 1.5).

Lake Ngami, and the possibilities of trade with the baTawana, were an early attraction to Kalahari travel. The more difficult route from the Atlantic seaboard via Ghanzi and Kobe Pan was pioneered by Charles Andersson in 1851, who also became the first European to travel into the Okavango Delta. Andersson also reached Etosha Pan in 1858, and travelled extensively in the area of the upper Nossop and Auob Rivers. Andersson's Ngami route was extended by later travellers such as Thomas Baines and James Chapman as far as the Zambezi at Victoria Falls, while others, such as Schulz and Hammar, passed through the Linyanti and Kwando areas.

These early travellers inevitably embodied the ideals of Victorian science, and were indefatigable documentalists of every aspect of topography, geology, natural history and ethnology; their books (e.g. Andersson, 1857; Livingstone, 1858a; Baines, 1864; Chapman, 1886; Oswell, 1900) are invaluable studies of the Kalahari at that time. Livingstone, in particular, covers a wide range of subjects, from the detailed engraving of the tsetse fly (*Glossina moritsans*) on the title page of *Missionary Travels*, to accurate descriptions of the form and interrelationships of the Middle Kalahari drainage which were not improved until the Du Toit expedition some 70 years later.

However, by the 1880s interest in the Kalahari and Ngamiland had waned, as valuable animal species, particularly elephant, declined, and the costs of travel in these regions became apparent. Some travellers, such as Dolman and Wahiberg (Tabler, 1973) paid with their lives; others, bankrupted by the heavy costs and uncertain profits, moved on to other schemes. The trade frontier moved northwards, propelled by gold strikes at Tati in 1868, and was extended as far as Barotseland, on the upper Zambezi, via the road through Mpandamatenga and Kazungula by George Westbeech in 1871 (Sampson, 1972). The Kalahari core came to be regarded as an obstacle, in which travel was hindered by lack of water and difficult terrain, while in the northern woodlands between the Boteti and Chobe Rivers intransigent natives, malaria and the tsetse fly proved to be major hindrances. This view prevailed through the later part of the nineteenth century. For example, Selous remarks on his narrow escape from death by drought in the area between Sua and Mababe in 1879 (Selous, 1893: p. 377), while lack of game encouraged him to move northeast into the Ndebele country of present day Zimbabwe, itself much overhunted by the 1870s (Tabler, 1960). When larger social groups attempted to traverse the sandveld, as with the 'Dorsland Trek' of 1877, the results were disastrous: 31 trekkers and thousands of cattle died before the group even reached Ngamiland (Tabler, 1973). A low point was reached at the turn of the century as the rinderpest epizootic swept through the region, and geological surveys, of which Passarge's study forms an integral part, failed to reveal the presence of viable mineral deposits.

1.3.3 Scientific perceptions during the colonial era

The political backdrop had also changed by the beginning of the twentieth century. In 1885 the British Government, at the repeated request of the Batswana chiefs, created the Bechuanaland Protectorate (sometimes abbreviated to 'BP'), with the land south of the Molopo River being attached to the northern Cape Province. South West Africa became a German colony, and, after World War I, a League of Nations Mandate administered by the Union of South Africa. For a

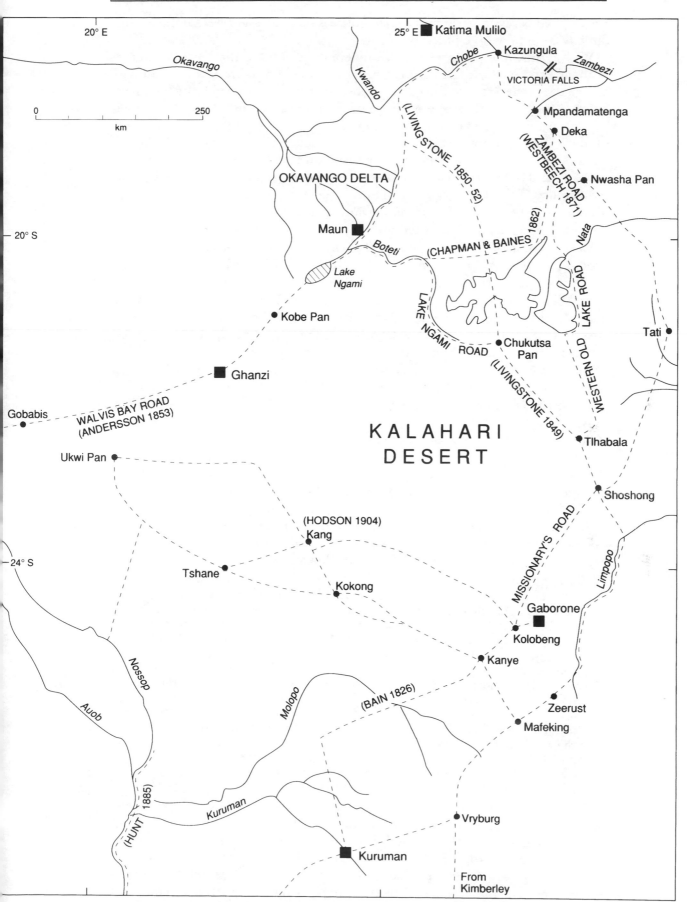

Figure 1.5. The routes of David Livingstone and some other nineteenth- and early twentieth-century European travellers in the Kalahari.

brief period between 1919 and 1940 the entire Kalahari Desert was administered, directly or indirectly, by Britain.

Within this multiplicity of rule lay different attitudes to the Kalahari. In the Bechuanaland Protectorate the British, reluctant landlords from the beginning, assumed that the Protectorate would eventually be assimilated into the Union of South Africa, regardless of the wishes of the populace. This led to decades of administrative neglect and underdevelopment which was to make Botswana one of the world's poorest countries at independence in 1966. Some Resident Commissioners, more ambitious or energetic than their colleagues, attempted development schemes, but they were, for the most part, thwarted by the indifference of the Treasury and Colonial or Dominion Offices (e.g. Rey, 1988). Within the Protectorate the Kalahari region, with its low populations and lack of resources, received the least attention.

Different conditions prevailed in the Union of South Africa. Land with agricultural potential along the Molopo River and in Gordonia District was rapidly taken up in the late nineteenth century, and in the aftermath of the Anglo-Boer War most of the northern Cape was divided into farm units, based on market centres such as Upington. Afrikaner settlers also moved into the Ghanzi area of the Protectorate in the 1890s, utilising the limestone aquifers as a water source for their stock. In South Africa the Kalahari was viewed as a valuable piece of real estate which was to be used for the benefit of the Union as a whole.

This is best illustrated by the Kalahari Scheme of Professor E.H.L. Schwarz, proposed in a number of newspaper articles in the Johannesburg *Star* from 1918 onwards, and a book entitled *The Kalahari, or Thirstland Redemption* (Schwarz, 1920). Schwarz's proposal, based on the premiss that the lakes of the Kalahari had dried out since 1820, entailed the restoration of the fluvial and lacustrine links in the region by the construction of dams on the Zambezi and Cunene Rivers, going as far as the transfer of water out of the Okavango–Makgadikgadi system to the Molopo and Orange Rivers by a supposed fossil network (Figure 1.6). The existence of standing water would then lead to an amelioration of the climate in the interior of southern Africa. The grandiosity of this scheme is contained in Schwarz's description (Schwarz, 1920: Introduction, p. v):

> When the gaps are blocked up and the old Kalahari lakes are once more there to supply the air of South Africa with moisture, the old central river of the Kalahari will once more flow. All down the course from Ngami to the Orange River, below the Falls, settlements will arise and agriculture, of the name [sic] nature and on the same scale as that in Egypt, will spring up. Yet it will be a better Egypt that will result from the occupation of this wilderness, for there will not be desert around. The Kalahari today is covered with bush and grass; when the lakes and rivers are reconstituted the vegetation will be that of the Soudan; rich pasturage and forest. Only the more important crops will need irrigation, and harvest will not be followed by the clean-sweep of every sign of vegetation. Away from the central river there will be stock farms, contributing to the wealth of the country, and every acre of the land will become habitable . . .
> . . . on the Kalahari Project, everyone in South Africa, whether he wants it or no, will receive additional rain, will see his land rendered more fertile, and all his difficulties from drought, famine and pestilence disappear.

The proposals caught the popular imagination, and the Union Government quickly commissioned the Du Toit expedition of 1925 to test the validity of the hypothesis. Du Toit (1926) concluded that the climatic scenario was unsubstantiated, and that the topographic relationships between the various hydrological links were not as favourable as surmised. However, the ideas put forward by Schwarz persisted in various forms through subsequent government enquiries into water use in the Kalahari (e.g. Jeffares, 1938; Mackenzie, 1946; Kokot, 1948; Brind, 1955). As the industrial heartland of South Africa faces water deficits in

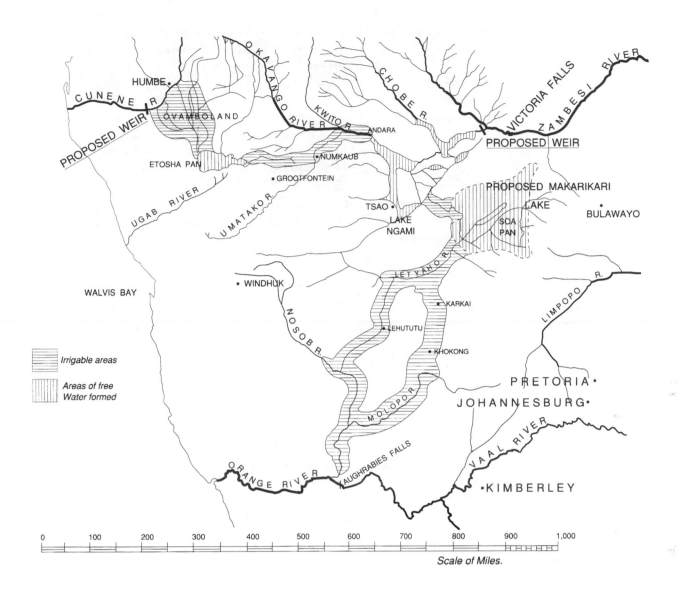

Figure 1.6. Schwarz's scheme to irrigate the Kalahari. After Schwarz (1920), Map 1).

the twenty-first century, transfer schemes from the Zambezi or Okavango to the Vaal triangle are often mooted (e.g. Borchert and Kempe, 1985; Nugent, 1987), an echo of the original transfer proposal.

The phase of environmental degradation ushered in by European hunting practices in the nineteenth century (Campbell and Child, 1971) was followed by rapid expansion in the cattle industry, with consequent demand for grazing land and water sources. The history of the cattle industry (section 9.2) is a complex issue, involving not only environmental concerns, but also land tenure systems, inter-territory relations and the changing attitude of the metropolitan power, and has been admirably summarised elsewhere (Hitchcock, 1985). In the Bechuanaland Protectorate the expansion was fuelled by economic demand, together with the increasing pressure from the British Government that the territory justify future investment (Pim, 1933). Scientists played a major part in this process; both Du Toit and Brind, the Government surveyor, played an active part in the expansion of the borehole network in the 1930s, Pole Evans (1950), a botanist, assessed the grassland resources, while Debenham (1952), on his post-war whistle-stop tour of the African colonies, enthused over the pastoral potential of the Kalahari. Others, with greater experience of the region, were not convinced (e.g. Wayland, 1953), but the general trend was for science to underwrite the cattle industry until as late as 1971, when concern was expressed about environmental degradation (Botswana Society, 1971). The Thirstland of the nineteenth century thus became the pastoral paradise of the twentieth.

In other fields scientific advances were slower. The description and mapping of natural resources continued in the Victorian tradition, largely in the hands of colonial administrators (e.g. Stigand, 1912, 1923; Clifford, 1928, 1931), and in 1933 it was possible to produce the first 1:500 000 map of the Bechuanaland Protectorate (Raffle, 1984). Other data accumulated from the work of government geologists, surveyors, tsetse control experts and agricultural advisers. In 1930 the first multidisciplinary research team took to the field. The Vernay-Lang Expedition, comprising 14 scientists, including such well known names as A.W. Rogers (geology), Austin Roberts (ornithology) and Vivian Fitz Simons (herpetology), collected 90 species of mammals, 330 species of birds, 600 fish, 2000 lower invertebrates and 21 000 insects (Vernay, 1931), as well as considerable botanical, archaeological and ethnological data. A solid scientific base had been achieved.

1.3.4 Current trends

Over the past quarter of a century great changes have occurred in the Kalahari landscape, accompanying political independence in Botswana, a rapid demographic increase and economic growth. Although population densities in Botswana and Namibia are among the lowest in the world, particularly in the Kalahari sectors (Figure 1.4), no longer is 'new' land available for expansion. The Kalahari as a whole is under pressure from a range of conflicting land uses.

The most important of these, from an economic viewpoint, is the exploitation of mineral resources. Though the sand itself is not of commercial value – and Du Toit's tongue-in-cheek suggestion that it be used for the manufacture of hourglasses and egg boilers appears not to have been heeded (Du Toit, 1927: p. 89) – significant finds of kimberlite pipe diamonds have been made, including one of the world's largest diamond pipes at Jwaneng, and they are now being exploited commercially at three mines located within the desert. Large-scale extraction of soda ash is about to commence at Sua Pan, in the Makgadikgadi Depression, while the world's eighth largest coal reserves lie only tens of metres below the surface in the Karoo Beds on the eastern edge of the Kalahari. At Sishen, south of the Molopo River, iron ore is mined. Elsewhere the search continues for hydrocarbons, uranium and a host of other minerals.

One of the most significant anthropogenic impacts upon the Kalahari environment has been its growing use as a grazing resource for cattle, both through traditional Batswana livestock systems and the expansion of organised ranching. The growth in use of the Kalahari as a rangeland resource has been based on the exploitation of ground-water aquifers through borehole construction, which has provided previously absent year-round water, together with improved veterinary services. This process has been accelerated in the past 20 years by the support of external funding agencies and links to world markets.

Whether the large-scale introduction of cattle to the Kalahari is economically or ecologically desirable is a matter for considerable debate (e.g. Kalahari Conservation Society (KCS), 1983; Cooke, 1985), but it can not be denied that its impact on indigenous human and wildlife populations has been considerable and its effect upon the character of the Kalahari environment substantial.

The Kalahari within the boundaries of South Africa and Namibia is largely fenced rangeland, with the exceptions of the Etosha National Park, proclaimed in 1907, and the Gemsbok National Park on the Nossop River, shared by South Africa and Botswana, created in 1931. Botswana, on the other hand, has devoted some 17 per cent of the total area of the country to wildlife reserves and management areas, with major parks, such as the Chobe National Park and Moremi Game Reserve, established from 1968 onwards. Unfortunately, none of these Botswana reserves is, in itself, an ecological unit. A conservation policy, backed by the National Parks Act and Controlled Hunting Areas, attempts to protect the wildlife asset, but, as Campbell (1980: p. 45) notes, unless the future expansion of cattle is limited, further diminution of this resource is likely to occur.

Inevitably, economic and population growth and the concomitant exploitation of resources raise issues of utilisation and conservation. The problems, discussed in Chapter 9, are most acute in the Kalahari Desert, but they also have repercussions within the regional and international communities.

Part One

THE PHYSICAL BACKGROUND

2 Tectonic and geological setting

MANY EARLY GEOMORPHOLOGICAL DESCRIPTIONS of the southern African landscape, with the notable exception of the work of Alex Du Toit (e.g. 1933) incorporated the premise that the African continent has been tectonically quiescent since the early Mesozoic, except for intermittent episodes of uniform and dramatic uplift. More recent research has led to a general revision of this view, suggesting that the tectonic history of the southern part of Africa has been more dynamic, complex and geomorphologically significant than previously assumed.

The initial development of the Kalahari as a geological entity was closely linked to the evolution of the African continent from the division of the supercontinent Gondwanaland (Figure 2.1) in the Mesozoic (Table 2.1). The broad tectonic

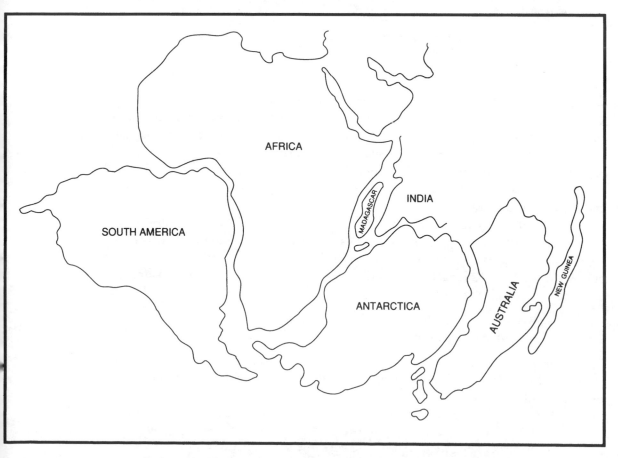

Figure 2.1. The probable configuration of Gondwanaland, showing the relationship of the modern continents prior to their separation.

Table 2.1. *Chronology of major structural events relevant to the history of the Kalahari*

Era	Period	Epoch	Ma	
		Holocene	0.01	
C	QUATERNARY			
A		Pleistocene	2	⎫ Establishment of modern course of
I				⎭ Zambezi River
N		Pliocene	5	
O				⎫ Deposition of Kalahari sediments
Z	TERTIARY	Miocene	22	⎪ throughout Tertiary; progressive
O		Oligocene	38	⎬ capture of endoreic rivers;
I		Eocene	55	⎭ further uplift of Escarpment
C				
		Palaeocene	65	⎫ Deposition of lower Kalahari Group
				⎬ sediments in interior basin by
				⎭ endoreic rivers?
			80 Ma	**Full marine conditions in South Atlantic**
M	CRETACEOUS		130 Ma	**Initial opening of the South Atlantic**
E				**Major phase of rifting**
S			136	**and initiation of the**
O				**hingeline and Great Escarpment**
Z	JURASSIC		180 Ma	**Separation of Antarctica from eastern**
O				**margin of southern Africa**
I			195	
C			200 Ma	**Beginning of Gondwanaland break-up**
	TRIASSIC			
			225	

development of Africa created the framework which led to the deposition of the Kalahari Group of sediments, and contributed to the nature of the sedimentary environment in the continental interior and to drainage evolution during the deposition of the sediments. These factors have, in turn, contributed to some of the notable characteristics of the present Kalahari environment, particularly the existence of vast expanses of flat terrain and unconsolidated sediments, and the paucity of surface drainage.

In this chapter we will examine this tectonic background, its influence upon the broad characteristics of the southern African landscape, particularly its effect on long-term drainage development in the interior, and we describe the geology present beneath the Kalahari Group sediments. The nature of the Kalahari Group sediments themselves is considered in Chapter 3.

2.1 Tectonic and structural framework

The African continent can be viewed as a series of large sedimentary basins separated by broad upwarps (Figure 2.2), with smaller, peripheral, basins also present around the coastal margins (Holmes, 1965; Summerfield, 1985a, b). The interior basins have developed by subsidence, mainly gentle downwarping, but in some instances by rifting, within a shield area or craton of very ancient Precambrian rocks (Burollet, 1984). Most of the African intercratonic basins were sites of significant or intermittent sedimentation throughout the Phanerozoic (the total period of geological time since the Precambrian) and therefore

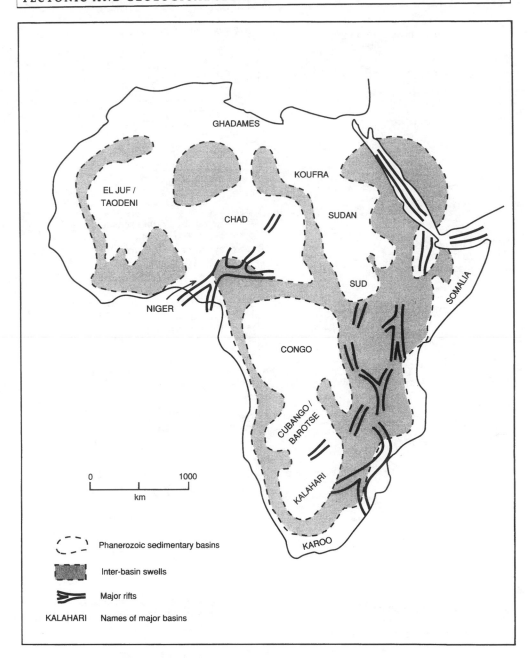

*Figure 2.2. The major structural elements of Africa. After Holmes (1965),
De Swardt and Bennet (1974), Burollet (1984) and Summerfield (1985a).*

contain great thicknesses of deposits. The Congo basin was probably not a centre
of sedimentation prior to the Mesozoic, but the southern part of the Kalahari
basin appears to have been in existence from the Palaeozoic (Figure 2.3*b*), as
evidenced by Permo-Carboniferous glacial tillites resting unconformably on the
Precambrian basement rocks (Visser, 1983).

The marginal, or pericratonic (Burollet, 1984) basins around the coastline of
Africa are, by contrast, all much younger features (Figure 2.3*c*), owing their
development to the break up of Gondwanaland (Summerfield, 1985*a*) which
commenced approximately 200 million years ago at the end of the Triassic–early
Jurassic. Tectonic activity at the commencement of the rupture of Gondwanaland involved some rifting, both in southeastern Africa (Rust, 1975) and offshore
(Dingle, Siesser and Newton, 1983) which influenced the nature of sedimentation
in parts of southern Africa, sometimes referred to as 'Zambezian tectono-

(*a*) Late Cambrian to early Permian

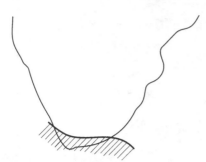

Cape sediments
Upto 9 km of sediments
deposited in fold trough

(*b*) Carboniferous to Triassic

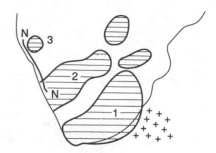

Karoo supergroup sediments.
Deposition in intercratonic
syneclise downwarps.
1 Karoo basin
2 Southern Kalahari basin
3 Etjo basin

Edges of Karoo terrain defined by
Precambrian massif [++] in east and
Namibia seismic zone N ⌒ N in west.

(*c*) Late Triassic to lower Cretaceous

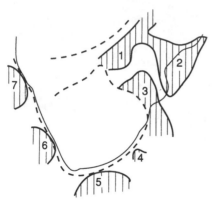

|||| Upper Karoo sediments
Deposition in rift zones
(Zambezian-taphrogenic style)
and continental margins

– – – General boundary of
Zambezian zone.

1 Mid Zambezi basin
2 Zambezi graben
3 Natal / Mozambique basin
4 St John's basin
5 Outeniqua basin
6 Orange basin
7 Walvis basin

(*d*) Upper Cretaceous to recent

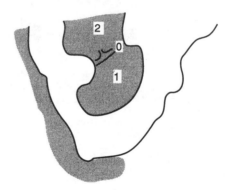

Kalahari Group sediments
and offshore counterparts.
Deposition in downwarps and,
locally, rifts (e.g. in 0, the
Okavango graben).
1 Kalahari basin
2 Cubango / Barotse basin

Figure 2.3. Sedimentary basins and deposits in southern Africa, at different periods of the Phanerozoic. Modified after Dingle, Siesser and Newton (1983).

sedimentary terrain' (Rust, 1975). Later earth movements coincident with the continued and final division of Gondwanaland (mid-Jurassic to early Cretaceous) were generally gentler, with downwarping in both the coastal zone and the African interior, the latter being responsible for the development of the contiguous Kalahari–Cubango–Congo basin, in which the Kalahari Group sediments have accumulated (Thomas, 1988*a*). Within this basin, rifting continued in the Okavango graben (Mallick *et al.*, 1981; Dingle *et al.*, 1983), locally enhancing sedimentation rates.

2.2 The development of the Kalahari basin

The preceding summary indicates that the development of the interior basin into which Kalahari Group sediments were deposited cannot be viewed as a 'one off' event. The existence of older basin sediments beneath the Kalahari deposits attests to earlier sedimentation in this region (sections 2.4 and 2.5), while the post-Jurassic structural evolution of the subcontinent (Summerfield, 1985*a*) has played an important role in later sedimentation in the region (Dingle, 1982). Nonetheless, the regional geomorphological setting induced by the final separation of the African continent from the rest of Gondwanaland can undoubtedly be viewed as the dominant factor in the evolution of the sedimentary history of the Kalahari.

2.2.1 The division of Gondwanaland

The rifting associated with the division of Gondwanaland involved the horizontal separation of lithospheric crustal plates, and is termed passive rifting (e.g. Summerfield, 1988). Although incompletely understood, the evolution of the passive margin of a plate involves not only subsidence in the rift zone, but also adjacent uplift, due to the thermal expansion of the crust during rifting (Sleep, 1971; Streckler and Watts, 1982). In southern Africa (Figure 2.4) uplift generated the hingeline or flexural bulge (De Swardt and Bennet, 1974; Summerfield, 1985*a*), which ultimately developed into the Great Escarpment of southern Africa, while subsidence was responsible for the development of both pericratonic offshore basins and the extensive intercratonic basin in the subcontinental interior (Dingle, 1982; Summerfield, 1985*a*; Thomas and Shaw, 1990).

2.2.2 Post-Gondwanaland morphotectonic evolution and sedimentation

The intercratonic Kalahari basin and the pericratonic coastal basins, which occupy 70 per cent of the offshore margins of southern Africa (Dingle, 1982) naturally became foci of post-Jurassic sedimentation, receiving, at least in part,

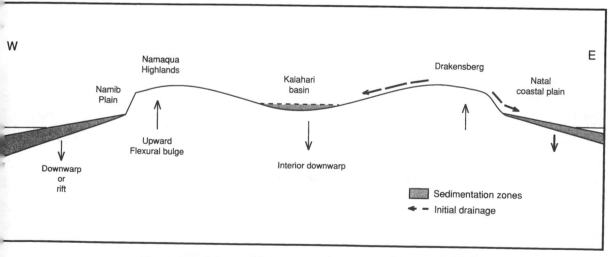

Figure 2.4. Schematic cross-section of southern Africa following the splitting of Gondwanaland.

material eroded from the uplifted hingeline. The rate of sedimentation varied from basin to basin, and is unknown for the interior, but was as great as 147 m Ma^{-1} in the offshore Walvis and Orange basins during the late Jurassic – early Cretaceous (Dingle, 1982).

Continued tectonic impetus for sedimentation during the period postdating both rifting and thermal uplift of the hingeline was derived from isostatic adjustments of the crust. These movements were essentially compensation for the transference of material from the hingeline to the basins through erosion (Summerfield, 1985a), generating further uplift of the flexural bulge and thereby contributing to continued erosion in the areas of higher ground.

Progressive cooling of the lithosphere following the phase of rifting and thermal expansion resulted in a greater crustal rigidity, causing a migration inland of the flexural bulge. Consequently, the later uplift, important for maintaining the impetus of sedimentation in the Kalahari basin through to the Pleistocene (Thomas and Shaw, 1990), was centred further inland from the initial hingeline zone (Summerfield, 1985a), approximately coincident with the present position of the Great Escarpment.

Seismic studies have identified discontinuities in the sedimentary fills of the offshore basins, suggesting that during the period from the Jurassic to the Pleistocene, hiatuses occurred in sedimentation (Siesser and Dingle, 1981), though the lack of studies precludes such an assessment being made for the Kalahari basin. Four breaks in offshore sedimentation have been identified, the last two of which, during the Oligocene–Miocene and late Pliocene–early Pleistocene, correlate with episodes of hingeline uplift (Summerfield, 1985a). Following a review of the geophysical, sedimentary and geomorphological evidence, Summerfield concludes that the overall picture is not sufficiently clear, however, to support a general concept of the post-Gondwanaland geomorphic history of southern Africa consisting of distinct periods of basin sedimentation separated by episodes of uplift, although such an arrangement persists in the most recent geomorphological assessment of landscape development in southern Africa since the Mesozoic (Partridge and Maud, 1987; Table 2.2).

2.2.3 Neotectonics in the Kalahari

While the general occurrence of sedimentation in the Kalahari–Cubango–Congo basin was intrinsically linked to the subcontinent-wide tectonic events outlined

Table 2.2. *Post-Jurassic landscape chronology of Southern Africa*

Period	Landscape changes
Late Jurassic/ early Cretaceous	Gondwanaland break-up and initiation of Great Escarpment
Late Jurassic to end of Early Miocene	'African' cycle of erosion; some minor tectonic interludes
End of Early Miocene	Moderate uplift; approx. 150–300 m
Early mid-Miocene to Late Pliocene	'Post African I' cycle of erosion
Late Pliocene	Major uplift of up to 900 m
Late Pliocene to Holocene	'Post African II' cycle of erosion

According to Partridge and Maud (1987).

above, sedimentation was locally enhanced by smaller scale tectonic activity within the basin itself, which continues to the present day (Reeves, 1972). Following the usage of Hancock and Williams (1986), these earth movements, occurring after the major plate configurations and structural framework of the region were established, can be termed neotectonic activity, though the application of this term is far from uniform or agreed (see discussion by Summerfield, 1987).

As long ago as 1926, Du Toit proposed that the Okavango Delta occupied a downwarped zone bounded by faults which were an extension of the East African Rift system. More recently, geophysical studies (e.g. Greenwood and Carruthers, 1973; Reeves and Hutchins, 1975; Hutchins, Hutton and Jones, 1976a) have confirmed this, identifying the role of faulting and rifting in the evolution of the Delta. The Delta is bounded abruptly along its distal margin by *en échelon* northeast–southwest trending faults (Figure 2.5) including the major Thamalak-ane fault, which follow the strike of sub-surface Precambrian rocks (Reeves, 1972; Mallick *et al.*, 1981). Further faults, at right angles to the main structural trend, probably also control the orientation of the 'panhandle' by which the Okavango River enters the Delta from the northwest (Reeves, 1974a, b; Mallick *et al.*, 1981).

Although this structure existed prior to the division of Gondwanaland, contributing to Upper Karoo sedimentation (Figure 2.2c; Rust, 1975), post-Gondwanaland (and therefore neotectonic in the regional context) subsidence in the rift has been estimated at between 300 m (Greenwood and Carruthers, 1973) and 1000 m (Hutchins *et al.*, 1976a), favouring the accumulation of great thicknesses of Okavango Delta alluvial sediments.

Overall, the Okavango Delta can be regarded as one of three massive alluvial fans occupying a lowly subsiding rift structure some 400 km long, which extends northeastwards and is linked to the East African Rift system (Reeves, 1972; Scholz, Koczynski and Hutchins, 1976). West of the Delta the graben is occupied by the now fossil fan of the Groot Laagte system, while to the east of the Delta the southeastward passage of the Kwando River is diverted by subsidence on the northern side of the Gomare/Linyanti Fault (Figure 2.5), contributing to the development of the Linyanti–Chobe swamps along the fault line. This swamp zone, now only partly active, continues northeastwards to the Zambezi Trough and Kafue Flats in Zambia. Traces of the ancient alluvial fan extending southeast from the Kwando can be identified from satellite imagery to the west of the Mababe Depression, where they merge at an oblique angle with the older deposits of the Okavango Delta.

Elsewhere in the Kalahari, it is probable that faulting with the same northeast–southwest trend has contributed to the development of the Makgadikgadi pan complex (Baillieul, 1979) and the small Nata drainage network which enters the pans from Zimbabwe in the east. Mallick *et al.* (1981) suggest that renewed movement of the Nata Fault has enabled the Nata River to capture the Tegwani and Maitengwe streams by altering its course to truncate their lower reaches.

While it is not possible to establish the precise timing of the earth movements described above, two lines of evidence strongly support their occurrence and continuance in the period since the establishment of the interior Kalahari Basin. First, to the northwest of the Okavango Delta, interpretation of Landsat imagery by Mallick *et al.* (1981) revealed linear sand dune ridges which have been disrupted by fault lineaments, while Cooke (1975a, 1980) described faulting in the nearby Gcwihaba Hills which has disrupted calcretes and cave deposits dated to the late Pleistocene. Second, a study of modern earthquake epicentres in Botswana (Reeves, 1972) has shown them to be distributed in the vicinity of the

Okavango Delta and about a northeast–southwest 'Kalahari seismicity axis'
which passes through the Makgadikgadi Basin (Figure 2.5). Very recent tectonic
activity has also been implicated in hydrological shifts in the Okavango Delta
within the past century (UNDP/FAO, 1977; Shaw, 1984).

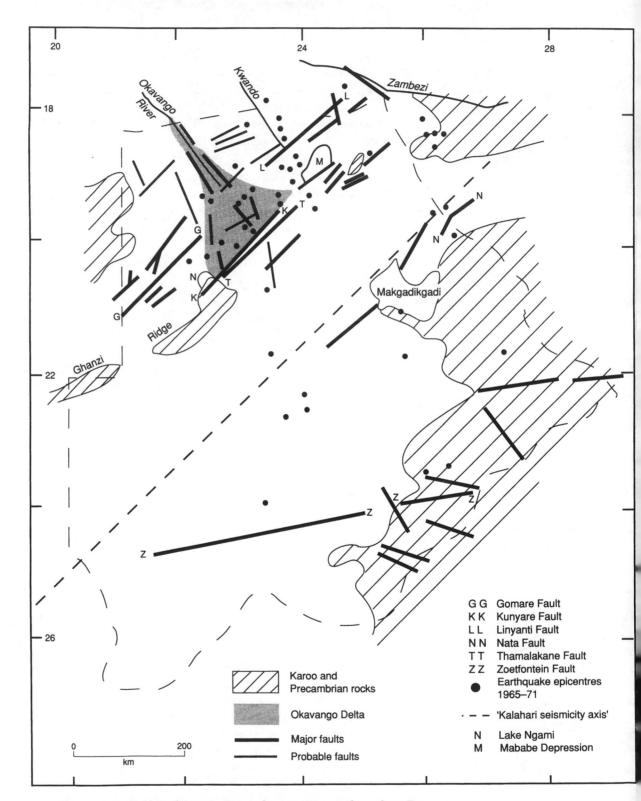

Figure 2.5. Earthquake epicentres and major structural trends in Botswana.
Data from Reeves (1972), Hutchins et al. (1976a) and Mallick et al. (1981).

2.3 Landscape responses to tectonic development

The tectonic history of southern Africa outlined above provided the framework for the structural and sedimentary development of the Kalahari. This cannot, however, be divorced from other major aspects of landscape development in southern Africa as a whole, particularly the evolution of the Great Escarpment and drainage development in the subcontinent, which has contributed significantly to the character of the Kalahari. These issues are examined below, while the landforms of the Kalahari itself are considered in detail in a later chapter.

2.3.1 The Great Escarpment

The Great Escarpment extends more or less continuously around the rim of southern Africa, at a mean distance of about 150–200 km from the present coastline (Figure 2.6: Rogers, 1920; Wellington, 1955; De Swardt and Bennet,

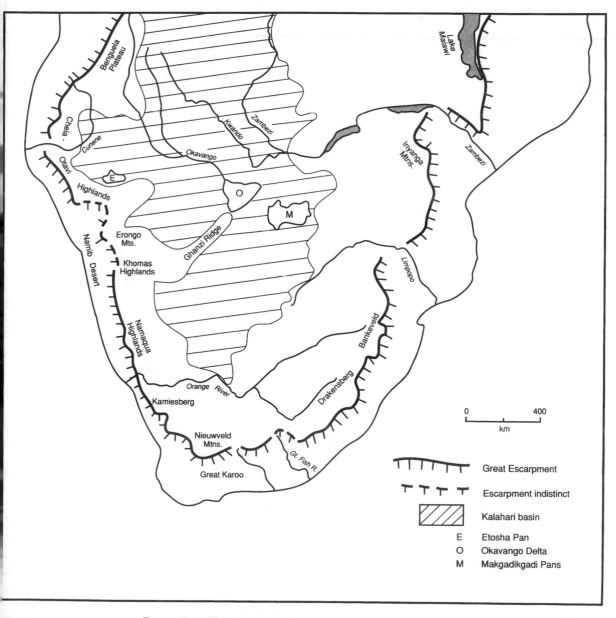

Figure 2.6. The Great Escarpment of southern Africa.

1974; Ollier, 1985). Its origins lie, as described in sections 2.2.1 and 2.2.2, in the uplift at the flexural bulge, the first of which accompanied the rifting responsible for the splitting of Gondwanaland, and the second which occurred due to subsequent isostatic crustal adjustment (Summerfield, 1985a). One mechanism which explains the location of the modern escarpment inland from the initial coastal hingeline is the progressively greater crustal rigidity which existed when later uplift occurred (Summerfield, 1985a, section 2.2.2). A second explanation, which does not need to be seen as an alternative, is that subaerial erosion since uplift has driven the escarpment inland (De Swardt and Bennet, 1974; Ollier, 1985).

Given that a variety of lithologies are present in the Great Escarpment, it is not suprising that erosion has imparted variations in its morphology and character from place to place, as described in the excellent account of Wellington (1955). The eastern sections, represented by the Drakensberg (Figure 2.7a) and Inyanga Mountains, are capped respectively by resistant basalt lavas and dolerite: consequently, they are distinct, upstanding and little dissected, attaining maximum altitudes of 3482 m and 2596 m asl respectively.

In the west, the Khomas and Namaqua Highlands (Hochland) of Namibia are developed in thick schists, grits and quartzites, and likewise form an abrupt escarpment edge, in which underlying granites are widely exposed (Figure 2.7b). This contrasts notably with the section immediately to the north where Karoo sediments and granites have been heavily dissected to leave the escarpment zone represented by the broken Erongo Mountains and isolated inselbergs such as the Spitzkoppe (1759 m asl).

2.3.2 The Interior Basin

From the Great Escarpment, the land surface drops abruptly to the coastal plain of southern Africa, while inland it is backed by the Highveld before generally declining, in many areas almost imperceptibly, to the interior, Kalahari, basin.

It is necessary here to note that the interior basin has sometimes been described as an elevated plateau, and though it does indeed attain a mean altitude of about 1000 m asl and can be extremely flat, the concept of interior elevation does not readily sit comfortably with the account of African tectonic evolution discussed above.

Early ideas concerning the development of an interior plateau credited its existence to episodic uplift, separated by periods of planation, during the long-term evolution of the African landscape (e.g. King, 1955, 1962). This explanation has generally been rejected, or at least substantially revised (e.g. Partridge and Maud, 1987), with the advent of a greater understanding of global tectonics and regional morphotectonics (see summaries by De Swardt and Bennet, 1974, and Summerfield, 1985b). Consequently, the altitude of the interior is now essentially regarded as a legacy of its former position as the central, elevated, part of Gondwanaland (Summerfield, 1985a) prior to its division and the subsequent elevation of the Great Escarpment. The flatness of the basin surface is mainly due to its sedimentary infill (Chapter 3), while in some places, such as the eastern Kalahari rim in western Zimbabwe, the apparent appearance of an elevated plateau has been due to geologically recent incision into the basin surface by tributaries of the Zambezi drainage network (Thomas and Shaw, 1988, and see below).

Figure 2.7. Contrasts in the character of the Great Escarpment. (a) Part of the
upstanding and distinct Drakensberg of Natal. (b) The dissected but nonetheless
abrupt escarpment of the Namaqua Highlands, Namibia. The flat-topped
mountain in this southward-looking view is the Gamsberg, some 2347 m asl.

2.3.3 Drainage evolution

The tectonic framework of southern Africa after the division of Gondwanaland
had a major impact upon the development of river systems within the
subcontinent, and, as a consequence, patterns of sedimentation in Kalahari
Group sediments. A dual system developed, whereby relatively short rivers with
steep gradients drained from the Great Escarpment to the coast (an exoreic
system), while longer but lower gradient endoreic rivers drained from the
opposite side of the escarpment into the interior Kalahari basin (Obst and Kayser,

1949; De Swardt and Bennet, 1974; Thomas and Shaw, 1988). This early endoreic system, of which the Okavango is the only major active river remaining today, is of great significance in the development of the Kalahari, as it is perhaps the only feasible mechanism which can account for the initial deposition of the Kalahari Group sediments in the interior basin (Thomas and Shaw, 1990).

Drainage evolution in southern Africa has evoked interest since the travels of David Livingstone (Livingstone, 1858a). Since the establishment of the initial dual drainage systems, major changes have involved the progressive capture of the internally draining rivers by components of the more aggressive exoreic system (De Heinzelin, 1963). Evidence for this process has been derived from various geomorphological and sedimentary studies, particularly those of the Zambezi and Limpopo systems (Du Toit, 1926, 1933; Bond, 1963; Moore, 1988; Thomas and Shaw, 1988), the Cunene River in Angola (Kanthack, 1921; Beetz, 1933; Wellington, 1938), the Molopo–Orange–Vaal system in South Africa (Schmitz, 1968; Smit, 1977; McCarthy, 1983) and the Fish River in Namibia (Wellington, 1955). Given the spatial dispersal and relative paucity of the evidence, especially within the Kalahari basin where many of the former traces of ancient rivers are likely to have been blanketed by reworkings of the Kalahari Sand, details of both the nature and chronology of the changes must remain in doubt, though the general principles involved accord well with many aspects of the tectonic and sedimentary history of the region. The development of the Zambezi system will serve to illustrate these issues and their significance to the development of the Kalahari.

2.3.4 The Upper Zambezi system

It is widely regarded that the modern course of the Zambezi River is relatively young, dating from the Pliocene (e.g. Lister, 1979) or possibly as recently as the mid-Pleistocene (Bond, 1975). Relatively abrupt changes in the direction and characteristics of the course of the modern Zambezi suggest that the Upper and Middle Zambezi evolved as components of separate river systems only to be linked relatively recently (Figure 2.8a); a theory which has gained additional support from studies of the distributions of fish populations in the rivers of central Africa (Jackson, 1961; Jubb, 1964; Balon, 1971, 1974; Bell-Cross, 1975).

From its source the modern Zambezi flows generally south to southeasterly until, at Katima Mulilo, near the border between Zambia and the Caprivi Strip of Namibia, it changes direction to flow almost due eastwards across the 'swamp and alluvium belt' (sections 2.2.3 and 5.2) that extends from the Okavango Delta to the Kafue River. About 70 km to the east, at the Mambova Falls, the river then enters the Zambezi fault zone, which continues until the river passes through the Great Escarpment to reach the coastal flats of Mozambique. The section between Katima Mulilo and the Mambova Falls provides the link between the Upper Zambezi and Middle Zambezi, which evolved as separate systems until the Pliocene to mid-Pleistocene.

Many investigators, including Du Toit (1927, 1933), Wellington (1955), Bond (1963) and Grove (1969) have inferred that the Upper Zambezi once continued its southerly course, joining either the Orange (Lister, 1979) or, more probably, the

Figure 2.8 (opposite). Drainage development in southern Africa. (a) Major modern drainage lines. The development of the Zambezi drainage system is shown in (b) and (c): (b) postdates the division of Gondwanaland and shows the proto-Upper Zambezi which may have joined the Limpopo; (c) is the situation prior to the union of the Middle and Upper Zambezi in the Pleistocene early Pleistocene. After Thomas and Shaw (1988).

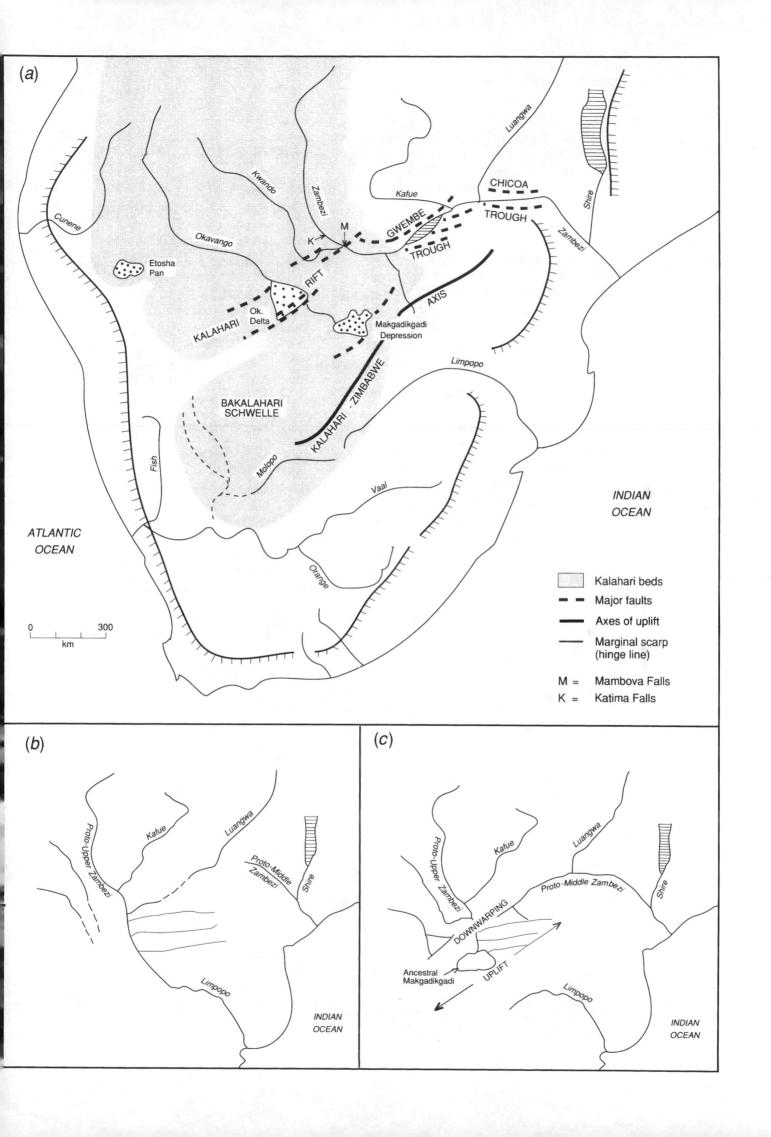

(a)

Cunene

Kwando

Zambezi

Luanga

Kafue

CHICOA

Shire

GWEMBE

M

K

TROUGH

TROUGH

Okavango

Etosha
Pan

RIFT

AXIS

Ok.
Delta

Makgadikgadi
Depression

KALAHARI

Limpopo

BAKALAHARI
SCHWELLE

KALAHARI - ZIMBABWE

INDIAN
OCEAN

Fish

Molopo

Vaal

ATLANTIC
OCEAN

Orange

Kalahari beds

Major faults

Axes of uplift

Marginal scarp
(hinge line)

M = Mambova Falls

K = Katima Falls

0   300

km

(b)

Proto-Upper Zambezi

Kafue

Luangwa

Proto-Middle
Zambezi

Shire

Limpopo

INDIAN
OCEAN

(c)

Proto-Upper Zambezi

Kafue

Luangwa

Proto-Middle Zambezi

Shire

DOWNWARPING

Ancestral
Makgadikgadi

UPLIFT

Limpopo

INDIAN
OCEAN

Limpopo (Figure 2.8b; Wellington, 1955; Bond, 1963). Given that distinct endoreic and exoreic river systems are likely to have existed following the initial development of a circum-southern African hingeline (Obst and Kayser, 1949; De Swardt and Bennet, 1974), it can be postulated that even before a link with the Limpopo was established, the proto-Upper Zambezi must itself have been an endoreic river, depositing sediment in the Kalahari basin in much the same way as Smit (1977) proposes for an ancestral north-flowing Orange River. Verboom (1974) has even suggested that the surface sands of western Zambia and neighbouring areas of the Kalahari basin owe their deposition primarily to fluvial processes.

The southward course of the proto-Upper Zambezi was disrupted by gentle earth movements in the Middle Kalahari, probably a combination of uplift in the vicinity of the Makgadikgadi basin, along the 'Kalahari–Zimbabwe Axis' and downwarping in the Okavango region (Du Toit, 1954; Bond, 1963). This may have caused renewed endorism in the Upper Zambezi system (Du Toit, 1926), prior to the link with the Middle Zambezi being established through the Livingstone syncline. This may have provided a source for a pre-Pleistocene ancestral Lake Palaeo-Makgadikgadi (Thomas and Shaw, 1988; Figure 2.8c).

The proto-Upper Zambezi had a number of left-bank tributaries flowing from the east and northeast, including the Kafue (Bond, 1963; Williams, 1975) and Luangwa (Wellington, 1955) Rivers (Figure 2.8b), together with numerous rivers with their sources in what is now Zimbabwe (Lister, 1979; Thomas and Shaw, 1988). These were progressively captured by the aggressive proto-Middle Zambezi system, the Luangwa by the proto-Zambezi itself and the others by the development of its extensive tributary network (Williams, 1975; Thomas and Shaw, 1988). The section of the modern Zambezi course between Katima Mulilo and the Mambova Falls is also assumed to have been a former left-bank tributary of the proto-Upper Zambezi, which subsequently suffered drainage reversal during the integration of the Upper and Middle Zambezi systems.

Prior to being captured by a tributary of the proto-Middle Zambezi, the Kafue flowed in a southwesterly direction to join the proto-Upper Zambezi, passing through what is now a 2 km wide gap at 1070 m asl in the modern Kafue–Zambezi watershed (Bond, 1963). Today this is a poorly drained alluvial flat, but the alluvium is underlain by coarse, current-bedded fluvial gravels (Clark, 1950).

In western Zimbabwe, a series of remarkably straight, sub-parallel westerly flowing rivers were probably also proto-Upper Zambezi tributaries prior to their capture by the Gwayi River, which drains into the Middle Zambezi (Figure 2.9). Evidence for this lies in the presence of fluvial gravel deposits and terraces to the west of the Gwayi and in northeastern Botswana, and the Kazuma and Mpandamatenga lacustrine features (Figure 2.9; Mallick et al., 1981), the existence of which cannot be explained in terms of either the present drainage or topographical setting of the eastern Kalahari rim (Thomas and Shaw, 1988).

2.3.5 The proto-Middle Zambezi and the modern Zambezi River

The precursor of the Middle Zambezi was an exoreic river on the coastward side of the Great Escarpment (Figure 2.8b). Downwarping along the ancient Gwembe and Chicoa troughs, part of the link between the Okavango graben and the East African Rift system, contributed to its considerable headward erosion and ultimately the capture of the proto-Upper Zambezi (Bond, 1963; Williams, 1975), in the Pliocene or Upper Pleistocene (Dixey, 1950; Bond, 1975; Lister, 1979).

After the establishment of the present Zambezi course, enhanced downwarping along the Gwembe trough led to further incision of the Middle Zambezi

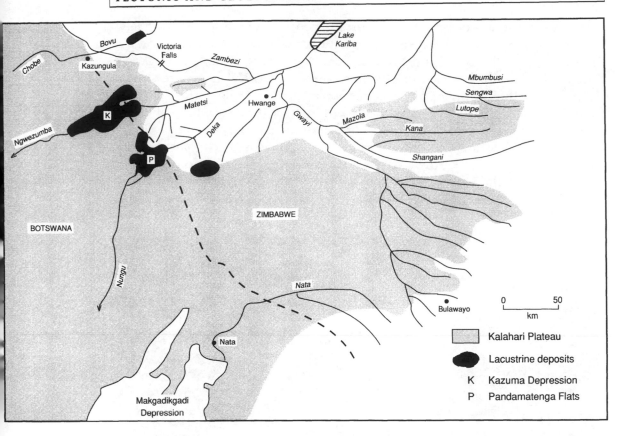

Figure 2.9 Drainage of the eastern Kalahari rim. The Matetsi, Deka and Gwayi are tributaries of the Middle Zambezi which have incised into the Kalahari rim and captured west-flowing tributaries of the proto-Upper Zambezi. After Thomas and Shaw (1988).

(Dixey, 1945). This was linked to renewed uplift at the flexural bulge, which contributed to a general rejuvenation of coastward flowing rivers in the subcontinent (De Swardt and Bennet, 1974). This allowed the Middle Zambezi to achieve considerable erosion during the Pleistocene (Dixey, 1950), resulting in the rapid development of the Victoria Falls (Bond, 1975) and the formation of the incised Middle Zambezi Gorge (Figure 2.10), in the space of no more than 250 000–500 000 years. Linked to this, the Middle Zambezi tributaries have achieved considerable erosion and incision into the eastern rim of the Kalahari Basin in Zimbabwe, exposing underlying Karoo basalts and sediments (Figure 2.9; Thomas and Shaw, 1988).

2.3.6 Drainage changes in the southern and eastern Kalahari

Equally significant changes in drainage patterns took place in the southern Kalahari as a result of the flexure of the Kalahari rim, particularly along the alignment of the Kalahari–Zimbabwe Axis to the east of the basin. Wellington (1955) suggested that the essentially 'dry' Molopo–Nossop–Auob valley network, considered by some to be a pre-Kalahari valley system (e.g. Smit, 1977; Rathbone and Gould, 1982), has suffered the capture of headwaters both by the Fish River in Namibia and by the Limpopo system to the east (Figure 2.8a), thus contributing to its current misfit appearance. Today the Molopo Valley network is part of the coastally oriented Orange River system, but Smit (1977) cites the presence of basal Kalahari Group conglomerates in valleys cut into the sub-Kalahari landsurface to infer that the original post-Gondwanaland drainage of the southern Kalahari was directed towards the sub-continental interior. Conversely, McCarthy (1983) has suggested, from alluvial deposits and valley forms below the Orange–Vaal confluence, that the Kalahari was drained southwards by a 'Trans-Tswana' River prior to the uplift of the present southern Kalahari watershed.

Apart from the beheading of the upper Molopo, the rise along the Kalahari–Zimbabwe Axis, probably since the late-Tertiary (Cooke, 1980), has led to the strong development of the left bank tributaries of the Limpopo network (Shaw, 1989a) and gradual incision of the headwaters of the fossil Mmone network draining to the Makgadikgadi Basin (Shaw and De Vries, 1988; section 5.4.1). As already noted, in the Makgadikgadi area itself Du Toit (1927) proposed that such an uplift was responsible for the isolation of the Makgadikgadi Basin by severing

Figure 2.10. The Victoria Falls (a) and Batoka Gorge (b) on the course of the Middle Zambezi have evolved rapidly under the influence of uplift during the Pleistocene.

the connection between the Upper Zambezi–Okavango network and the Limpopo to create the proto-Makgadikgadi lake, the mechanics of which have been discussed by Cooke (1980). A possible relict of this Tertiary coastwise network is a large wind-gap, up to 10 km wide, which crosses the present Kalahari watershed from Mea Pan in the southwest corner of the Makgadikgadi Basin to the headwaters of the Motloutse River, a tributary of the Limpopo. Although the geology and geomorphology of this feature have never been examined, it is significant that the discovery of diamond-bearing kimberlite pipes at Orapa on the southern side of the Makgadikgadi followed the prospecting of kimberlite-related garnet and limenite in the Motloutse River, in accordance with Du Toit's theory.

2.4 Pre-Kalahari geology

With the Kalahari owing its structural origin to events which began in the Jurassic (Table 2.1) and the Kalahari Group sediments therefore being Cretaceous at the oldest, it could be argued that the pre-Kalahari geology of the southern African interior is irrelevant in a consideration of the Kalahari environment. It is, however, the pre-Kalahari geology which is yielding mineral wealth within the desert today, while these older rocks also provide a number of dispersed outcrops which, given the Kalahari's notable flatness, are significant topographical features of parts of the region. Without discussing the lithological minutiae, which are considered in other sources (e.g. Rogers, 1936; Du Toit, 1954; Coates *et al.*, 1979; Lockett, 1979; Smith, 1984) an understanding of the geology beneath the Kalahari is therefore of value, though in this chapter it is largely restricted to the area of the Kalahari Desert.

The pre-Kalahari geology of central southern Africa can be considered as consisting of two major elements (Table 2.3): first, ancient Precambrian rocks, which form the geological basement, outcropping around the Kalahari margins and also as dispersed inliers within the desert, such as the Ghanzi Ridge; and second, members of the extensive Carboniferous to Triassic Karoo Supergroup, which contains both complex sedimentary sequences and extensive basaltic lava outpourings. Karoo lithologies occur unconformably between the Precambrian rocks and Kalahari sediments beneath most of the Kalahari Desert, though Precambrian rocks lie directly beneath the Kalahari Group to the southwest, northwest and east of the Okavango graben (Figure 2.11). Post-Karoo sills and dykes complete the sub-Kalahari geology.

2.4.1 Precambrian rocks

Surface exposures of Precambrian rocks are limited and both Karoo and Kalahari lithologies provide an extremely thick cover throughout much of the Kalahari Desert, in excess of 1000 m in central areas. However, borehole records, magnetic surveys (Reeves, 1978) and information from beyond the margins of the Kalahari have yielded invaluable data on the sub-Kalaharian/Karoo geology, which is composed of rocks which are frequently mineralifically rich as a result of a high degree of alteration.

Though the precise stratigraphical affinities are often doubtful, the Precambrian geology beneath the Kalahari can be considered (Reeves, 1978) to comprise two areas of basement or platform rocks (so called because they provide the base beneath subsequent formations): the southeastern zone of very ancient Archaean cratonic rocks (predating 2600 Ma) and the northwestern zone of Proterozoic

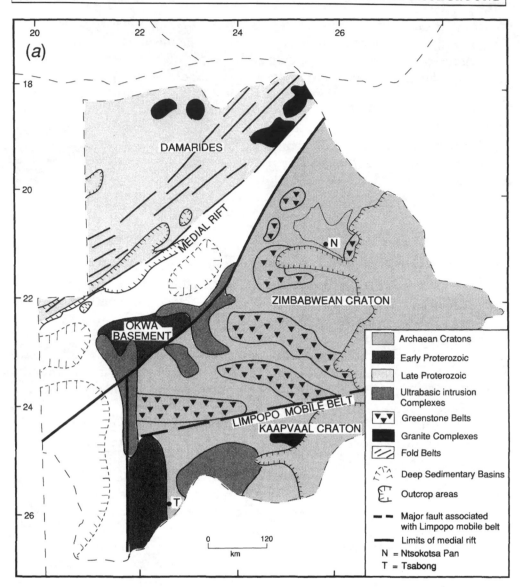

rocks, separated by the southwest–northeast Medial Rift which extends from the Nama Basin of Namibia to the Zambezi Metamorphic Belt in western Zimbabwe (Reeves, 1978; Lockett, 1979; Jones, 1982). In the ancient rift, which probably dates from about 1000 Ma, the oldest Precambrian basement geology has been depressed by over 15 000 m in places (Reeves, 1978) and blanketed by late Precambrian and possibly early Proterozoic rocks. These have in turn been covered by a minimum of 1000 m of Karoo sediments throughout most of the rift's length.

2.4.2 South of the Medial Rift

South of the Medial Rift, the Archaean Basement geology that is well exposed in the eastern Botswana hardveld (Figure 2.11 *a* and *b*), and, to a lesser extent, in a southerly direction (Mallick *et al.*, 1981), continues beneath the Kalahari sandveld. Sediments and basalts of the Central Karoo Basin separate the basement from the Kalahari Group sediments throughout most of the central Kalahari, confirmed by Coates *et al.* (1979), who encountered only one Archaean-age outcrop, at Ntsokotsa Pan in the Makgadikgadi Basin, on the Kalatraverse project. In the Southern Kalahari, however, south of approximately 25° 30′ S and

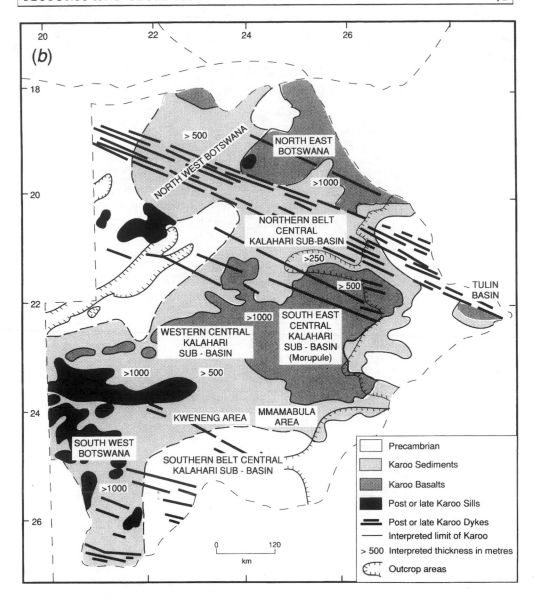

Figure 2.11. (a) Precambrian geology and structure and (b) Karoo geology
beneath the Kalahari in Botswana. Modified after Jones (1982), Smith (1984)
and other sources.

east of 22° E, Archaean rocks lie directly beneath between 50 and 200 m of
Kalahari sediments in places. This is not reflected in the geological surveys, for
the small and scattered surface outcrops of Precambrian rocks shown by Mallick
et al. (1981) in this area, usually in the vicinity of pans, are largely placed in the
Proterozoic Waterberg Group and Transvaal Sequence, confirming Jones's
(1982) view that even to the south of the Medial Rift, basement rocks possess a
fairly extensive sub-Kalahari cover which dates from the mid-Precambrian.

The Archaean Basement represents many periods of sedimentation, volcanism
and tectonic activity before approximately 2600 Ma. In the Middle and Southern
Kalahari the basement comprises rocks of the ancient Kaapvaal and Zimbabwean
cratons (Figure 2.11a). Though these are dominated by (often pink) gneisses,
small relicts of the quartzitic and magnesian sediments from which the gneisses
were, in part, derived by metamorphism still survive (Coates et al., 1979). In
places, the gneiss has been affected by granitisation to produce granodioritic
gneiss with post-granitisation granitic intrusions and complexes also evident. In

Table 2.3. *Simplified column of the solid geology beneath the Kalahari south of the Zambezi*

Era	Period	Ma	Geological representation beneath/in Kalahari*
CAINOZOIC			See Table 2.1 and Chapter 3
M E S O Z O I C	Cretaceous	136	
	Jurassic	195	Late or post Karoo dolerite dykes and kimberlite pipes associated with early Gondwanaland break-up
	Triassic	225	*KAROO SEQUENCE* 4. Stormberg Lavas 3. Lebung Group 2. Ecca Group 1. Dwyka Group (See Figure 2.11)
U L P A P E P A E E O R Z O I C	Permian	286	
	Carboniferous	360	
	(Devonian)	415	Unconformity
LOWER PALAEOZOIC	(Silurian) (Ordovician) (Cambrian***)	570	
P R E C A M B R I A N**	Proterozoic**	2600	Ultrabasic lavas of Bushveld Igneous Complex SOUTH AND EAST 3. Waterberg Group sandstones 2. Transvaal Sequence dolomites and sandstones 1. Ventersdoorp Supergroup lavas NORTH 3. Damara Sequence sediments 2. Ghanzi Group metamorphosed sediments 1. Kgwebe Formation volcanics
	Archaean**		3. Greenstone schists 2. Post-gneiss granite bodies and intrusions 1. Gneisses of Southern Kalahari basement Bulawayo Group lavas in extreme east

*   Terminology used in table and text is wherever possible consistent with SACS (1980), except in areas beyond its jurisdiction. There is inevitably debate about some names, for example, Transvaal Supergroup or Transvaal Sequence.

**   Terms not strictly analogous with normal use of Era and Period. Periods in brackets are not represented in the geological record beneath the Kalahari.

*** Cambrian Nama Group sediments may underlie parts of the western Kalahari.

the Zimbabwean craton, belts of greenstone (schist) reflect more limited metamorphism affecting the original sediments and lavas, with deformation under relatively low temperatures and pressures allowing the structures of the parent rocks to remain. These belts, which include the Bulawayo Group, regarded as some of the oldest, or Primitive, Archaean rocks (e.g. Du Toit, 1954), beneath parts of the eastern extremity of the Kalahari Group in central Zimbabwe (Lockett, 1979), are of considerable economic importance, yielding a range of metal ores including gold. By way of contrast to the Greenstone belts, the contact between the two cratons, known as the Limpopo Mobile Belt (Key and Hutton, 1976), is a zone displaying strong metamorphism and fracturing of the cratonic material.

Parts of the Archaean geology of the Kaapvaal craton, and less extensively, the Zimbabwean craton, have a cover of Proterozoic rocks which include members of the Ventersdorp Supergroup, dominated by acid lavas, Transvaal Sequence comprising members of the component Griquatown and Dolomite Groups and Blackreef Formation (Mallick *et al.*, 1981) and Waterberg Group consisting almost entirely of thickly bedded dark red sandstones and conglomerates (e.g. Wellington, 1955), the whole spanning the period 2600 to 1700 Ma (Jones, 1982). The westernmost of these deposits lie adjacent to the rift-truncated limit of the Kaapvaal craton at the so-called 'Kalahari line' (Figure 2.11*a*) in a belt which is heavily folded (Reeves, 1978) and well represented by outcrops which appear through the Kalahari Sand in the vicinity of Tshabong and Maralaleng in southern Botswana and in the nearby Molopo valley (Mallick *et al.*, 1981). The outcrops have been described as exposing maroon, glassy sandstones and hard, ripple-bedded grey and white glassy quartzites near Tshabong (Boocock and Van Straten, 1962; Mallick *et al.*, 1981) and red, blue and grey quartzites some 30 km to the northeast at Maralaleng which Mallick *et al.* (1981) now believe to be undifferentiated deposits of the Transvaal Sequence. At Khakhea, further northeast, both Transvaal and Waterberg sediments are exposed.

In places the Proterozoic sediments of the Southern Kalahari, but not the overlying Karoo deposits, have been intruded by ultrabasic lavas, which present themselves in the form of diabase and dolerite sill, dyke and sheet complexes (Jones, 1973; Mallick *et al.*, 1981). Jones (1982) equates the major intrusive complex in the Southern Kalahari (Figure 2.11) to the Bushveld igneous complex of Zimbabwe and the Transvaal, dated to 2100 Ma by Hamilton (1977). Comparable dolerite intrusions are found in the Proterozoic rocks beneath the Kalahari Sand in western Zimbabwe and are dated on stratigraphical grounds to in excess of 1000 Ma (Lockett, 1979).

2.4.3 North of the Medial Rift

To the north of the Medial Rift, the Precambrian rocks beneath the Kalahari have a thick intermediary blanket of Karoo sediments in the Mesozoic–Cainozoic Okavango Rift (see section 2.2.3), but north and west of the Ghanzi Ridge and in the Mababe Depression they generally lie unconformably beneath the Kalahari sediments themselves. With the possible exception of gneiss found in association with the Aha Hills of northwestern Botswana (e.g. Mallick *et al.*, 1981), the Precambrian rocks in this area belong to Proterozoic times.

The later Precambrian rocks of this area are represented by three groups (Table 2.3): the Kgwebe Formation volcanics, Ghanzi Group weakly metamorphosed sediments and Damara Sequence (Katanga Group in Zambia) sediments. The Kgwebe volcanics, comprising both acid and basic igneous rocks with associated sediments, were regarded as the oldest rocks in this area by Thomas (1973) and have subsequently been dated to c.980 Ma (Key and Rundle, 1988). They have limited surface outcrops but are exposed on the southeastern side of the Kgwebe Hills and at the northeastern end of the Ghanzi Ridge, to the south of the Ngami Basin.

The Ghanzi Group outcrops along the southwest–northeast trending Ghanzi Ridge, which is the major pre-Kalaharian outcrop found in the Kalahari Desert, and in the Shinamba Hills which maintain the trend to the east of the Mababe Depression. The trend of the ridge represents the axis of tight folding to which the various sedimentary formations of the group have been subjected after the sediments were deposited in a volcanic trough dated to about 1000 Ma (Key and Rundle, 1988). The group contains sediments ranging from coarse gritstones to mudstones and limestones, all displaying evidence of weak metamorphism.

Though having a more widespread distribution than the Ghanzi Group, the younger sediments of the Damara Sequence have fewer outcrops within the Kalahari, with major exposures confined to the Aha Hills and Nxaunxau Valley to the northwest of the Okavango Delta and more minor outcrops forming the inselbergs of the Koanakhe, Gcwihaba and Tsodilo Hills, and flanks of short stretches of the Okavango River valley (Mallick et al., 1981). The Damara rocks do, however, continue both westwards to be widely exposed in Namibia where they form the Windhoek and Khomas Highlands (e.g. Du Toit, 1954), and northeastwards, where their Katangan equivalents are found in western Zambia (Money, 1972). Damara sediments are economically important, containing base metal ores (Jones, 1982), and consist of quartzites, sandstones, siltstones, limestones and massive dolomites which are more deformed than the Ghanzi sedimentary sequence and which pose considerable problems of stratigraphical correlation in the northwestern Kalahari (see discussion in Mallick et al., 1981) due to their great thickness, exceeding 10 500 m (Du Toit, 1954), deformation and poor exposure. Kroner (1977) has however been able to suggest that deposition occurred between 700 and 630 Ma, in a trough resulting from deformation of the basement at about 890 Ma (Key and Rundle, 1988). The Damara rocks themselves experienced major deformation during the early Palaeozoic Damara Orogeny.

2.5 The Karoo Sequence

Karoo rocks, for which SACS (1980) prefer the term Sequence to Supergroup, date from the Upper Palaeozoic (Table 2.3) and therefore rest unconformably upon Precambrian lithologies, for the most part intervening between the latter and the Kalahari Group sediments. Karoo rocks are widespread throughout southern Africa so it is no surprise that considerable Karoo deposits are found beneath the Kalahari (Figure 2.11), but their surface exposure within the Kalahari is extremely poor (Boocock and Van Straten, 1962; Mallick et al., 1981).

The geology of the Karoo Sequence can be divided simply into lower sediments ranging from the Carboniferous to the Triassic and upper volcanics. More accurately, five series or groups are widely identified, particularly in the main Karoo Basin: Dwyka (properly called a Formation rather than a Group according to SACS, 1980), Ecca, Beaufort, Stormberg and Drakensberg. The Stormberg and Drakensberg rocks have sometimes been linked together as the upper and lower Stormberg, but are now more widely considered as separate groups, though SACS (1980: p. 539) argue that there is no lithological justification to link the various units collectively known as the Stormberg into a Group. The terminology proposed for the upper Karoo strata in Botswana confuses the issue further in that Smith (1984; see Figure 2.1) terms the Stormberg Group sediments the Lebung Group and the Drakensberg Group the Stormberg lavas.

The Karoo sediments consist largely of shallow freshwater and terrestrial deposits, though marine and glacial facies are present in the oldest deposits. The fact that deposition occurred over a very wide area and in a number of structural basins means that the stratigraphical sequence displays notable regional and sub-regional facies variations. Recent developments in Botswana, particularly extensive mineral prospecting beneath the Kalahari sediments and the use of regional gravity surveys, have enabled lithological interpretations and correlations to be carried out for the Kalahari region in its own right (Smith, 1984) rather than simply by reference to the classical succession of South Africa (e.g. Green, 1966). This has led not only to the formal division of the Stormberg rocks described above, but also to the determination that the Beaufort Group of South Africa is probably not represented beneath the Kalahari, with the Tlhabala/Kwetla sediments currently being given only formation status and only tentatively correlated with the Beaufort Group rocks found further to the south (Smith, 1984). Despite the advances made in recent years, it cannot be assumed that further significant revisions will not occur as more data are collected from beneath the Kalahari.

2.5.1 Karoo depositional sequence and setting

The lengthy period of Karoo deposition was one 'beginning with an ice age and ending with a volcanic episode' (Maufe, 1935). The depositional sequence of the sedimentary rocks from basal Karoo Dwyka sediments, with a strong glacial imprint, to upper Karoo Lebung Group sediments, with considerable desertic 'red bed' development, with intervening Ecca times being cool temperate, is a reflection of the equatorward drift which Gondwanaland experienced during the Upper Palaeozoic (Van Eeden, 1973). Deposition itself occurred in a number of basins, including the main Karoo basin of South Africa, and the Kalahari Basin centred on what is now Botswana, where the deposits have a general thickness of 1000–1500 m (Smith, 1984). These basins are to some extent descriptional conveniences but their existence, to varying degrees, has been confirmed by geophysical surveys and facies changes. The nomenclature used for them has also changed over time and from author to author, which adds confusion. Many authors, however, refer to the existence of the Kalahari Basin (e.g. Boocock and Van Straten, 1962; Green, 1966; Jones, 1982; Visser, 1983; Smith, 1984) of Karoo sedimentation, also variously calling it the Karroo (sic) Kalahari Basin and Central Karoo Basin.

Facies variation led Green (1966) to regard Karoo deposition to have occurred in three basins-within-basin in the Kalahari region, but with the basins being continuous over gentle rises in Precambrian basement. In the terminology of Smith (1984), these are the Southwest Botswana Basin, the Central Kalahari Basin and the Northeast Botswana Basin (Figure 2.11b). The Central Kalahari Basin has also been further divided into sub-basins on stratigraphical grounds, with only tentative correlations currently possible between individual sub-basin stratigraphies (Coates, 1980; Smith, 1984). Karoo deposition in the separate Northwest Botswana Basin of Smith (1984) represents the earliest sediment accumulation in the Okavango Rift, which was from the mid-late Karoo (Lebung/Stormberg) or possibly Ecca, onwards, and possibly earlier, in response to the tectonic 'emergence' of the Ghanzi Ridge. Rifting and faulting also affected sedimentation in the Northeast Botswana Basin and its continuation into western Zimbabwe

(Lockett, 1979; Smith, 1984), to the extent that Ellis (1978) regarded sedimentation in the southern area to have occurred in a 'Nata Sub-basin'.

Overall, early Karoo (Dwyka) sedimentation can be said to have been influenced considerably by the pre-Karoo topography, which in the Southwest Botswana Basin and its continuation into Namibia and the northern Cape Province, for example, consisted of a high but strongly dissected upland which exerted a major control on Dwyka glaciation (Visser, 1983). The length and degree of sedimentation meant that this influence diminished, particularly where centred over stable basement cratons, as erosion lowered the topography during the progression of the Karoo, but with notable disturbances occurring in the tectonically active zones which emerged in the Northern Kalahari as the beginning of the division of Gondwanaland approached at around 200 Ma.

2.5.2 Dwyka Group

The Dwyka Group sediments reflect late Carboniferous to Permian glaciation and subsequent deglaciation (Visser, 1983), though it should, however, be noted that the South African Committee for Stratigraphy (SACS, 1980) argues that there are no sound reasons for grouping the Dwyka tillites with overlying shales. The term Dwyka Group is nonetheless retained here due to its recent usage in a detailed study of the Karoo deposits beneath the Kalahari (Smith, 1984). Limited outcrops of glacial sediments were reported and then described from the southwestern Kalahari and Molopo Valley by Rogers (1907) and Du Toit (1916), they have subsequently been described in detail by, for example, Frakes and Crowell (1970) and analysed, with the additional advantage of data from borehole records, by Visser (1983).

The sedimentary sequence in southwestern Botswana (Smith, 1984) consists of massive basal tillites (the Malogong Formation: Figure 2.12), overlain by the Khuis Formation of pebbly mudstones and shales and the siltstones of the Middlepits Formation. The Dukwi Formation is found elsewhere at the base of the sub-Kalahari Karoo sequence, displaying an overall marked thinning to the northeast, and consisting of patchy basal tillite overlain by glaciofluvial deposits. Analysis of striations on exposures of basement rocks in the Molopo valley (Du Toit, 1916) and detailed fabric analyses of the tillites (Visser, 1983) have allowed the direction of ice movement and even a tentative glaciation history to be constructed for the southwestern Kalahari basin in early Dwyka times.

At its maximum extent a massive ice sheet covered this part of Gondwanaland, including the main Karoo Basin to the south, which was isostatically depressed below sea level. Deposition of the basal tillite, which is interdigitated with fluvial sandstones, did not occur until the break-up of the ice sheet allowed successive phases of ice movement to affect the basin in which deposition occurred, with ice transporting material derived from the surrounding highlands (Visser, 1983; Figure 2.13). Ice generally moved in a southwesterly direction, dominated by the Botswana Ice Lobe (Frakes and Crowell, 1970). The glacio-fluvial, glacio-lacustrine and mudstone deposits which form the overlying formations are generally deglaciation facies (Smith, 1984) which accumulated during the demise of the last, early Permian, Tses glaciation (Visser, 1983). In the Southwestern Basin, the Dwyka glacial deposits may have accumulated in shallow marine conditions due to isostatic depression of the land mass during glaciation; the upper Dwyka deposits in this area also include marine fossils (Heath, 1972). There is no evidence, however, that the marine incursions extended into the Central Kalahari Basin (Smith, 1984). Following the Dwyka glacial periods, the

BASIN / GROUPS	SOUTHWEST BOTSWANA	KWENENG AND WESTERN CENTRAL KALAHARI	MMAMABULA	MORUPULE + S.E. KALAHARI	NORTHEAST BOTSWANA AND NORTHERN BELT	NORTHWEST BOTSWANA
STORMBERG LAVA [1]		STORMBERG 105	LAVA 150	GROUP 285	(undivided) 375	BASALTS
LEBUNG [2]	NAKALATLOU SST. 60	NTANE 64	SANDSTONE 120	130	FORMATION 95	BODIBENG SST. FM. 60
	DONDONG FM.	MOSOLOTSANE 115	23	FORMATION 60	(north east only) NGWASHA FM. 80 / PANDAMATENGA FM.	SAVUTI FM.
TLHABALA FM (= Beaufort Series)	KULE FM.	KWETLA FM.	TLHABALA 41	90	FORMATION 92	?
ECCA	OTSHE FM. 220	BORITSE FM. ?	KOROTLO FM. 73	SEROWE FM. 50	TLAPANA FM. 120	MARAKWENA FM.
			MMAMABULA FM. 100	MORUPULE FM. 70		TALE FM.
		KWENENG FM.	MOSOMANE FM. 72	KAMOTAKA FM. 40	MEA ARKOSE FM. 25	?
	KOBE FM. 135	BORI FM. 135	SORI FM. 80	MAKORO FM. 108	TSWANE FM. 40	?
DWYKA	MIDDLEPITS FM. 73		DUKWI		FORMATION	?
	KHUIS FM. 200					
	MALOGONG FM. 175	258	166	37	25	

1 Drakensberg Group of others (e.g. Dingle *et al*., 1983)
2 Stormberg Group of others

90 = maximum depth encountered in boreholes

———— unconformity

– – – partial unconformity

Figure 2.12. Formations of the Karoo Sequence of the Botswana Kalahari. Nomenclature of Smith (1984).

overlying Karoo strata represent the infilling of a shallow basin (Lockett, 1979), first by waterlain deposits and finally by terrestrial sediments.

2.5.3 Ecca Group

The Ecca Group contains carbonaceous sediments of economic value which are today being exploited, not only in the well established coalfields of South Africa and the eastern Kalahari margin at Hwange in Zimbabwe, but also within the Kalahari itself. Although Clark-Lowes and Yeats (1977) and Jones (1982) considered Ecca sediments to be absent from the Southwestern Botswana Basin, Smith (1984) has demonstrated their presence, while they may also occur at the base of the Karoo sequence in the Okavango graben as the Ngami Beds, Ngwako Formation (Clark-Lowes and Yeats, 1977; Coates, 1980), or Tale and Marakwena Formations (Smith 1984), representing local lacustrine infilling of the subsiding graben.

Ecca sediments were deposited during the Permian in a shallow basin which may have been open to the sea in the west. Boocock and Van Straten (1962) divided the Ecca deposits beneath the central Kalahari into lower arenaceous and upper, dominantly argillaceous, phases. Though the division into upper and lower formations is also used by Smith (1984), the presence of argillaceous facies at the base of the lower formations in the Northeast and Central Kalahari Basins, and sandstones only at the bottom of the sequence in the Southwest Basin

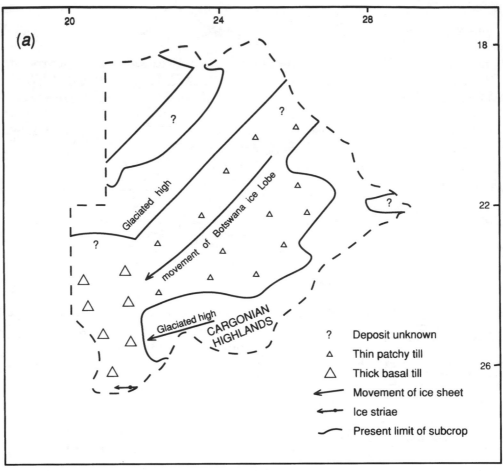

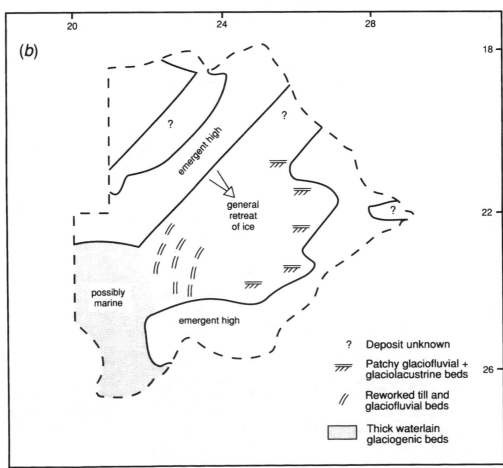

(Ncojane/Nossop sandstone member at the base of the Kobe siltstone formation) suggests that the division into upper and lower members would be better served by the descriptors Lower, Pre-carbonaceous and Upper Carbonaceous, Ecca formations are more appropriate. The sequence in western Zimbabwe, where the Upper and Lower Hwange sandstones bracket the major coal-bearing shales (Lockett, 1979), does not indicate a simple 'fining up' of Ecca deposits in this area either.

The lower Ecca sediments represent generally conformable deposition of shallow-water sediments upon the Dwyka Group lithologies. The earliest muddy deposits probably accumulated in shallow lacustrine conditions, though marine bivalve remains in the basal member of the Southwest Basin indicate that the 'lake' initially maintained the link with the sea which existed during Dwyka times. These mainly argillaceous sediments grade up into sandstone formations which have characteristics of deltaic deposition (Smith, 1984).

The upper carbonaceous Ecca formations generally represent a continuation of deltaic aggradation allowing the accumulation of peat on flood plains, under climatic conditions which were predominantly cool temperate or tundra-like (Clark-Lowes and Yeats, 1977). The thickness of the coal seams varies significantly within both the upper Ecca sediments of the central Kalahari (Smith, 1984) and the deposits to the northeast in western Zimbabwe (Watson, 1960; Lockett, 1979).

2.5.4 Tlhabala Formation and Beaufort Group

The Tlhabala Formation outcrops to the south of the Makgadikgadi basin in the Mosu Escarpment (Stansfield, 1973; Coates *et al.*, 1979), but is not clearly represented in borehole records from many parts of the Kalahari. The Permian–Triassic was a time of rapid faunal evolution, resulting in palaeontological zonation of Beaufort Group strata (Seeley, 1892; Du Toit, 1954; Dingle *et al.*, 1983), but the absence of reptilian remains within the Tlhabala Formation of the central Kalahari Karoo succession, and equivalent formations in other Botswana basins (Figure 2.12), makes its correlation with the Beaufort Group of the main Karoo basin very tentative (Smith, 1984). The abrupt transition from the carbonaceous Ecca sediments to the mudstones of the Tlhabala Formation, formerly the upper non-calcareous component of the Ecca Tlapana Formation of Stansfield (1973), has been interpreted as indicative of a significant environmental change in Karoo times from the preceding fluvial landscape dominated by floodplains and deltaic conditions to open-water lacustrine conditions (Smith, 1984) in which silts and muds accumulated.

2.5.5 Lebung or Stormberg Group

The Lebung Group of Botswana and its Stormberg equivalents elsewhere represent the final (mid- to late Triassic) phase of Karoo sedimentation prior to the episodes of igneous activity linked to the commencement of Gondwanaland's division. This group is well represented in the sediments of the Central Kalahari basin but is less well developed in the southwest. The sediments are important in that the transition from lower 'red bed' formations containing red to mauve

Figure 2.13 (opposite). Possible palaeoenvironments during times of (a) the deposition of the basal Dwyka till deposits and (b) subsequent glaciofluvial/ glaciomarine sediments. After Smith (1984).

siltstones, pebbly beds and sandstones (the Bushveld Mudstone of earlier authors) to upper red, pink or white sandstone beds with marked bedding structures is suggestive of a progression from predominant water deposition, under seasonal or ephemeral conditions, to one where, despite indications of alluvial fan deposition in some basin marginal areas, aeolian activity predominated (Stansfield, 1973; Smith, 1984).

These generally massive sandstones, widely referred to as the Cave or sometimes Bushveld Sandstone throughout southern Africa (e.g. Du Toit, 1954; Green, 1966), the Etjo Sandstone in Namibia, Forest Sandstone in Zimbabwe (e.g. Sutton, 1979), Ssake, Samkoto, Bodibeng or Ntante Sandstone in Botswana (Passarge, 1904; Wright, 1958; Boocock and Van Straten, 1962; Smith, 1984) and now called *in toto* the Clarens Formation (Beukes, 1970; SACS, 1980), are the 'most widely developed of all the Stormberg Group sediments in the extra Karoo areas' (Dingle *et al.*, 1983: p. 48). In the Kalahari they are well exposed to the south of Lake Ngami and where subsequent Karoo volcanics are absent, at the contact with the Kalahari Group sediments in parts of eastern Botswana and western Zimbabwe. Their frequent pink-red coloration, and that of underlying marls and shales, is often interpreted as indicative of warm conditions at the time of deposition, while the weathering of the upper sandstone to an unconsolidated sandy regolith make the determination of the boundary of the Kalahari Sand difficult in places.

2.5.6 Upper Karoo basalt

Triassic sedimentation in the Kalahari basin was ended by volcanic activity which resulted in the formation of the basalts that are found extensively beneath the Kalahari Group sediments in the eastern Middle and central Kalahari (Figure 2.11b). Potassium-argon dating of basalts from the Kalahari (Coates *et al.*, 1979) indicates that vulcanicity marked the commencement of the Jurassic, with dates clustered at 180 ± 10 Ma comparing well with those obtained from the main Karoo Basin (Smith, 1984).

The Drakensberg (or late Stormberg) basalts in Botswana have been interpreted as indicating five separate centres of outpouring (Green, 1966) of which four – central Botswana, Kazungula–Victoria Falls, Maitengwe River and Okavango – lie within or on the margins of the Kalahari. Surface exposures of all have been identified and in the case of the Kazungula and Maitengwe basalts are considerable, being exposed where the eastern Kalahari rift has retreated or been dissected in western Zimbabwe (Figure 2.14). Okavango basalts are represented by limited outcrops at the intersection between the rift and the Ghanzi Ridge, while central Botswana basalts are represented in the Orapa–Letlhakane area on the southern margins of the Makgadikgadi Basin.

Borehole records indicate considerable basalt thicknesses of up to 375 m (Smith, 1984). The nature of the basalt flows indicate quiet, fluid, subaerial eruptions which blanketed the irregular desert topography of the Lebung/Stormberg sedimentary facies. Individual subsequent eruptions were quite closely spaced as subaerial weathering on the surfaces of individual flows is relatively limited; the overall duration of the volcanic episode is estimated to have lasted no more than 20 Ma (Smith, 1984).

2.5.7 Late or post Karoo intrusions

Dolerite dykes have been identified in many parts of the Kalahari by field survey, aeromagnetic survey, borehole analysis and vegetation patterning (e.g. Green,

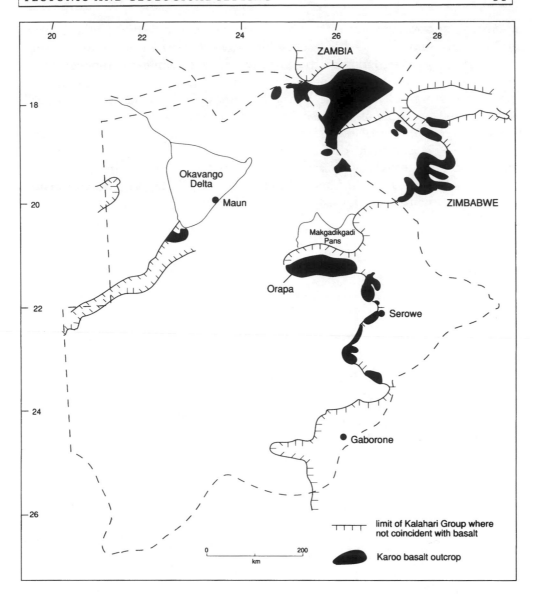

Figure 2.14. *Surface outcrops of Karoo (Drakensberg/Stormberg) basalt.*

1966; Reeves, 1978; Coates *et al.*, 1979; Mallick *et al.*, 1981), cutting across both Precambrian rocks and Karoo sediments. Sills have also been identified, primarily but not exclusively in the southwestern Kalahari (Gidskehaug, Creer and Mitchell, 1975; Figure 2.11*b*). The dykes occupy a main swarm, which has been equated to a failed Gondwanaland spreading axis by Reeves (1978), up to 100 km wide, which trends northwest–southeast from the extreme east of Botswana through the Okavango Delta. More dispersed forms occur, with the same general trend, to the south of this swarm, but not to the north of it (Mallick *et al.*, 1981; Figure 2.11*b*).

In southwestern Botswana dykes with more variable trends have also been identified. Major dolerite sills have also been mapped (Kingston *et al.*, 1961; Figure 2.11*b*), extending into the Keetmanshoop area of Namibia where they intrude into Dwyka sediments (Dingle *et al.*, 1983). The relationship of the dykes and sills to the main basalt bodies has proved debatable. Some authors have considered the dykes to have been feeders to the basalt bodies, giving them a late Karoo age, but a K/Ar date of 140 Ma (reported in Coates *et al.*, 1979) from a dyke in the Serowe region suggests that they are younger. Basalts near Mariental in central Namibia have been dated to the mid-Jurassic (161–173 Ma: Gidskehaug *et* *al.*, 1975), but dolerite sills further south at Keetmanshoop dated by the same authors fall within the 178–199 Ma bracket, suggesting late Karoo emplacement.

While the precise dating of these igneous events remains equivocal, their association with the break-up of Gondwanaland is not doubted. The emplacement of economically valuable kimberlite pipes, found beneath the eastern Kalahari and probably also present in the southwest (Baldock, Hepworth and Marengwa, 1976) is also associated with crust weaknesses linked to this event (Baldock, 1977). The main kimberlite pipe at Orapa has been dated to 93.1 Ma, and the crater created by its eruption provided suitable conditions for the preservation, in the following one million years, of a diverse range of Cretaceous insects and flora (Rayner and McKay, 1986).

3 Kalahari sediments

THE TECTONIC FRAMEWORK discussed in the previous chapter created the setting for the deposition of the extensive Kalahari sediments, which occupy the Mega Kalahari, as defined in Figure 1.2.

These sediments, now known as the Kalahari Group (SACS, 1980), first received serious geological consideration in the early years of this century with the publication of *Die Kalahari* (Passarge, 1904), the culmination of 2 years of rigorous fieldwork centred on Ngamiland in northern-central Botswana. Subsequent investigations have occurred more widely within the area occupied by the sediments: for example, in western Zimbabwe (e.g. Maufe, 1920; Bond, 1948), western Zambia (Money, 1972), Zaïre (e.g. Claeys, 1947; Cahen and Lepersonne, 1952) and in Namibia (e.g. Mabbutt, 1957), but undoubtedly the greatest volume of data has been derived from within Botswana.

A major limiting factor on investigations, noted by many authors (e.g. Du Toit, 1954: p. 456) has been the lack of natural exposures which reveal the stratigraphy of the Kalahari Group, especially its lower units. This is a consequence of the flat terrain of the Kalahari, its position within the downwarped African interior and the relative lack of surface drainage lines which might create natural sections. However, though there are few exposures within the Kalahari Desert, stratigraphical sections are exposed around its perimeter where river valleys do exist and where erosion has cut into the basin rim, creating small escarpments. These locations provided fruitful sources of information to earlier investigators such as H.B. Maufe, who examined the sequences exposed in the Kalahari's eastern margins between Bulawayo and Victoria Falls in Zimbabwe (Maufe, 1920, 1939), and Range (1912a, b), Wagner (1916) and Mabbutt (1957), who investigated the stratigraphy exposed in the Urinanib Escarpment, part of the Kalahari's southwestern margin in Namibia.

Given the virtual absence of exposures within the Kalahari Desert itself, it is somewhat paradoxical that the largest body of data, especially that gathered most recently, should have been acquired from within Botswana. This, however, is largely because the data have focused upon surficial and immediate sub-surface units, primarily the Kalahari Sand and duricrust suites. Valuable data have nevertheless been gained concerning the thickness of Kalahari deposits, utilising information collected from the sinking of boreholes for sub-surface water supplies and mineral prospecting in the hard-rock geology beneath the Kalahari.

3.1 Depositional setting, distribution and thickness

The structural history of southern Africa (Table 2.1) indicates that the Kalahari Group sediments are at their oldest of Cretaceous age, with local deposition,

53

reworking and post-depositional modifications continuing up to the present day. These extensive, exclusively continental, sediments have accumulated in the downwarped sub-continental interior of southern Africa, as outlined in Chapter 2. Some authors (e.g. Cahen and Lepersonne, 1952; Wright, 1978) have suggested that sedimentation occurred in a series of component basins within the interior, though Jones (1982) has disputed this. Nevertheless, it is apparent that the Ghanzi Ridge (Figure 2.6) does effectively divide the Kalahari Desert from more northern parts of the Kalahari physiographic region: it has possibly existed as an upstanding ridge since the earth movements responsible for adjacent continental rifting in Triassic times (e.g. Rust, 1975). Additionally, the tectonically controlled Okavango Rift and Makgadikgadi and Etosha depressions are essentially basins within the downwarp where sedimentation has been a response to local rather than continental-scale tectonic factors.

It is apparent today that erosion of some sections of the rim of the interior basin is reducing the spatial extent of the Kalahari sediments and creating outliers in some areas, for example in western Zimbabwe (Thomas and Shaw, 1988). It has even been proposed that the surface sand unit of the Kalahari Group was once continuous with the Namib Sand Sea in the west (Du Toit, 1954) and the Mozambique coastal plain in the east. Such a relationship would have entailed the Kalahari Sand extending outwards from the southern African interior, across the Great Escarpment in both easterly and westerly directions, blanketing the whole of southern Africa, approximately between latitudes 20° S and 26° S. Such a situation is highly improbable, especially given the structural and drainage evolution of the subcontinent. Instead, it seems more likely that the sands of the Namib Sand Sea and on the Mozambique Plain are a consequence of the accumulation and aeolian reworking of materials brought to the coast by rivers draining outwards from the interior, such as the Limpopo, Zambezi and Orange. In the case of the Namib, sedimentary and mineralogical studies by Lancaster and Ollier (1983) have indicated that much of the sand has been derived from material brought to the Atlantic Ocean by the Orange River system and transported northwards and onshore by tides and ocean currents. Additionally, smaller streams such as the Kuiseb and Tsauchab, flowing westwards from the Great Escarpment, are likely to have contributed locally to the sands of the Namib, particularly during times of increased effective precipitation (Besler and Marker, 1979; Besler, 1980).

Borehole stratigraphical records, supplemented by data from geophysical surveys and a limited number of other published sources have permitted the thickness of the Kalahari sediments to be ascertained for the area south of latitude 15° S (Thomas, 1988a; Figure 3.1). Given the flat surface of the Kalahari, variations in surface relief cannot explain the considerable range of thicknesses which are indicated. Instead, three factors account for them: the relief of the pre-Kalaharian topography which the sediments blanket; the existence of sub-basins within the overall interior downwarp; and the deposits thinning towards the rim of the original continental downwarp.

The Kalahari Group sediments are generally thicker to the north of the Ghanzi Ridge than to the south (Figure 3.1). The greatest thicknesses, in excess of 300 m, are attained in northern Namibia, including the Etosha basin, and in the Okavango graben, the latter, and possibly the former, being subjected to neotectonic subsidence (see section 2.3.2) since the initial continental downwarping occurred. The sub-Kalahari floor also locally exceeds 200 m in depth in the Southern Kalahari, in an area where Visser (1983) also identified an earlier, pre-Karoo centre of sedimentation.

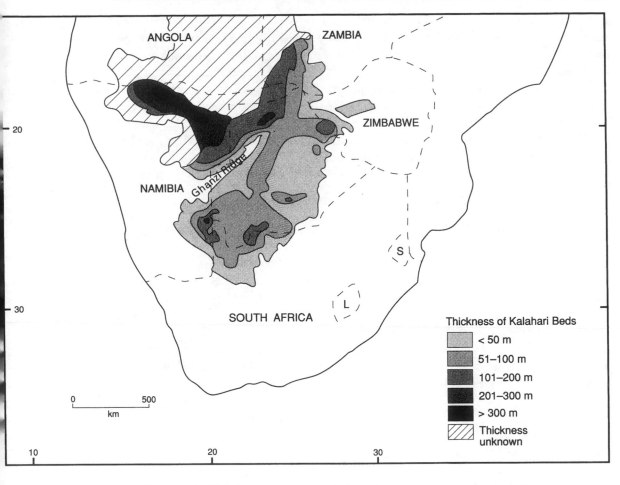

Figure 3.1. The thickness of the Kalahari Group sediments, constructed largely from borehole data. After Thomas (1988a).

3.2 Classification of the Kalahari Group

Despite the great thicknesses which the Kalahari sediments attain and their considerable spatial extent, only a limited number of lithological units and facies have been recognised, most of which were described by Siegfried Passarge in the early part of this century, and placed in a stratigraphical framework (Passarge, 1904). Although the German's investigations were spatially extensive, especially given the age in which he was working and the difficulties of travel which the Kalahari presents, the observations upon which he based his stratigraphy of the Kalahari sediments were restricted to the deposits of what is now northern Botswana, together with some brief comments concerning the deposits of the Victoria Falls area.

Subsequent investigations have added lithological and stratigraphical information from other locations within the considerable distribution of the Kalahari sediments, notably by Maufe (1920, 1939), Veatch (1935), Rogers (1936), Cahen and Lepersonne (1952) and Money (1972). One consequence of the seemingly limited lithological variations which these and other investigators described has been a preoccupation with establishing broad stratigraphical correlations of units from locations separated by considerable distances, though some notable exceptions have erred on the side of caution in this respect (e.g. Rogers, 1936; Boocock and Van Straten, 1962). For, as they and others such as Farr et al., (1981) have noted, there are marked vertical and lateral variations within the units into which the sediments are generally grouped, which is a common characteristic of terrestrially deposited sediments. These variations, however subtle, may be indicative of different ages and origins, as well as reworking and diagenetic replacement of the sediments, which may make some of the broad correlations and the palaeoenvironmental conclusions drawn from them of doubtful value, especially in the absence of satisfactory means of dating the sediments. In the remainder of this chapter we examine the issues concerning the classification of the Kalahari Group sediments, before examining the characteristics, distribution and significance of the various lithological units.

3.2.1 Passarge's classification

Passarge (1904) divided the sediments of the Kalahari into five groups, with lithological variations within them represented by a number of subdivisions or units (Figure 3.2). The lowest, and therefore oldest group in Passarge's succession was the 'Botletle Beds,' named after the seasonal channel (now Boteti) which links the Okavango Delta with the Makgadikgadi basin. The major characteristic of the units in this group is chalcedonic (silica) cementation, though Passarge (1901, 1904) also included laterites (iron cementation) in his *Botletleschichten*. The three main units identified by Passarge in this group were a basal transitional unit or conglomerate comprising the cemented weathering rubble of underlying, pre-Kalaharian, lithologies; a chalcedonic sandstone, and a pan sandstone which was frequently calcareous but displayed signs of silica replacement. The latter unit gained its name not because of a suggested environment of deposition, but because it was frequently exposed in the floors and margins of pan depressions. Passarge recorded that the Botletle Beds were frequently thin and often absent from the base of the Kalahari succession. Although he interpreted this as evidence that they had undergone erosion before overlying units were deposited, it was subsequently suggested that this might be a consequence of their deposition being local and sporadic rather than pan-Kalaharian.

Above the Botletle Beds the succession continued with the more widely distributed Kalahari Limestone (*Kalaharikalk*). The lower unit of this group was divided into calc-sinter, present in the hardveld (i.e. not overlain by Kalahari Sand) and, in the sandveld, poorly consolidated calcareous sandstone, in which the quartz grains were rarely in contact, floating in the $CaCO_3$ matrix. The upper, younger beds of the limestone group consisted of marls, frequently saline, and pan tufa or limestone. Passarge describes the younger beds as being frequently pierced by hollow tubes and also possessing inclusions of weathered chalcedony, from which an erosional hiatus was inferred between the deposition of the Botletle Beds and the Kalahari Limestone (Passarge, 1904: p. 600).

The three remaining groups in Passarge's scheme all consisted of unlithified deposits. The most extensive of these, the Kalahari Sand, was subdivided into four varieties largely on the basis of colour. It was also noted that in places the base of the sand contained fluviatile gravels which included pebbles inferred as being derived from the Botletle Beds and Kalahari Limestone. Above the Kalahari Sand, reworked surface or *Decksand*, mixed with rock fragments, and alluvial deposits (*Alluviale Bildungen*) of the Okavango–Zambezi swamp zone and from pans and channels in the sandveld, formed the final two groups of the stratigraphical sequence.

3.2.2 Stratigraphic successions from other locations

Passarge's (1904) stratigraphical succession provided the 'master framework' against which findings from other areas within the distribution of the Kalahari

Passarge (1904) Kalahari (Botswana)	Maufe (1939) Victoria Falls	Cahen & Lepersonne (1952) Zaïre	Money (1972) Zambia
ALLUVIALLE BILDUNGEN (Alluvium) 　1. In swamp zone 　2. In sandveld DECKSAND			ZAMBEZI FORMATION (3. Limestone and clays on pan floors) (2. Various interspersed duricrusts)
KALAHARI SAND (Divided into 4 subgroups) Basal river gravels	KALAHARI SAND CARSTONE NODULE BED	SABLE OCHRES	1. Mongu sand member (of Zambezi formation)
KALAHARI KALK (Limestone) 　2. Young marls and pan deposits 　1. Calc-sinter and calcareous sandstone	PIPE SANDSTONE		(UPPER BAROTSE FORMATION) (MIDDLE BAROTSE FORMATION)
BOTLETLESCHICTEN (Botletle Beds) 　3. Pan sandstone and sandy limestone 　2. Chalcedonic sandstone 　1. Cemented weathering debris of underlying geology	KALAHARI - CHALCEDONY	GRÈS POLYMORPHES Silicified sandstone Chalcedonic limestone	LOWER BAROTSE FORMATION 2. Cemented sandstone with pipes
		KAMINA SERIES Gravels and sandstones	1. Basal conglomerates

NOTES

Horizontal lines indicate implied correlation between areas	Alternative correlations from literature ( )　No direct correlations implied	Pipe sandstone has been correlated with the Gres Polymorphes, and therefore the Botletle Beds, by Cahen and Lepersonne (1952), but with the lower unit of the Kalaharikalk by Maufe (1939)

Figure 3.2. Passarge's (1904) Kalahari stratigraphical scheme, and attempted correlations from other areas.

sediments were commonly compared and correlated (Figure 3.2), with the latter being achieved through comparison of lithological characteristics (e.g. Maufe, 1939) or the continent-wide erosion surfaces upon which deposits supposedly rested (e.g. Cahen and Lepersonne, 1952; Mabbutt, 1957). Notwithstanding the problems bedevilling such bases for correlation (see below), it is of value to examine some of the major schemes produced for other areas.

Zimbabwe

Prior to the 1950s, the Kalahari beds had probably been no more thoroughly described than from parts of Zimbabwe (then Southern Rhodesia), especially from the northwestern part of the country and in the vicinity of Victoria Falls in particular. To some extent this was probably because the succession is fairly well displayed (in Kalaharian terms) in this well-dissected landscape, but it was also a consequence of the large volume of stratigraphical work which was conducted in conjunction with Stone Age archaeological investigations in the area in the 1930s and 1940s (e.g. Jones, 1944; Bond, 1946, 1948; Dixey, 1950). However, other investigations were purely geological in focus, with MacGregor (1916), Lamplugh (1902, 1907: whose underrated work included the first ever use of the term 'silcrete') and Maufe (1920, 1930, 1939) making notable contributions. Maufe's (1939) stratigraphical framework (Figure 3.2), derived from exposures at Victoria

Falls, summarises the ideas emanating from this area and the attempts to correlate them with Passarge's (1904) scheme.

Though the lowest unit of Maufe's sequence was the 'Kalahari Chalcedony', basal gravels and conglomerates have been identified at the bottom of the Kalahari sequence elsewhere in Zimbabwe (e.g. Cochran, 1969). The chalcedony, interpreted as silicified limestone by Lamplugh (1907) and Dixey (1950), was described as a 'greyish to yellow brown, frequently mottled, translucent to opaque, flint like rock' (Maufe, 1939: p. 212) occurring in tabular layers or nodular masses rarely exceeding 10 cm in thickness. Overlying the chalcedony, Maufe identified a unit up to 4 cm thick which he named the 'Pipe Sandstone', due to the numerous tubes penetrating the mass of the poorly cemented white, pink or red rock. Though silica cemented, Maufe correlated this unit with the Kalahari Limestone of Passarge (1904) and implied (Maufe, 1939: p. 219) that it too had undergone silica replacement of carbonate. He attributed the presence of pipes to deposition in 'a reed or sedge bed'.

Beneath the blanketing Kalahari Sand, but above the Pipe Sandstone, Maufe described a weathered layer of the latter containing pebbles and up to 2 m thick, and a carstone rubble bed, which can be interpreted as a pisolithic ferricrete, ranging from 10 to 60 cm in thickness. Both Maufe (1939) and Dixey (1950) noted that, apart from the presence of quartz and chalcedony pebbles, the pebbly sand was identical to the Kalahari Sand, while Dixey (1950) suggested that the siliceous component of the carstone layer was derived from the weathering of the Pipe Sandstone. Maufe correlated this unit with the gravels which Passarge reported at the base of the Kalahari Sand in some exposures in Botswana.

Zaïre and Angola

Maufe (1939) attempted to correlate his sequence from the Victoria Falls area with that of Veatch (1935) from the Belgian Congo (Zaïre) and Angola. Veatch's investigations had largely concentrated upon the post-Miocene surficial deposits (sands and river gravels), using their relation to erosion surfaces to link deposits from different locations. Subsequently, Cahen and Lepersonne (1952) used the same approach to classify the whole sequence of the 'Kalahari System' of the Belgian Congo, dividing the sediments into three 'Series'.

The lower Kamina Series, consisting of gravels and sandstones, was regarded as older than any of the units identified by Passarge (1904), being truncated by a Cretaceous land surface. The *Grès Polymorphes* (silicified sandstones and chalcedonic limestones) comprised the middle series and were correlated with the Botletle Beds, while the upper plateau sands or *Sables Ochres*, given a late Tertiary age, were seen as equivalent to the Kalahari Sand.

Zambia

Although Dixey (1950) gave some consideration to the Kalahari deposits on the northern side of the Zambezi near Victoria Falls, the most thorough assessment of the deposits in Zambia was provided by Money (1972), who utilised borehole stratigraphical logs as well as natural exposures in the Zambezi valley to produce his interpretation. This detailed scheme divided the sediments into the Barotse Formation, with lower, upper and middle divisions, overlain by the Zambezi Formation (Figures 3.2 and 3.3).

The lower Barotse Formation is comprised of patches of basal conglomerates containing Karoo basalt and agate pebbles, which Money (1972) correlated with the Kamina Series of Cahen and Lepersonne (1952) in Zaïre, overlain by a fine- to medium-grained sandstone, equated with the Botletle Beds and Pipe Sandstone. The middle Barotse represents units of ferruginous sandstones and quartzites

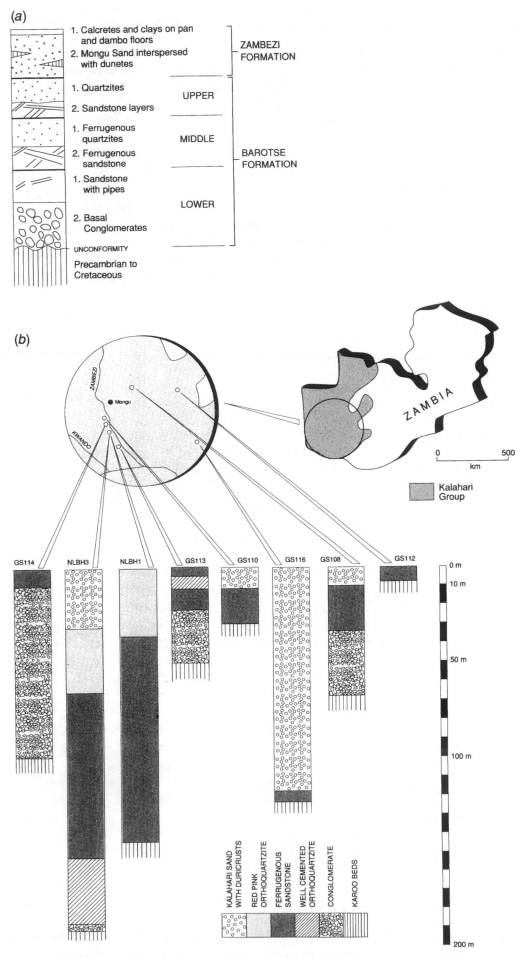

Figure 3.3. *Stratigraphical column for the Kalahari Group in western Zambia and local variations in stratigraphy. After Thomas (1988a) with additional data from Money (1972).*

displaying aeolian bedding, while the upper Barotse consists of units of massive pink quartzites, sandstones and feeble $CaCO_3$ cemented conglomerates displaying evidence of silica replacement. Money (1972) noted that as the upper units were mainly exposed in valley sides and were only poorly represented in borehole records, it was difficult to ascertain their stratigraphical position. Finally, the Zambezi Formation, though interspersed with duricrust units, primarily comprised the Mongu Sand Member, yet another name for the Kalahari Sand.

Southwestern Kalahari

Although Mabbutt (1957) examined the relationship of Kalahari deposits in Namaqualand (southeastern Namibia) to erosion sufaces, the most useful investigation from the southwestern area has been that of Smit (1977). His examination, employing the records of in excess of 2000 borehole logs, divided the Kalahari succession into four units: basal gravels and clays, frequently mixed as gravelly clays or clayey gravels; calcareous clay (marl), containing beds of sandstone, sands and gravels; calcareous sandstones, calcretes and silcretes; and aeolian (Kalahari) sand.

3.2.3 Problems of correlation, stratigraphy and age

It is clear from the preceding section that many investigators have regarded the Kalahari sediments as comprising a relatively clear-cut stratigraphical succession which can also be used to correlate the deposits from different locations within their total distribution. Figure 3.4 shows the schematic stratigraphical succession proposed by Dingle *et al.* (1983), after Mallick *et al.* (1981). Duricrusts and the Kalahari Sand are divided into 'older' and 'younger' units, while alluvial deposits are clearly regarded as younger than those of lacustrine origin. This succession is based on morphostratigraphical and lithostratigraphical criteria, for division of the Kalahari Sand is achieved according to its association with different dune systems (see section 3.4 and Chapter 6).

Some authorities, like Maufe (1939), made further deductions from the lithological characteristics of the sequence, interpreting it climatostratigraphically. Maufe considered the deposition of the Pipe Sandstone in a reed bed to indicate humid conditions, the subsequent replacement of the calcium carbonate by silica to be the product of arid conditions, as was the deposition of the Kalahari Sand. Debenham (1952) and Wayland (1953) built on Maufe's idea to suggest that silcrete in the Kalahari was everywhere the product of the replacement of calcium carbonate, implying a chronological interpretation for these deposits.

Others urged greater caution in interpreting the stratigraphy of the Kalahari. Although it can be argued that a general stratigraphical succession can be developed for the Kalahari Group, a major problem is that not all units are present everywhere, and it is impossible to assign a deposit to a stratigraphical position on lithological characteristics alone (Figures 3.3. and 3.5). As far back as 1936 Rogers noted that particular lithological characteristics were not confined to individual beds or horizons. This, for example, may have contributed to the term 'Botletle Beds' being used in a variety of ways not coincident with Passarge's (1904) original application. King (1947) suggested that the Botletle Beds may in fact be of multiple ages, while Du Toit (1954) employed the name to describe all the Kalahari beds beneath the surficial sand. By 1962 it was regarded as no more than a 'sack term' used to label silcretes, calcretes and sandstones (Boocock and Van Straten, 1962). Today the term is not used.

One of the major problems in the chronostratigraphical interpretation of lithological units of the Kalahari sediments is that the cemented sandstones,

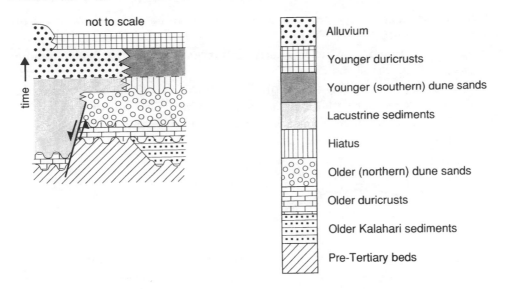

Figure 3.4. Schematic sedimentary succession for the Kalahari, as proposed by
Dingle et al. (1983).

chalcedony, quartzites and sandy limestones described by many authors (see
sections 3.2.1 and 3.2.2) are duricrusts, which are not necessarily the product of
primary deposition, but commonly develop through the diagenesis and
replacement of pre-existing sediments and consolidated rocks (see below).
Coupled with the fact that, for a variety of reasons, it is extremely difficult to
determine the age of duricrusts, and that these processes can occur on a very local
as well as regional scale, the chronological interpretation of lithological units
within the Kalahari sediments becomes extremely hazardous. Reflecting this, and
in an attempt to remove the chronostratigraphic connotations which it implies,
Boocock and Van Straten (1962) preferred the term 'Kalahari Beds' to Roger's
(1936) 'Kalahari System'. More recently, following usage recommended by the
International Subcommission on Stratigraphic Terminology, the alternative
'Kalahari Group' has been used (SACS, 1980).

Recognising the problems of employing lithological characteristics in relating
the Kalahari Group deposits of different areas, King (1947) noted that correlation
might best be achieved using the erosion surfaces upon which different units
rested. This method was employed by, amongst others, Cahen and Lepersonne
(1952) and Mabbutt (1957), but given advances in knowledge of the tectonic
evolution of southern Africa (Chapter 2), this approach to inter-regional
correlation is now also regarded as essentially untenable (DeSwardt and Bennet,
1974; Thomas, 1988a). Correlation is not assisted by the absence of diagnostic
fossils. Although gastropod shells and Chara seed fossils were reported from the
Pipe Sandstone by Maufe (1939) and from silicified limestone (Veatch, 1935),
such remains have made a valuable contribution only to studies of surface
lacustrine deposits of relatively recent origin (e.g. Shaw, 1985a). Although the
onset of sedimentation in the Cretaceous has been confirmed by the recent
discovery of fossil wood from this period in the basal beds in the Southern
Kalahari (T. Partridge, pers. comm.), the paucity of the fossil record between the
Cretaceous and the period covered by archaeological remains and radiocarbon
dating creates a chronostratigraphic gap the length of the Tertiary. Within this
timespan a number of questions remain to be answered, including the spatial and
temporal continuity of sedimentation within the Kalahari as a whole. The
assumption that the bulk of the Kalahari Group is of Tertiary age has not, as yet,
been proved in any respect.

A third approach to the correlation problem may be the identification of
unconformities within the Kalahari Group, rather than correlation of the units
themselves. Recent work on the Kalahari overburden at the Orapa diamond mine
(N. Lock, pers. comm.) indicates three unconformities in the Kalahari Sand,
including some palaeosol development. The application of this technique over
short distances, together with detailed lithological investigations of some units,
particularly the upper members and duricrusts, may be the best route to an
understanding of the regional sequence as a whole.

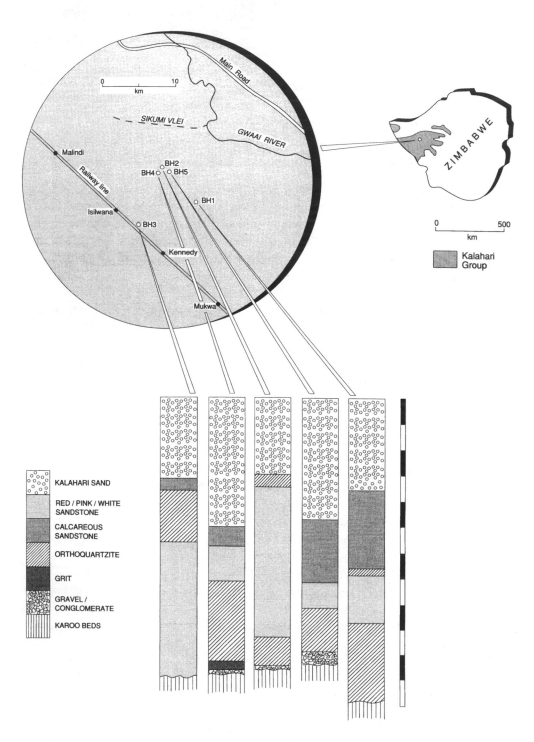

Figure 3.5. Variations in Kalahari Group stratigraphy in the Kennedy area,
western Zimbabwe (Thomas, 1984b; data from Cochran, 1969).

3.2.4 Post depositional modifications

Chemical activity within the Kalahari Group sediments is greatly favoured by a number of environmental factors, including long-term tectonic stability throughout most of the region, climatic variations around a semi-arid mean, endoreic surface and sub-surface drainage, an active groundwater regime, and the availability of silica, calcium and low-valency solutes. The chemical modification of sediments has led to the formation and further alteration of a suite of duricrusts, mostly in the calcrete–silcrete spectrum, which are among the most extensive in the world; calcrete profiles of 60 m have been reported from the Molopo valley (Goudie, 1973) and of 100 m at Khakaea in the Southern Kalahari (Bruno, 1985).

The processes of alteration are complex, and include gradual diagenesis of the Kalahari strata, chemical weathering of underlying rock, particularly the removal in solution of base elements, the addition of aeolian silica and calcium, and the continual precipitation and re-solution of calcium carbonate and silica under a wide pH range. Although much of the duricrust is pedogenic and has formed within the range of soil moisture fluctuation, others have formed, and continue to form, as a result of ground water activity. As groundwater is inconsistent in both quality and distribution (section 3.7), and is controlled by the geohydrological properties of the host rock, particularly in the determination of flowpaths, and the incidence of both regional and local recharge events, there is considerable variation in the extent of modification. As already noted, the assumption that duricrusts have stratigraphic context (e.g. Heine, 1978a) has led to problems with correlation. Likewise, the assumption that the duricrusts are of Tertiary age (e.g. Netterberg, 1969a; Goudie, 1973) is likely to be challenged as local and regional variations are distinguished. The nature and distribution of the duricrust suite is examined in section 3.5.

3.3 Lithologies of the Kalahari Group

Six major lithological components can be identified within the Kalahari Group: conglomerate and gravel; marl; sandstone; alluvium and lacustrine deposits; Kalahari Sand; and duricrusts. Additionally, several more minor units can be recognised. Of all the units, the Kalahari Sand and the duricrusts (silcrete, calcrete and ferricrete) are most significant in terms of their extent, the importance attached to them in environmental reconstructions, and what is known about them: to this end, each is dealt with in a separate, later section. In this section we examine each of the other major units.

3.3.1 Conglomerate and gravel

Conglomerates and gravels of varying characteristics are found sporadically at the base of the Kalahari Group throughout its distribution (Figure 3.6a), in a clay (Smit, 1977), calcrete (e.g. Rogers, 1936; Mallick et al., 1981) or silcrete (Shaw and De Vries, 1988) matrix. Although these basal deposits range widely in composition, thickness and degree of cementation, their widespread distribution to some extent reflects the importance of the post-Gondwanaland endoreic drainage systems (Smit, 1977; Thomas, 1988a; Thomas and Shaw, 1990). Gravels were also placed higher up Passarge's (1904) Kalahari succession, at the base of the Kalahari Sand, and they have also been widely reported from valley terrace sequences in a variety of relationships with other lithological units of the Kalahari Group (e.g. Dixey, 1950; Thomas and Shaw, 1988).

(a)

(b)

(c)

(d)

Figure 3.6. Examples of lithologies within the Kalahari Group sediments.

(a) Horizontally bedded Karoo sandstones near Letlhakeng, southeast Kalahari, are overlain by a thin veneer of basal Kalahari conglomerate (foreground). In this location, one of the few exposures of this unit within the Kalahari, the conglomerate matrix has largely been removed by weathering resulting in the presence of an unconsolidated gravel spread. At adjacent sites the conglomerate matrix is calcretised or silcretised.

(b) Massive sil-calcrete derived from Karoo siltstone exposed as a cliff at the side of one of the Letlhakeng dry valleys. The free face is about 3 m high and shows clear signs of karstic micro-features.

(c) Diatomaceous earths exposed in the banks of the Boteti River at Moremaoto. They were probably deposited in still-water conditions as water ponded to the west of the Gidikwe Ridge during the 920 m asl lacustrine phase in the Makgadikgadi Basin (see section 5.3.4).

(d) Recent saline silts and clays on the surface of Sua Pan in the Makgadikgadi Basin. These sediments are subject to frequent cycles of inundation and desiccation.

(e) Laminar to massive calcrete beneath Kalahari Sand, Matetsi area, western Zimbabwe.

(f) Pisolithic ferricrete containing some quartz pebbles, Gokwe area, central Zimbabwe.

The best developed basal deposits have been reported from the southwestern area of the Mega Kalahari. Both Du Toit (1954) and Mabbutt (1955) described the deposits exposed in the Urananib Escarpment in southeastern Namibia, where they attain a thickness of up to 90 m (Du Toit, 1954; Figure 3.7). Mabbutt (1955) noted that, at their base, a coarse conglomerate rested on deeply weathered Karoo sandstone, with the gravels themselves being interbedded with sands and displaying evidence of current bedding. Du Toit (1954: p. 458) regarded the basal grits and gravels to be an integral component of the base of the 'Kalahari limestone'; although this is, in fact, a calcrete, he considered the whole grit–gravel–limestone unit to be of fluviatile origin. Smit's (1977) study of deposits of the northern Cape Province, South Africa, found the gravels often to be mixed with clays and to reach a thickness of 100 m, and in one instance 180 m, infilling old drainage lines (Smit, 1977; Tankard et al. 1982).

Most of the occurrences reported from elsewhere are of thin deposits of only a few centimetres or tens of centimetres, for example in Zambia (Money, 1972) and in Zimbabwe (Cochran, 1969), though a basal gravel deposit of almost 100 m was found in a borehole in the vicinity of the Zambezi River (Money, 1972; Figure 3.3).

The lithologies of the gravels reflect local and regional geology. In the Urananib Plateau deposits the gravels comprise granite, gneiss, quartz, Nama Sandstone quartzite and Fish River Sandstone pebbles (Range, 1912a; Du Toit, 1954); in the lower Molopo area local Dwyka tillite pebbles are present (Rogers, 1936). In Zambia, basalt and agate pebbles form an important component of the basal conglomerate (Money, 1972), while the coarse current-bedded fluvial gravels identified by Clark (1950) beneath the alluvium of the former southwesterly course of the Kafue (see section 2.3.5) may also be inferred as basal Kalahari gravels.

Little is known of the basal gravels and conglomerates from Botswana. Although Passarge (1904) mentioned their occurrence from the Middle Kalahari, Boocock and Van Straten (1962: p. 134) considered conglomerates to be only a minor component of their Kalahari succession. However, boreholes sunk during the course of the Kalatraverse One project (Coates et al., 1979) revealed between 4 and 40 m of basal Kalahari sediments that included units of unspecified sandstone conglomerates in the area to the southwest of the Makgadikgadi basin. Further south, Mallick et al. (1981) mention gravels in the Kweneng and Letlhakeng areas; Shaw and De Vries (1988) have described these deposits as beds of rounded quartz, quartzite and jasper pebbles of Precambrian Waterberg and Karoo lithologies at the base of the Kalahari Group.

3.3.2 Marl

Marls are an important component of the Kalahari Group in southwestern areas of the Mega Kalahari, but in some areas they may simply be the weathering remnants of underlying Karoo beds (Farr et al., 1981). Boocock and Van Straten (1962), despite their limited consideration of basal gravels and conglomerates, implied that marls were the most important basal unit of the Kalahari succession. However, marls are found both overlying such deposits (as 'red calcareous clay' in Smit's (1977) succession) and resting directly 'in hollows on the pre-Kalahari surface' (Boocock and Van Straten, 1962: p. 153), even beneath Kalahari gravels, as pointed out by Du Toit (1907, 1954). Although impersistent in their distribution, the pink to red marls described by both Rogers (1936) and Boocock and Van Straten (1962) can reach 65 m in thickness, and those from the northern Cape reach 100 m (Smit, 1977). The Kalahari marl is generally fine-grained, homogeneous and without stratification (Du Toit, 1954) and while commonly

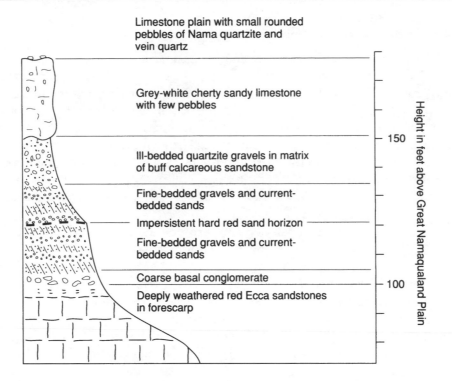

Figure 3.7. Kalahari Group gravels and conglomerates, Urinanab Plateau, Namibia. After Mabbutt (1957).

occurring as one bed, it has been noted to inter-digitate with one or more silcrete layers in some locations.

Records of marls from the central Kalahari are sparse – they are mentioned only briefly by Boocock and Van Straten (1962) – nor have they been reported from Zimbabwe, Zaïre or Angola. They were not included in Money's (1972) Zambian succession, but more recently 22 m of red to dark red marls were encountered by Martinelli and associates (1980) at the base of the Kalahari Group near Mongu, western Zambia. They are also present in the Grootfontein area of northern Namibia (Thomas, 1988a), while marl can additionally form as local deposits within pan depressions (e.g. Mallick et al., 1981).

3.3.3 Sandstone

Sandstones of varying characteristics have been described from throughout the distribution of the Kalahari Group, as discussed in sections 3.2.1 and 3.2.2. Many of these deposits are, however, better considered as duricrusts (see section 3.5), especially silcrete, being the result of diagenesis and cementation of pre-existing unconsolidated deposits. Summerfield (1978) for example, recognised that the Pipe Sandstone of Maufe (1939) and others, and the various quartzites which appear in the Kalahari literature, were in fact silcretes, developed through 'physico-chemical processes active in the zone of weathering' and not the deep burial sedimentary processes which are responsible for the development of sandstone proper.

There are, however, one or two instances where it is conceivable that sandstones *per se* are being described, with indications being provided of their initial origin. Perhaps most significant is Money's (1972) account of his Middle Barotse formation, where both the lower ferruginous sandstones and upper ferruginous quartzites are described as possessing well developed, steeply dipping cross strata taken as indicative of an aeolian origin.

3.3.4 Alluvium and lacustrine deposits

Alluvial and lacustrine deposits are important and spatially extensive surface members of the Kalahari Group within the context of the Okavango–Zambezi swamp zone and the lake basins of the Middle Kalahari, including the Makgadikgadi, Ngami and Mababe depressions (see Chapter 5). Locally, alluvium occurs on the floors of some of the larger 'dry' valley systems present in the central Kalahari, notably the Mmone, and is, of course, found in conjunction with the perennial rivers of the Northern Kalahari. The deposits on the floors of numerous pan depressions (Chapter 6), particularly in southern Botswana and western Zambia, also include sediments which can be regarded to be of lacustrine origin.

The most extensive alluvial deposits are found in association with the Okavango Delta. Due to its setting in a subsiding graben (section 2.2.3), the Okavango alluvium is up to 300 m thick (Hutchins *et al.* 1976a); laterally, the deposits extend beyond the limits of the current distributary system (Figures 2.5 and 5.1) and merge with those associated with the Chobe and Zambezi rivers to the east (Figure 5.2). A second, but significantly smaller, suite of alluvial fan deposits is associated with the Nata River system which enters the Makgadikgadi depression from Zimbabwe in the east: these, too, have been affected by fault control (Mallick *et al.*, 1981).

While both alluvial and lake deposits largely comprise silts and clays, those of lacustrine origin also contain evaporites, notably in the Makgadikgadi basin and in some cases significant admixtures of Kalahari Sand (Thomas, 1987a), due to reworking processes and the introduction of material along aeolian pathways. Diatomaceous deposits, duricrusts and shell beds are also found as components of Kalahari alluvial and lacustrine deposits, the last two particularly in conjunction with strandline features (sections 3.5 and 3.6 and Chapter 5).

3.4 Kalahari Sand

Because of its enormous spatial extent (section 1.2.2) and its surface location, the unconsolidated Kalahari Sand is perhaps the most studied unit of the Kalahari Group, yet in many respects it remains as poorly understood and as controversial as the others. The origin, mode of deposition, age and environmental significance have all remained elusive and have been debated only in terms of a general explanation, yet if the information which is available is applied at the appropriate scale, then much of value is in fact known. Correct interpretations are not only based on sedimentological grounds, considered below, but also through the context of the morphological expressions in which the sand is found, especially the extensive dune systems, which are considered in Chapter 6.

The term Kalahari Sand is not applied to a homogeneous deposit but to one which varies markedly in colour, composition, thickness and possibly even age. Towards the fringes of its distribution it becomes mixed with and indistinguishable from the weathering products of underlying Karoo and Precambrian rocks (Baillieul, 1975). In other locations, such as in the southwestern Kalahari, it appears as a distinct deposit, 20 to 30 m thick, of aeolian origin because of its landform association (Thomas, 1984a) and grain size characteristics (Lancaster, 1986a), resting on an extensive limestone (calcrete) surface which has sometimes been interpreted as mantling an old landsurface (Mabbutt, 1955). A further contrast is proved by the Kalahari Sand in northern Namibia, where between 200 and 300 metres of sand lie uninterrupted on lower Kalahari gravels and marl or directly on Karoo bedrock (Thomas, 1988a).

One of the most often mentioned attributes of the Kalahari Sand has been its colour. It has commonly been described as red (e.g. Maufe, 1930; Bond, 1957; Cooke, 1957; Lancaster, 1976) though it is in fact frequently ocherous (Cahen and Lepersonne, 1952; Cooke, 1957) and surface layers are sometimes bleached (Wright, 1978), particularly where the sand has been water worked (Baillieul, 1975). Sand reddening is the result of individual sand particles being coated by a pellicle of chemically precipitated iron oxide, resulting from the weathering of iron-rich minerals within the sand matrix (e.g. Norris, 1969; Walker, 1979; Gardner, 1981); in the case of dune sand, its presence has been interpreted as evidence of sand stability since the iron coating is easily removed during transport. However, as the presence of sand reddening is affected by many variables, including subsurface lithologies (Mallick *et al.*, 1981), climate, mineral availability and mobilisation, and there are notable local scale variations in the colour of the Kalahari Sand (Thomas, 1984*b*), little of environmental significance can currently be gained from attempting to interpret the regional scale colour variations which are implied by Poldervaart (1957) and Petrov (1976).

3.4.1 Provenance and age of the Kalahari Sand

A number of authors have conducted sedimentological (mineralogy, particle size and particle shape) studies of Kalahari Sand, largely, but not exclusively, from within Botswana, Zimbabwe, Zambia and parts of the Southern Kalahari dunefield in South Africa. Broadly, mineralogical studies can yield data concerning the origin of material, and particle size and shape can give information concerning transport and depositional processes, though as these characteristics can also be inherited from parent materials their interpretation can be complicated.

The Kalahari Sand is largely composed of quartz grains (90 per cent or more by weight) with accessory heavy minerals. MacGregor (1947) considered this material to have come from a 'vast seafloor sand area exposed during the upper Pleistocene ice age . . . blown by trade winds from the east'. Such a dramatic and distant origin does not accord well with our understanding of the tectonic history of southern Africa, and more realistic interpretations have been derived through analysis of the non-quartz fraction of the sand.

Poldervaart (1957) suggested that the Botswana Kalahari Sand was derived from a westerly direction as both heavy minerals (especially kyanite) and mean particle size decreased in a southeasterly direction. Later analysis by Baillieul (1975) refuted this idea, painting a more complex picture and dividing the sand into four zones based on mineralogical and grain size composition (Figure 3.8). While noting the importance of wind-transported sand in the northern half of the country, he regarded *in situ* weathering of sub-surface Karoo bedrock (also noted by Boocock and Van Straten, 1962), and the agency of bioturbation, to have made significant contributions to the sand in other areas.

Investigations from Zambia by Trapnell and Clothier (1957), Savory (1965), Verboom (1974) and Musonda (1987) have also proposed that locally derived weathering deposits, together with detrital material carried along aeolian and fluvial pathways, account for the origin of the Kalahari Sand in the Western Province. In Zimbabwe, analysis of staurolite mineral particle sizes led Bond (1948) to suggest that Kalahari Sand has been derived from rock outcrops in the northwest of the country. Subsequent investigations (Sutton, 1979; Lockett, 1979; Thomas, 1984*b*), which employed analysis of the total heavy mineral suites from Kalahari Sand and Karoo and Cretaceous rocks in the vicinity, point to much of the sand being of local origin with additions being derived from sources in an easterly direction.

Though of limited detail, the overall picture which has been built up from

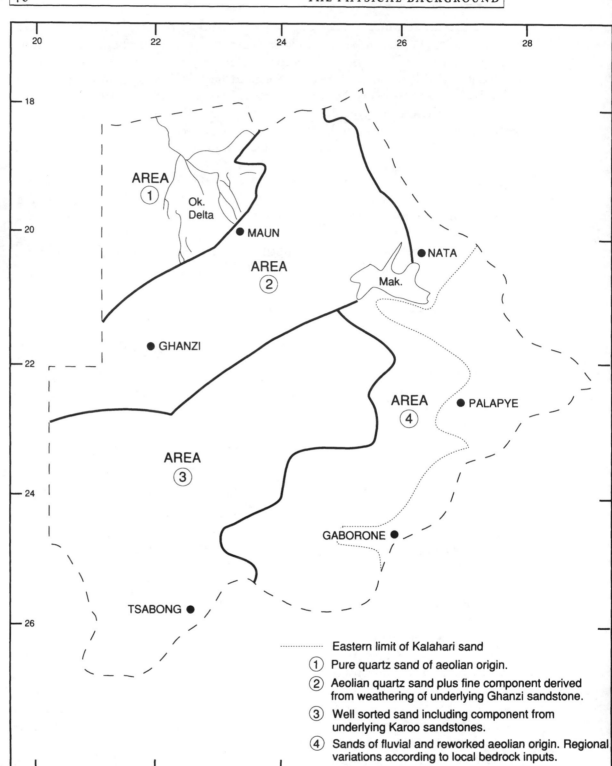

Figure 3.8. Surface sand types in Botswana, as defined by Baillieul (1975).

mineralogical studies is therefore that the Kalahari Sand represents the accumulation of material derived from local sources, including accumulated *in situ* weathering products of pre-Kalahari lithologies, supplemented by material transported to the interior over relatively short distances in a manner which is consistent with endoreic depositional basin infilling (Chapter 2).

A number of authors have attempted to provide an age for the Kalahari Sand. Cahen and Lepersonne (1952) gave the sand a Miocene age, Maufe (1939) and

King (1962) assigned it to the late Pliocene or early Pleistocene, Mabbutt (1957) to the Pliocene and Bond (1948) and Dixey (1956a) to the mid-Pleistocene. Wayland (1954) used archaeological grounds to date the 'true' Kalahari Sand in Botswana as post-Acheulean, Acheulean being the later part of the Early Stone Age.

A number of authors have recognised that the sand has been reworked on a number of occasions (e.g. Dixey, 1938), and attempts have been made to distinguish reworked sand from 'true' Kalahari Sand. Poldervaart (1957: p. 114) noted that 'there are Kalahari sands and there are Kalahari sands, their ages extending into recent times'. Baillieul (1975) felt that Tertiary sand found between datable cave deposits in the Gcwihaba (Kwihabe) Hills in northwestern Botswana might be significant in dating the original Kalahari Sand, though he added that it was unknown whether these were simply a local deposit or a remnant of that subsequently redistributed over the Kalahari region. Cooke (1975a) has, however, argued that the cave deposits are not stratigraphically significant and are distinguishable from the sand found beyond the cave only because of the addition of cave detrital material.

Overall, we are no clearer today in determining the age of the Kalahari Sand than were earlier workers. However, it is doubtful whether the concept of an 'original' age is of any consequence given that the sand comprises both *in situ* weathering and transported components. Conversely, an understanding of the mechanisms and ages of sand reworking, through its construction into definable landforms, is of more significance in terms of Kalahari environments and history (see subsequent chapters).

3.4.2 Sedimentary characteristics

The sedimentological characteristics of the Kalahari Sand have been described by many authors and from many locations within the total distribution. Table 3.1 gives examples of summary grain size data, expressed on the logarithmic phi (Φ) scale (Folk and Ward, 1957), for samples taken from Baillieul's (1975) four sand types in Botswana, showing samples designated as aeolian and those considered to be derived from *in situ* bedrock weathering, while Table 3.2 gives values obtained from Kalahari Sand samples collected from a number of different landform associations (Thomas, 1987a).

Aeolian characteristics have been widely cited for the Kalahari Sand. It has frequently been described as fine to medium with rounded to sub-rounded grains (e.g. Bond, 1948; Savory, 1965), the larger grains commonly being better rounded and more spherical than those which are smaller (Thomas, 1987b), while the total cumulative grain size distribution may be comparable with that from desert dune fields in other localities (e.g. Binda and Hindred, 1973). In the southwestern Kalahari, the sand is found in association with a distinct dune system and displays evidence of fining and increased sorting in the direction of sand transport (Lancaster, 1986a), inputs of material from local sources within the system (Thomas and Martin, 1987), as well as small-scale sedimentary variations due to the processes operating over individual dunes.

Despite the existence of sedimentological studies which demonstrate the importance of aeolian activity in the 'history' of the Kalahari Sand (e.g. Flint and Bond, 1968; Binda and Hindred, 1973; Baillieul, 1975), both locally and regionally, this must not be taken as evidence for an aeolian origin for this material. In some cases, the sand may have aeolian attributes because that was the origin of the parent Karoo rock from which it was weathered (Wright, 1978), whereas in others it may be the consequence of aeolian reworking rather than initial deposition. That there has been more than one episode of aeolian activity

Table 3.1. *Typical Folk and Ward (1957) descriptive parameters for Kalahari Sand samples*

Area*	Sample	Mean	Sorting	Skewness	Kurtosis
			Expressed in phi (ϕ) values**		
I	From dunefield	2.06	0.75	0.002	0.05
		2.23	0.76	0.16	0.55
	Reworked sand	2.27	0.43	−0.09	0.52
		2.19	0.81	0.08	0.49
	From swamp area	2.53	0.60	−0.05	0.34
II	Unspecified	2.80	0.61	−0.09	0.62
		2.33	0.68	0.04	0.54
		2.66	0.76	−0.01	0.64
		2.49	0.73	0.12	0.55
III	Unspecified	2.57	0.69	−0.04	0.40
		2.38	0.76	0.08	0.45
		2.33	0.74	−0.16	0.49
		2.36	0.61	−0.15	0.49
IV	Over basalt	2.36	1.00	−0.23	0.49
	Over Ecca S'stone	1.86	1.22	−0.07	0.46
	Over Cave S'stone	2.74	0.68	0.21	0.53

After Baillieul (1975).
* From Baillieul (1975), see Figure 3.8.
** The phi scale is an inverse log scale, where 0.0 $\phi = 1.0$ mm, 1.0 $\phi = 0.5$ mm, 2.0 $\phi = 0.25$ mm, 3.0 $\phi = 0.125$ mm, etc.

affecting the Kalahari Sand is clear from morphological studies of the Kalahari dune systems (see Chapters 6 and 7).

The complex and polygenetic history of the Kalahari Sand has been demonstrated by a study of the total sedimentary characteristics (mineralogy, grain size, grain shape and individual particle surface textures) of Kalahari Sand samples found in conjunction with a range of depositional landforms (Thomas, 1987*b*). In this study sedimentological characteristics were not sufficiently distinct within any particular landform group to allow its statistical differentiation from the other groups. However, scanning electron microscope (SEM) studies (see Krinsley and Doornkamp, 1973; Bull, 1978) did provide a means for group separation, though particles from all landform categories possessed evidence of being affected by wind activity, suggesting the great importance of aeolian processes in the environmental history and reworking of the Kalahari Sand.

3.5 Duricrusts: calcrete, silcrete and ferricrete

A *duricrust* has been defined (Goudie, 1973: p. 5) as follows:
A product of terrestrial processes within the zone of weathering in which either iron and aluminum sesquioxides (ferricrete, alcrete), or silica (silcrete), or calcium carbonate (calcrete) or other compounds have dominantly accumulated in and/or replaced a pre-existing soil, rock, or weathered material, to give a substance which may ultimately develop into an indurated mass.

Such materials have a widespread distribution in tropical and sub-tropical landscapes, and have received considerable scientific attention over the past 20 years. Major texts which discuss their genesis, distribution, form, classification and impact on the landscape include Goudie (1973), McFarlane (1976), and Goudie and Pye (1983). In the context of the Kalahari there have been

Table 3.2. *Characteristics of Kalahari Sand samples from different landform associations*

Landform element	n	% of sample in size category sand grade*					Graphical moment (ϕ) (Folk and Ward method)			
		Coarse	Med.	Fine	Silt	Clay	Mean	Sort	Skew	Kurt
Dune crest	46	7.87	35.78	50.78	4.59	1.04	2.13	1.01	0.16	1.18
Interdune	25	9.81	30.64	45.08	11.42	3.52	2.43	1.31	0.15	1.53
Pan	21	10.79	27.07	33.94	22.36	6.42	3.09	2.49	0.51	1.12
Palaeolake	18	13.19	25.45	32.59	19.60	10.40	3.32	2.46	0.48	1.16
Fossil channel	21	7.60	30.90	43.63	14.71	2.24	2.58	1.72	0.44	1.40

After Thomas (1987a).
* Particle diameters: coarse sand $= -1.0$ to 1.0 ϕ; medium sand $= > 1.0$ to 2.0 ϕ; fine sand $= > 2.0$ to 4.0 ϕ.

innumerable references to duricrusts since Passarge's (1904) observations, but few studies of duricrusts *per se*; the most comprehensive include Goudie (1971), Netterberg (1969a, 1978, 1980) and Watts (1980) on calcretes, and Smale (1973) and Summerfield (1982, 1983a) on silcretes.

As already noted (section 3.2.4) conditions within the Kalahari favour the accumulation of duricrust suites (Figure 3.6b, e and f). Such suites are not only well developed, but also extremely complex. In regional terms the distribution of duricrust types follow the precipitation gradient (Goudie, 1973), with ferricretes in the Northern Kalahari, and silcretes and calcretes dominating the Kalahari core. However, abrupt boundaries in the duricrust distribution, especially along the Kalahari–Limpopo divide, together with the co-existence of all three major types at some localities, confirm that site conditions are more important in their formation than gross climatic parameters. The assumption that specific duricrust types are representative of certain palaeoclimatic conditions, in particular the postulated relationship between calcrete and silcrete and semi-aridity (e.g. Heine, 1979, 1982), has been critically evaluated by Summerfield (1983a) and Rust, Schmidt and Dietz (1984), and found wanting.

The range of duricrust types is probably also greater in the Kalahari than elsewhere. Much difficulty is experienced in identifying and distinguishing between calcrete and silcrete without chemical or petrological analysis. This has led to much confusion in field identification, the more so in early reports; the substitution of the terms calcrete for silcrete, sandstone for silcrete and limestone for calcrete are common errors (e.g. Passarge, 1904).

The problem is compounded by the existence of a large range of intermediate forms, termed cal-silcretes and sil-calcretes dependent on their dominant constituent. Recognition of this variety led Mazor *et al.* (1977) to propose the use of the term 'crete' to cover the full range of duricrust in the Orapa area. Detailed chemical studies (e.g. Gwosdz and Modisi, 1983; Figure 3.9) also revealed considerable variation of duricrust profiles over short horizontal and vertical distances. Other forms which are known to occur include ferruginous calcretes and ferruginous silcretes (e.g. the Letlhakeng valley) and a range of ferricrete–silcrete–calcrete conglomerates, in which one material forms a matrix for pebbles of another, as in the Mpandamatenga and Nata areas.

Duricrusts, because of their indurated nature, are usually assumed to form prominent features in the landscape. In erosional landscapes this is generally true, but in a depositional setting, such as the Kalahari, such resistant features are encountered only where the duricrust has been exposed to sub-aerial processes, as in fossil valleys (Figure 3.6b), escarpments, pan rims and surface pavements. Although these probably represent the most prominent and significant duricrust landforms, in which the main tendency is towards gradual replacement of $CaCO_3$ by silica, duricrusts are common throughout the Kalahari Group sequence in general, and are related to variations in groundwater chemistry. In these cases the silica–$CaCO_3$ relationship is bidirectional and the duricrust far less stable. For this reason the morphology, characteristics and distribution of the duricrusts are discussed *en suite* in subsequent sections, rather than as individual entities as is common practice.

3.5.1 Classification and morphology

Calcretes and silcretes have developed from a wide variety of host materials in the Kalahari, including Karoo sedimentary rocks and basalts, Kalahari Sand, beach gravels, alluvial, pan and lacustrine sediments, including shell beds (section 3.6.1) and diatomaceous earths (section 3.6.2), as well as pre-existing duricrusts. The range of materials, therefore, is extensive. Netterberg (1969b, 1980) and Van Zuidam (1975) have proposed a series of classifications for calcretes on the basis of standard descriptors, secondary structures and other properties. This has been modified by Goudie (1983) to a more generalised morphological description (Table 3.3), and with the exception of the soil and powder categories, is applicable also to silcretes.

Silcrete classification on morphological grounds is more difficult. Summerfield (1982, 1983b) chose to differentiate silcretes on the presence or absence of

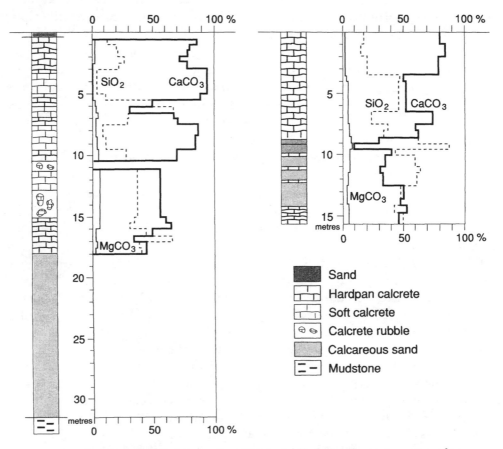

Figure 3.9. Borehole logs from a dry valley in Letlhakeng, showing stratigraphy and relationships of calcretes and silcretes. After Gwosdz and Modisi (1983).

Table 3.3. *A morphological classification of calcretes*

Calcrete type	Characteristics	Occurrence
Calcified soils	Weakly cemented soil	Soil horizons
Powder calcrete	Fine powder, some carbonate replacement	Pans and playas
Nodular calcrete	Concretions or nodules in a calcareous matrix	Various
Honeycomb calcrete	Honeycomb texture of coalesced nodules. May be conglomeritic	Various
Hardpan calcrete	Hard layer, often composed of cemented honeycomb or nodular horizons. Includes calcretised gravels	Above or between nodular or powder calcretes. Frequent as a surface horizon
Laminar calcrete	Laminated crust or layers < 25 cm thick	Frequent cap to hardpan exposures
Boulder calcrete	Discrete to coalesced boulders. Resolution often apparent	Secondary calcrete formed from other types

After Goudie (1983).

Table 3.4. *A micromorphological classification of silcrete*

Fabric type	Characteristics
GS (grain-supported) fabric	Skeletal grains form a self-supporting framework
Sub-types:	optically continuous overgrowths Chalcedonic overgrowths 'Microquartz' mix (microquartz, cryptocrystalline and opaline silica)
F (floating) fabric	skeletal grains float in the matrix and are not self-supporting Skeletal grains > 5% content
Sub-types:	Massive (glaebules absent) Glaebular (glaebules present)
M (matrix) fabric	Skeletal grains < 5% content
Sub-types:	Massive (glaebules absent) Glaebular (glaebules present)
G (conglomeratic) fabric	Detrital component Sediment > 4 mm diameter present

After Summerfield (1983*b*).

weathering profiles, an important factor in the Kalahari where weathering profiles are absent, with the possible exception of silcretes on the Serorome and Serowe escarpments. The silcretes are then subdivided on chemical and micromorphological criteria, and, in particular, on fabric morphology (Table 3.4). Smale's (1973) classification (Table 3.5) tends to ignore intermediate categories that can be identified only on micromorphological grounds, but is more useful for field identification. Again Summerfield's four fabric types are also found in calcrete samples, with a calcite matrix instead of silica. Goudie

Table 3.5. *A morphological classification of silcrete*

Silcrete type	Characteristics
Terrazo	GS fabric, roughly 60% quartz grains with solutional cavities. Chert, opaline or cryptocrystalline cement. Conchoidal fracture. Shatters on impact
Conglomeratic	Pebbles of terrazo silcrete or other material. C fabric
Albertina	Matrix of terrazo type, few skeletal grains. F or M fabric. Usually cryptocrystalline
Opaline	Massive silcretes of opaline, chalcedonic or cryptocrystalline quartz. Few grains. M fabric
Quartzitic	Orthoquartzite texture created by cementation and authigenic growths on quartz grains. GS fabric

After Smale (1973).

(1983) warns of the limitations of classification where composite profiles occur, citing the superimposition of six calcrete profiles in the Molopo valley.

Ferricretes can also be classified by a number of criteria (e.g. McFarlane, 1983), including mineralogy, structure and chemical content. In the Mega Kalahari all reported ferricretes are of pedogenic origin without weathering profiles, and have a limited range of characteristics. Most are limited to shallow depths of vermicular or pisolithic structures (Figure 3.6f) formed by the cementing of saprolite, gravel or sand, though spreads of pisolithic gravels occur. The most extensive duricrust of this type is the 1.5 m deep lateritic manganese deposit at Mombezhi in Zambia, covering some 3 km², and assaying 30–60 per cent MnO_2 (Johns, 1985).

3.5.2 Distribution and relation to landform

The general characteristics of duricrust distribution have already been noted, i.e. a tendency to follow the precipitation gradient, and an association with specific landforms, namely pans and drainage lines. These are obvious as they frequently allow observation of the profile as well as surface distribution. However, there is strong evidence to suggest that there is a genetic relationship between these landforms and duricrusts, and this is discussed in sections 5.4 and 6.4.

Summerfield (1982) notes four generalised modes of occurrence of silcrete (Figure 3.10), namely horizons of silicified sands, valley sides and terraces, pan rims and replacement of calcrete in profiles. To this list can be added silicified spring tufas, as in the Serorome and Letlhakeng Valleys (Shaw and De Vries, 1988), as bars in perennial rivers such as the Boteti (Snowy Mountains Engineering Corporation, 1987), and as recent accumulations in highly saline environments, where it often preserves desiccation cracks and 'crocodile-skin' features (Coates et al., 1979). There are also sites where silicification of bedrock has proceeded without an intermediate calcification stage, as in the basalts of the Chobe Escarpment.

Calcrete shows a similar tendency towards pan rims, drainage lines and localised horizons within the Kalahari strata. It is also apparent in association with present or relatively recent fluvial and lacustrine systems, often containing shell deposits.

Within the Kalahari there are also regional tendencies worthy of note. Calcretes are best developed on the Bakalahari Schwelle and in the vicinity of the

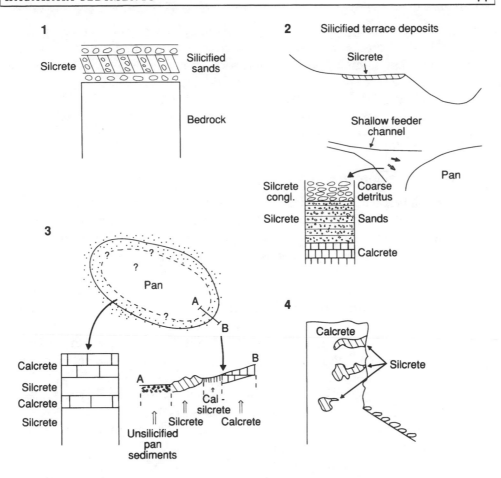

Figure 3.10. Schematic representation of the main types of silcrete occurrence, after Summerfield (1982). (1) Silicified sand horizons, (2) silicified terrace and fluvial deposits, (3) pan silcretes, (4) replacement of existing calcretes.

Molopo network, where, as already noted (section 3.2.4), depths of up to 100 m have been recorded. Calcrete thickness tends to decrease northwards. The fossil lake system of the Okavango–Makgadikgadi has a well developed recent alluvial calcrete suite formed c. 10 000 BP (Chapter 7) which attains thicknesses of up to 10 m, while older calcretes in pans are usually part silcretised in this area. North of the Makgadikgadi limited calcrete development is found in association with pans, and in the Northern Kalahari calcretes are rare. This accords to Netterberg's (1969b) view that hardpan calcrete is restricted to areas receiving more than 550 mm annual precipitation, nodular calcretes extend to 800 mm p.a. and calcification is absent beyond the 800 mm isohyet.

The Okavango–Makgadikgadi system offers useful insights into silcrete formation and distribution, as it forms a closed surface and groundwater basin with silica concentrations above the equilibrium solubility of quartz. Summerfield (1982) notes the increasing concentrations of SiO_2 in surface waters of the Okavango and Nata Rivers towards the Makgadikgadi Basin, with shifts of pH values above 9, the point of silica solubility, in the vicinity of the basin itself. This corresponds with an increasing visibility of silcrete in the river channels. Pisolithic silcretes are forming at present at depths of about 3 m in the lower Okavango (Snowy Mountains Engineering Corporation, 1987), and appear as cross-channel bars in the lower Thamalakane, Boteti and Nata Rivers. Within the Makgadikgadi Basin, particularly in Sua Pan, silcretes occur on and within the pan surface, while calcretes are restricted to the pan surrounds and old shorelines. The highly alkaline environment of Sua Pan, in the sump of the

Makgadikgadi Basin, appears to be one in which contemporary silica precipitation and dissolution occur, to form a distinctive suite of 'terrazo' and chert type silcretes containing uncommon mineral associations (see below). This progression may represent an analogue for other parts of the Kalahari, where, as Sumerfield (1982: p. 62) notes, temporal and spatial changes in groundwater pH have probably been a major cause of silica precipitation in the past, mostly in the vicinity of pans.

3.5.3 Chemistry and mineralogy

Within the range of materials described it is difficult to make generalisations about their chemistry. The antipathy of $CaCO_3$ and silica allows for the existence of the range of intermediate forms encountered in the field. Comprehensive analyses of Kalahari calcretes have been made by Gwosdz and Modisi (1983) in the search for carbonate resources, and indicates great variety at any given site; at Ghanzi, for example, $CaCO_3$ and SiO_2 varied from 44 to 89 per cent and 8 to 34 per cent respectively, while at Nata the comparative figures were 27–70 per cent and 20–70 per cent. The mean $CaCO_3$ contents fall well below the 79 per cent quoted for world calcrete means (Goudie, 1973: p. 18), and themselves probably represent high concentrations by virtue of the sampling exercise. Iron and aluminium sesquioxides fall into the ranges of 0.5–1 per cent and 1–3 per cent respectively, while MgO and $MgCO_3$ lie in the ranges 0.5–13 per cent and 1.5–20 per cent, with great disparity between sample sites.

Calcite is the dominant carbonate, mostly in the form of micrite, which precipitates in pores and exerts pressure by crystal growth on the surrounding fabric. Dolomite has also been recorded (Watts, 1977; Coates *et al.*, 1979) and uranium prospecting (Levin, Hambleton-Jones and Smit, 1985) has indicated low concentrations of carnotite in the Gordonia region.

Chemical analyses of silcrete (Summerfield, 1982) from the Kalahari indicates SiO_2 values of the order 92–98 per cent, with small fractions of iron, aluminium, calcium and magnesium oxides. Compared with weathering profile silcretes encountered in the northern Cape Province and in Australia, the Kalahari silcretes were low in TiO_2 (Summerfield, 1983*a*).

Thin section analysis and X-ray diffraction (Summerfield, 1982) suggest that the dominant form of authigenic silica is of quartz, with matrices formed from chalcedony, microquartz, cryptocrystalline or opaline silica. Lussatite has also been recorded. The green colouring found in pan silcretes in alkaline environments, originally believed to be chlorite (Smale, 1973), has been identified as glauconite–illite (Summerfield, 1982), accounting for the higher Al_2O_3, Fe_2O_3 and K_2O contents of these particular samples. Smale (1968, 1973) has recorded the presence of clinoptilolite, a mineral usually associated with volcanic ash, in pan sediments associated with silcrete at Nata, although not in the silcrete itself. Both glauconite and clinoptilolite have limited distribution in terrestrial sediments, having been recorded from a few salt lakes.

The main chemical pathways for the formation and transformation of calcretes and silcretes in Kalahari environments have been given by Watts (1980; Figure 3.11).

3.5.4 Models of genesis

A number of genetic models have been proposed for duricrust development, including *in situ* weathering (relative accumulation), lateral transfer by geomorphic agencies, vertical *per descensum* and *per ascensum* transfer by enriched soil and

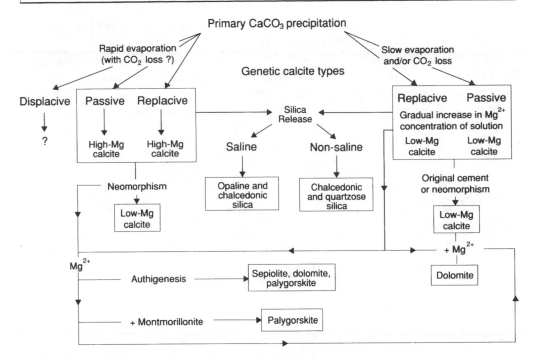

Figure 3.11. Schematic summary of the main pedogenic and diagenetic processes affecting Kalahari calcretes. After Watts (1980).

ground water (absolute accumulation), detrital origin (secondary duricrusts) and organic origin. These models are explored fully in Goudie (1973) and Goudie and Pye (1983).

There is no doubt that Kalahari duricrusts are polygenetic; all of these models are applicable to formation in different localities. *In situ* weathering, often to depths exceeding 100 m, is apparent in the Kalahari (e.g. Shaw and De Vries, 1988), and duricrusts derived from sub-Kalahari rocks are common. Coates *et al.* (1979: p. 92) note that it is difficult to differentiate between Karoo and Kalahari strata where alteration has occurred. Likewise, a range of materials has derived from the activity of surface, soil and groundwater movement. Organic pathways have been under-researched, with the exception of the roles of the Mopane tree (*Colophospermum mopane*) in calcium fixation (De Winter, De Winter and Killick, 1966) and diatoms (Du Toit, 1954) in silica fixation in pan sediments.

Within this range of processes the importance of some chemical mechanisms has been identified. Netterberg (1969*b*, reiterated by Coates *et al.*, 1979), suggests that soil suction (pore water pressure or pF) is an important mechanism in the solution and precipitation of $CaCO_3$, as increasing pF is encountered as ground water ascends, leading to loss of CO_2 and carbonate precipitation. The inverse applies to descending waters. The position of this process in the profile is dependent on evapotranspiration conditions and the pore status of the host material.

Summerfield (1983*b*) identifies evaporation and changes in pH as the major determinants of silica precipitation, with cooling, reaction with cations and organic processes as supernumary possibilities. The role of carbonates in raising pore water pH to above 9, at which point silica becomes soluble, has been identified as a possible factor in the bidirectional replacement process in existing calcretes and silcretes (Summerfield 1982: p. 56). On a larger scale, fluctuations in saturated ground water, as proposed by Senior and Senior (1972) and Smale (1973), is a probable mechanism of genesis. Although Summerfield (1983*b*) expresses reservations about the widespread availability of saline ground waters in the Kalahari, upward 'flushes' of saline ground water resulting from recharge and subsequent displacement by fresh water are common (section 3.7).

The comprehensive nature of the calcrete–silcrete suite also raises questions about the source of the component elements. Although calcium is available in the Karoo strata, and more so, in the Ghanzi limestones, it does not explain the quantity of calcrete that has formed. Goudie (1973) suggests that aeolian dust is a major agent in the recycling of carbonate, but little research has focused on this possibility. Grain abrasion has likewise been proposed as a major contributor to the pool of soluble silica (Summerfield, 1982).

3.5.5 The age of duricrusts

The dating of duricrusts has proved to be a difficult proposition. Already noted are the problems caused by the assumptions that duricrusts have stratigraphic (section 3.2.3) and palaeoclimatic (section 3.5) contexts. Netterberg (1978, 1980) also notes that calcrete formation is not the result of a discrete event, and may therefore have a composite age, while the radiometric dating of calcretes has yet to accommodate the problems of carbonate origin and contamination. No successful attempts have been made to date silcretes, although it is possible to distinguish contemporary silcrete from older forms on the basis of mineral stability.

Duricrusts, therefore, have tended to fall into the categories of 'older' and 'younger' deposits suggested for the Kalahari Group as a whole (e.g. Wright, 1978). Younger duricrusts include soft, largely unaltered calcretes associated with topographic depressions or drainage systems, often containing shells or archaeological artefacts, which have proved amenable to radiocarbon dating, and are thus less than 40 000 years old (Chapter 7). The older duricrusts have proved undatable, though in some cases ESA and MSA artefacts have provided a very approximate date. The general assumption has been that older duricrusts without artefacts are at least Pliocene in age, based on palaeoclimatic assumptions (e.g. Netterberg, 1980) or estimations of the rate of formation (Goudie, 1973). Both of these approaches are fraught with difficulties, for, as Summerfield (1983: p. 67) notes, duricrust may be associated with short-term departures from long-term climatic (and process) means. Although several authors (e.g. Wright, 1978) have pointed out the distinct unconformity which separates duricrust benches on valley sides from overlying sand, with alteration and karstification on the bench surface, there is, as yet, no justification for the assignment of these duricrusts to stratigraphic units (e.g. Malherbe, 1984).

There is, however, ample evidence of multiple phases of duricrust development during the Cainozoic. Two approaches to the issue may be of value in the future. The first is the application to duricrusts of the range of dating techniques now becoming established in other fields of earth science, such as fission track and thermoluminescence (TL) dating. The other is detailed site studies which will lead not only to a better understanding of duricrusts themselves, but perhaps to a recognition of local and regional phases in duricrust formation.

3.6 Minor lithological units

Lithological units with limited spatial distribution include sediments associated with lakes and rivers, such as diatomaceous earths, shell beds and evaporites, together with cave deposits. Despite their limited distribution, these units are important as palaeoenvironmental markers, and have been the foci of research activity.

3.6.1 Diatomaceous earths

Diatomaceous earths were first recorded in the Kalahari by Passarge (1904) in the banks of the Boteti River and in the bed of Lake Ngami. Subsequent investigations (Rogers, 1936; Kent and Rogers, 1947; Coates et al., 1979; Shaw, 1985a; Heine, 1987; Shaw and Thomas, 1988) have found that such earths are a frequent occurrence throughout the lake beds, terraces and pans of the Lake Palaeo-Makgadikgadi system (Figure 3.6c), and also occur in the vicinity of the Molopo network (Table 3.6; Figure 3.12). Diatoms have also been described from the Kalahari Sand itself (Rogers, 1936: p. 62), and from the channel peats of the Okavango Delta (Ellery, 1987).

Most of the diatomaceous earths have been formed by still-water sedimentation under neutral to slightly alkaline conditions in lake beds, usually in the lee of sand ridges and shorelines. Although the earths have not, as yet, been dated, their proximity to the surface suggests a relatively recent origin during a high lake phase, probably during the late Glacial, or, in the case of the extensive and distinctive beds of Ngami, even more recently. Others, such as the deposits at Moremaoto on the Boteti River, are clearly related to past stages in the evolution of Lake Palaeo-Makgadikgadi.

In the Molopo region diatomaceous earths have been located in the terraces of the Molopo River (Rogers, 1936) and at shallow depth in the Vryburg area related to old drainage alignments (Smit, 1977). One of these occurrences, at Rus-en-Vreede, has been found to be uraniferous (Levin et al., 1985).

3.6.2 Shell beds

The presence of deposits of freshwater gastropod and bivalve shells inevitably contained within a soft, sandy calcrete matrix was commented on by several nineteenth century travellers (e.g. Bradshaw, 1881). Passarge (1904) included these shell 'limestones' within his classification of Kalaharikalk and recorded 11 genera in the Boteti and Ngami regions, often coincident with deposits of diatomaceous earth. Subsequent investigations (Rogers, 1936; Heine, 1982, 1987; Shaw, 1985a; Shaw and Thomas, 1988) have confirmed a wide but localised distribution of shell beds throughout the Lake Palaeo-Makgadikgadi and Molopo systems, particularly in river terraces, and also in a number of pans in the Kalahari (Table 3.6; Figure 3.12).

Rogers (1936: p. 75), comparing his own findings with the species list of Boettger (Schulze, 1907) and of the Vernay–Lang Kalahari Expedition of 1930, suggests significant differences in the gastropod fauna on either side of the southern Kalahari divide, with an absence of the genera Lanistes, Ampullaria, Melania and Viviparus south of the Okavango–Makgadikgadi region, suggesting a long period of faunal separation.

Shell-bearing calcretes have been regarded as the youngest members of the calcrete corpus present in the Kalahari (Rogers, 1936), and the dating of both shells and their calcrete matrices have invariably yielded late Quaternary ages, particularly within the late Glacial (18 000–13 000 BP) and late Holocene (c. 2000 BP). As the shell beds are encountered above present lake and river levels they have provided useful evidence for past lacustral phases (e.g. Heine, 1982; Shaw and Cooke, 1986), and, within the limitations imposed by the hydrological and tectonic conditions, wetter climatic episodes (Chapter 7). Less convincing has been the climatic interpretation of the land snail Xerocerastus (Heine, 1982), often found in association with freshwater molluscs, as along the Molopo and Okwa Rivers, which has widespread distribution in the Kalahari under present conditions.

Table 3.6. *Distribution of fossil Gastropoda deposits and diatomaceous earths in the Kalahari (Key to Figure 3.12)*

Number	Location	Reference	Species
Fossil Gastropoda			
A	Serondela	Bradshaw, 1881; Shaw and Thomas, 1988	*Bellamya* spp. *Lymnaea natalensis*
B	Ngwezumba	Shaw, 1985a	*Afrogyrus* spp. *Bulinus* spp. *Ceratophallus* spp. *Lymnaea natalensis*
C	Matapa	Rogers, 1936	*Bulinus* spp. *Burnupia* spp. *Planorbis* spp.
D	Boteti and Nhabe Rivers	Passarge, 1904; Heine, 1987	*Corbicula africana* *Melanoides tuberculata* *Vivipara passargei*
E	Okwa Gorge	Shaw (unpublished)	*Bulinus* spp. *Corbicula africana* *Lymnaea natalensis* *Melanoides tuberculata*
F	Nata	Heine, 1987	*Bulinus* spp. *Corbicula* spp. *Melanoides* spp. *Unio* spp.
G	Gweta	Heine, 1987	As above
H	Molopo River	Heine, 1982	*Bulinus* spp. *Corbicula fluminalis* *Unio* spp. *Xerocerastus* spp.
Diatomaceous earths			
1	Parakarungu	Shaw and Thomas, 1988	
2	Savuti	Rogers, 1936	
3	Ngami	Passarge, 1904; Rogers, 1936; Shaw, 1985a; Snowy Mountains, 1986	
4	Moremaoto	Passarge, 1904	
5	Rakops	Kent and Rogers, 1947	
6	Tokotse Pan	Coates *et al.*, 1979	
7	Letlhakane	Kent and Rogers, 1947	
8	Nata	Rogers, 1936; Heine, 1986	
9	Witkop	Rogers, 1936	
10	Kuruman	Levin *et al.*, 1985	
11	Nqoga Channel	Ellery, 1987	

3.6.3 Pan sediments and evaporites

Pans form a specific sedimentary environment, and are discussed in detail as landforms in section 6.4. Sediments within pans have a complex relationship with geomorphological processes, including influx by aeolian activity, surface flow and lateral groundwater transfer, alteration by descending and ascending ground water, and aeolian deflation controlled by groundwater tables and surface moisture content. In addition, pans are frequently the foci of deep weathering, which permits the formation of tens of metres of duricrusts and laminated sands and clays, while repeated wetting and drying of the pans surface may lead to the burial and preservation of cracks and desiccation features. Organic sedimen-

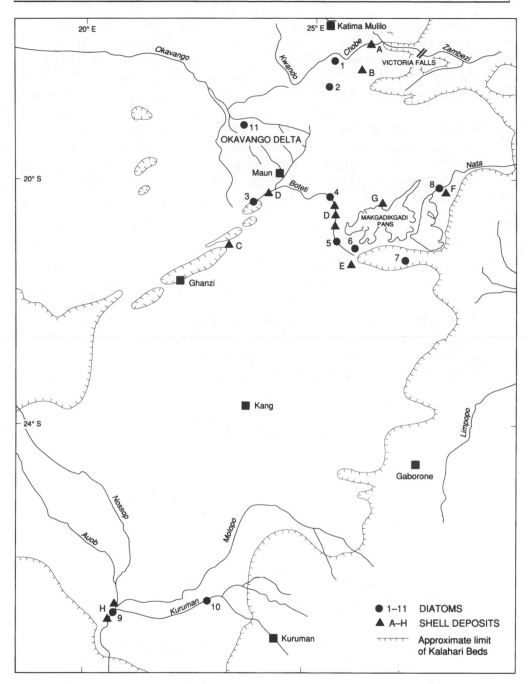

Figure 3.12. Location of shell deposits and diatomaceous earths in the Kalahari. (Legend in Table 3.6.)

tation and turbation also lead to near-surface alteration. A sample profile, from Njare Pan in the Middle Kalahari, is shown in Figure 3.13.

Evaporite deposits occur in closed basins where saline waters are found close to the surface, either as a result of horizontal transfer or by recharge forcing from below. The best known examples are the massive Sua Pan in the Makgadikgadi Basin, and the Soutspan complex south of the Molopo–Kuruman confluence, whose hydrology has been discussed by Arad (1984).

Sua Pan (see also section 5.3.4) is one of the world's few remaining pristine salt lakes (Figure 3.6d), although it is shortly to be disturbed by soda ash extraction. The bed consists of brine-saturated sand and clay layers, with surface efflorescence of trona (Na$_2$CO$_3$.NaHCO$_3$.2H$_2$O) and halite (NaCl), with an absence of calcium and magnesium carbonates, already precipitated out of the system. Silica precipitation and dissolution are frequent following influxes of fresh surface water, while the pan forms a major source of aeolian dust. The surface itself is constantly changing in response to alternating reduction/ oxidation and precipitation/solution of salts dependent on the inundation– desiccation cycle, haloturbation, bioturbation and deflation to form a sequence of minor ephemeral landforms. These complex mechanisms of evaporite accumulation and the microgeomorphology of salt lakes have been discussed in detail elsewhere (e.g. see Eugster and Kelts, 1983; Shaw and Thomas, 1989; for recent summaries).

3.6.4 Cave deposits

Cave deposits include sediments flooring present-day caves within the Kalahari, together with remnants from cavities now removed by erosion. They are limited to the Gcwihaba Hills in Ngamiland and the Gaap Escarpment to the south of the Kalahari, and are discussed in the context of their cave environments in section 6.5.2. The deposits consist of a range of materials including calcrete, breccias, flowstones, clastic material and organic matter, particularly bat dung. Microfaunal remains at available sites range in age from mid-Pliocene to Recent.

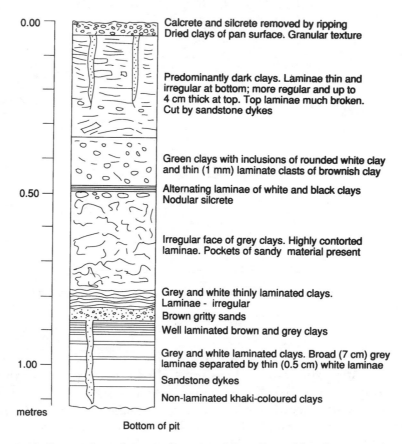

Figure 3.13. Section through pan sediments at Njare Pan. After Coates et al. *(1979).*

3.7 The hydrogeology of the Kalahari Group

Man in the Kalahari relies heavily upon groundwater resources. Until the advent of deep drilling and diesel pumps this inevitably meant the exploitation of shallow and frequently perched aquifers within the Kalahari Beds. In the twentieth century technological advances have allowed the widespread exploitation of deeper resources, particularly in the underlying Karoo strata.

The lithological units outlined in this chapter vary greatly in porosity and permeability, characteristics which affect their potential as aquifers. The sporadic distribution of both saline and freshwater bodies within the Kalahari Beds reflects these lithological variations. The Kalahari Sand, in particular, displays high transmissivity, which allows potential recharge with poor storage capacity. Duricrusts, on the other hand, form good, if localised, aquifers (e.g. Bruno, 1985).

Lively debate has been conducted over the past 80 years as to whether groundwater recharge takes place through the Kalahari Beds *in toto* under contemporary climatic conditions. Early judgements (Passarge, 1906; Debenham, 1948; Frommurze, 1953; Dixey, 1956b; McConnell, 1959; Boocock and Van Straten, 1961, 1962; Martin, 1968) were unanimous on the lack of recharge, assuming that high evaporation rates and effective extraction by vegetation would rapidly deplete any rainfall infiltrating into the sand, thus restricting ground water to a capillary fringe close to the ground surface. Van Straten (1955) suggested 6 m as the limit of recharge on the basis of soil moisture retention studies. The role of vegetation has been substantiated; Verhagen *et al.* (1978) comment on the extraction of ground water by trees in the Gamagara valley, Southern Kalahari, while Jennings (1974) has noted the presence of living tree roots at a depth of 68 m in an unused borehole, possibly reaching the water table at 141 m (De Vries and Von Hoyer, 1988). Recharge would thus be possible only where the sand cover was thin, as in the vicinity of rock outcrops and at the Kalahari rim, and in localities of substantial duricrust development, as around pans and along ephemeral drainage lines.

Subsequent studies, using the isotopes ^{3}H (tritium), ^{13}C, ^{14}C and ^{18}O, carried out in both the central Kalahari Basin (Verhagen, Mazor and Sellschop, 1974; Mazor *et al.*, 1974, 1977; Mazor, 1982) and the Gordonia region of the southern Kalahari (Verhagen *et al.*, 1978; Verhagen, 1983) have revised this view, suggesting that recharge occurs by accelerated infiltration during years of above-average rainfall, as occurred in the southern Kalahari in 1973–74. The tritium and ^{18}O profiles, in particular, suggest that vertical recharge is common along preferential lines (Mazor *et al.*, 1977; Verhagen, 1983). This process would be particularly important in the Middle Kalahari, where the rainfall mean is higher (Mazor *et al.*, 1974).

Foster *et al.* (1982) repeated the isotope studies in southeastern Botswana and sounded a note of caution over the earlier studies, pointing out that levels of tritium were very low and did not reflect the patterns of rainfall in the Kalahari in the post-bomb period, while the selection of pre-existing boreholes by Mazor *et al.* (1974, 1977) biased their results towards points of maximum recharge. Further, the utilisation of pumped samples from these boreholes did not allow for well-head contamination from adjacent aquifers. They concluded that, in an area with a mean annual rainfall of 450 mm, diffuse recharge was unlikely in sand cover greater than 4 m deep. Mazor (1982) has concurred that recharge in the Kalahari is far from uniform, and is most likely in zones of high infiltration (joints, zones of bioturbation, calcrete outcrops, drainage lines) during periods of prolonged excess rainfall, which itself may not be uniformly distributed.

De Vries (1984) has investigated the regional hydraulic gradient between the regional divide and the terminal sink of the Makgadikgadi Basin and suggests that either the hydraulic head has decayed since the last period of pluvial replenishment prior to c. 12 000 BP, or the present sub-surface hydrology is in equilibrium with a present-day recharge of less than 1 mm yr^{-1}. However, subsurface recharge at the eastern rim of the Kalahari is likely from Precambrian and Karoo outcrops, given the constant high yields, of the order of 5 million m^3 yr^{-1}, in the wellfield supplying the Jwaneng diamond mine over the past decade.

4 Climate, soils and vegetation of the Kalahari

IT IS NOT SURPRISING, given the great extent of the Mega Kalahari environment, that some of its fundamental characteristics vary considerably in spatial terms. Of these characteristics, perhaps of central significance, due to its influence upon other variables, including vegetation, soils and geomorphological activity, is climate, which ranges from the aridity of southwestern Botswana to the humid tropical conditions of Zaïre. The spatial dimensions of climatic variability can, to a large extent, be explained in terms of the components of the global atmospheric circulation which affect southern and central Africa and the continental setting of the Kalahari. Other influences are spatial and temporal embellishments upon the framework which these establish. In this chapter, we aim to outline this framework and to examine its influence on the major characteristics of Kalahari climates, soils and vegetation communities. Greatest reference will be made to the Middle and Southern Kalahari, the area for which most information is available, although more northerly locales will be considered where appropriate.

4.1 Climatic controls

The synoptic climatology of the Kalahari, which covers the major elements of the global atmospheric circulation influencing the climate of the region, have been discussed in detail in the context of the whole of southern Africa by Schulze (1972) and Tyson (1986). In southern Africa the global circulation patterns are, in turn, tempered by the influence of the southern part of the African continent upon atmospheric characteristics. The land mass can be regarded essentially as a southward-tapering meridional finger which disrupts the development of atmospheric high pressure, especially during the summer. Climatic characteristics are further influenced by the elevation of the Kalahari, and its interior position, which adds the dimension of continentality to the atmospheric factors contributing to aridity. Thus, while the Kalahari Desert lies astride the tropic of Capricorn, its altitude gives it a climate which in some respects is more temperate than tropical (e.g. Pearce and Smith, 1984).

The Kalahari Desert is located within the southern hemisphere subtropical high pressure belt and, while northern parts of the Mega Kalahari lie equatorward of this belt, the climate of this region *in toto* is affected by this feature more than by any other atmospheric factor (Torrance, 1972). The characteristics of this belt and the seasonal fluctuations in the position of the components of the tropical atmospheric system, particularly the equatorial trough or Inter-tropical Convergence Zone (ITCZ), result in three major near-surface air streams influencing

Kalahari climates: South Atlantic air, the northeast monsoon penetrating from East Africa, and Indian Ocean tropical easterlies (Tyson, 1986).

4.1.1 Atmospheric circulation: vertical components and anticyclones

The significant all-year-round receipt of solar radiation in the tropics causes atmospheric instability which drives the tropical atmospheric circulation in the vertical plane. This contributes to the two (one in each hemisphere) 'thermally direct' Hadley cells, which have their rising limbs in the equatorial zone, and their subsiding limbs in the subtropics. In the southern hemisphere, the subsiding components of the southern Hadley cell and southern temperate Ferrel cell are responsible for the nature and position of the subtropical high pressure belt which so influences the character of Kalahari climates (Figure 4.1).

Above the atmospheric boundary layer, southern Africa experiences a mean anticyclonic circulation as the southern hemisphere subtropical high pressure belt is centred on about latitude 30° S. This mean condition is modified by two factors which cause important seasonal fluctuations. First, the presence of the southern African land mass in this latitudinal zone creates differential sea–land heating that causes the high pressure belt to be split into two component cells: the South Atlantic and Indian Ocean anticyclones (Schulze, 1972). The splitting of the high pressure is greatest during the summer, when low pressure develops over the land, while in the winter, high pressure also occurs over the land mass, either as an independently developed high (Streten, 1980), or as an extension of the Indian Ocean anticyclone, which is probably a series of eastward moving composite cells (Trewartha, 1962). The second factor is that the high pressure cells fluctuate in position, in conjunction with the seasonal migration of the global pressure and climatic belts (Figure 4.1). On average, the South Atlantic and Indian Ocean anticyclones experience respective seasonal migrations covering 5° and 6° of latitude (McGee and Hastenrath, 1966; Tyson, 1986). The overall result of the two effects is that the centres of both cells appear to migrate northwest in winter months (Schulze, 1972), which has a significant influence upon seasonal winds and rainfall in the Kalahari. Even as far north as the southern half of Zaïre, the Mega Kalahari experiences a winter dry season under the influence of southeasterly winds associated with the South Atlantic anticyclone (Bultot and Griffiths, 1972).

4.1.2 Atmospheric circulation: air masses and the ITCZ

A further dimension of the synoptic climatology of the region is gained by looking at the significance of the ITCZ and the movements of the major air masses affecting southern Africa. The ITCZ or equatorial trough represents the zone of convergent trade winds and thermal uplift at the meeting of the two Hadley cells. It fluctuates in position north or south of the equator according to the seasons, and, as with the subtropical high pressure belt, land–sea thermal contrasts complicate the picture; in particular, the ITCZ tends to be less pronounced and less continuous over land.

The ITCZ has no direct influence on the Kalahari during the winter, but in the summer months the displacement and extension of the northern Hadley cell into the southern hemisphere (Figure 4.1) results in the ITCZ and its southern extension, the Zaïre Air Boundary (the ZAB) playing an important part in the region's climate (Tyson, 1986). The ITCZ and ZAB fluctuate in position and intensity from day to day and year to year, but as well as marking the convergent boundaries between the air masses which influence the region, they represent

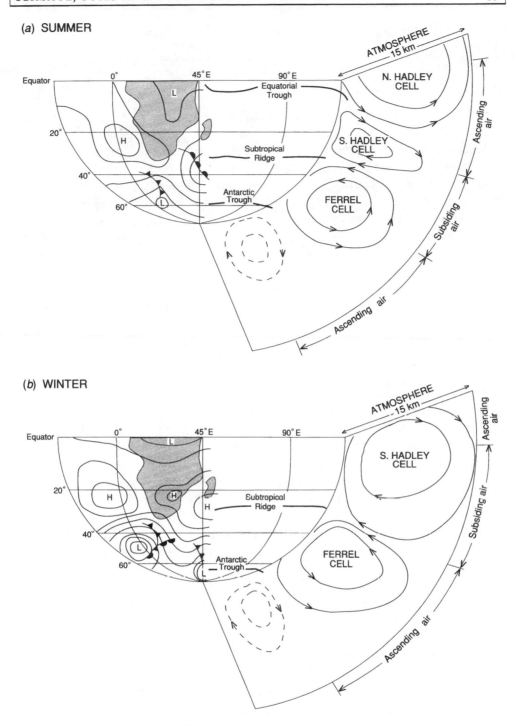

Figure 4.1. Vertical and horizontal components of the atmosphere affecting
southern and central Africa in (a) December–February (summer wet season) and
(b) June–August (winter dry season). After Tyson (1986: Fig. 5.4).

troughs of low pressure where convective activity enhances rainfall (Tyson, 1986;
Figure 4.1). In about January, the ZAB may link the thermal low which develops
over the southern African interior with the ITCZ (Taljaard, 1981).

In northern parts of the Kalahari, summer rainfall influenced by this situation
is likely to occur in conjunction with recurved South Atlantic air, which has a
strong westerly component and enters the region from Zaïre and Angola; in more
central and southerly areas easterlies blowing around the Indian Ocean high are
more likely to be responsible. In winter, especially before July, the southern
Kalahari may also occasionally receive rainfall in conjunction with the westerly

tracking depressions that influence the climate of the Cape winter rainfall zone (Tyson, 1986).

4.2 Climatic characteristics

As well as controlling surface winds in the Kalahari (Figure 4.2), the synoptic situation imparts a distinct seasonality upon the climate of the Kalahari. Within the Mega Kalahari, it is variations in precipitation rather than temperature which contribute most to the seasonal contrasts in climate, making it more appropriate to talk in terms of wet and dry seasons rather than summer and winter, though the latter terms are frequently used. As well as seasonal patterns, there are marked year-to-year variations which contribute significantly to the character of the Kalahari as a desert or thirstland (section 1.2.1) and make the availability of moisture a pertinent factor for plant and animal communities and human activities. With regard to temperature, it is perhaps the high diurnal ranges which can occur during the winter (dry season) months that are a major feature. In the following sections, we outline these and other characteristics of Kalahari climates.

4.2.1 Precipitation

Annual precipitation in the Kalahari increases in easterly and northerly directions (Figure 4.3a) in response to the varying spatial impacts of the different air masses affecting southern Africa and their seasonal fluctuations. The driest, southwestern areas of the Kalahari are truly arid in the context of accepted definitions (Table 1.1), receiving on average less than 200 mm of rainfall per annum, but virtually all of the Southern and Middle Kalahari receives on average less than 500 mm p.a. and is therefore at least semi-arid. The amounts received increase into the Northern Kalahari and the remaining areas of the Mega Kalahari so that, for example, Mongu in western Zambia has a mean value of 960 mm and Lubumbashi in the Katanga region of Zaïre receives 1243 mm (Figure 4.3b).

The whole of the Kalahari Desert and the Mega Kalahari as far north as the southern half of Zaïre is essentially a summer rainfall zone, with precipitation occurring in conjunction with the summer synoptic situation described above. The southwestern Kalahari is therefore the driest area because it is furthest away from the convective trough of low pressure linked to the Zaïre Air Boundary and the ITCZ and it is therefore more likely to be affected by the fringes of the descending air associated with the South Atlantic anticyclone during summer months. It is also most 'continental' in respect to the moisture-bearing air masses which penetrate the region from the east and north. Longley (1976) has developed a typology of the various weather patterns involved with the production of rain, which almost exclusively comes in the form of convective thunderstorms (Tyson, 1979), tending to occur in late afternoon and early evening. It is important to note that while most rainfall is derived from high-intensity showers (Cooke, 1979c), not every shower is a significant rainfall event. Analysis of Botswana data by Pike (1971) showed that over 80 per cent of rainfall events were of low intensity and rainfall amounts were less than 10 mm for about half of the total number of showers.

The significant seasonality of Kalahari rainfall is illustrated for selected weather stations in Figure 4.3b. For most of the region, on average over 80 per cent of annual rainfall occurs between October and April. Deviations from this situation occur in northern Zaïre, where increasing proximity to the equator results in a progressively greater equitability of rainfall, and in the extreme southwestern

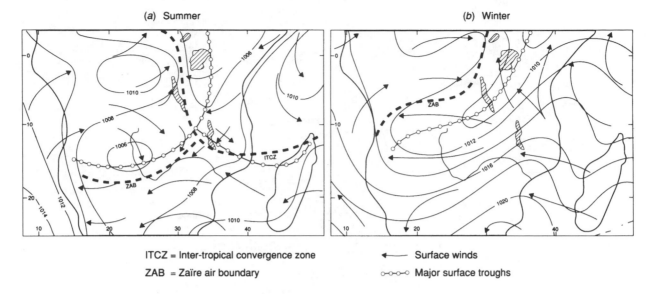

ITCZ = Inter-tropical convergence zone ← Surface winds
ZAB = Zaïre air boundary ○–○–○ Major surface troughs

Figure 4.2. Mean position of the ITCZ and ZAB over southern Africa in (a) summer and (b) winter, with related air masses and wind regimes. After Tyson (1986: Fig. 5.6).

part of the Kalahari Desert. In the latter, a not insignificant part of a year's rainfall can fall in winter months in conjunction with the passage of the depressions which bring rainfall to much of the Cape Province of South Africa (Tyson, 1986).

The length and onset of the wet season also vary spatially in conjunction with the synoptic developments which occur. Basically, the onset is earlier and the duration of the wet season longer in the north than the south and a similar gradient applies from east to west. During the period 1930–59 at Gandajika in Zaïre, just to the north of the Mega Kalahari, the average date for the onset of the wet season was 28 August and the start of the dry season was 9 May, resulting in the wet season lasting for over two-thirds of the year (Bultot and Griffiths, 1972). At Lubambashi, 650 km to the southeast, the respective mean dates were 22 October and 20 April, with the wet season lasting for just under half the year. In Botswana, the rains commonly do not begin until late November, lasting through to April (Cooke, 1979c). Within this period, the rains can be divided into three distinct spells centred on November, January–February and, in some areas, April (Vossen, 1986). As the length of the wet season and total precipitation amounts decrease in a south and westerly direction, so the interannual variability (mean deviation divided by the mean) increases (De Queiroz, 1955; Schulze, 1972; Bhalotra, 1985a; Tyson, 1986): in the southwestern Kalahari (Figure 4.3a), interannual variability exceeds 45 per cent.

4.2.2 Temperature and humidity

The relative altitude of the southern African interior results in mean surface temperatures being somewhat lower than those of other land masses at comparable latitudes. Mean daily temperatures range from 20 to 24°C throughout the area from Lubumbashi in the north to southeastern Namibia (Table 4.1), within which two trends can be seen: seasonal and diurnal. At all the locations in Table 4.1, both mean daily maximum and minimum temperatures are higher during the wet than during the dry season. The higher maximum values reflect the fact that the wet season occurs during the southern hemisphere summer, whereas the lower minimum values are a consequence of the

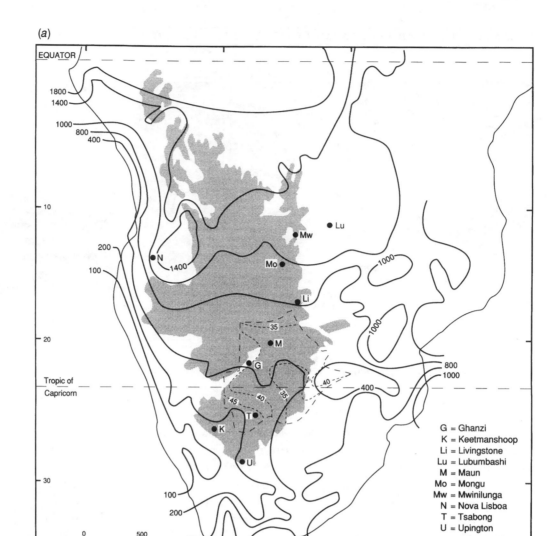

(a)

EQUATOR

G = Ghanzi
K = Keetmanshoop
Li = Livingstone
Lu = Lubumbashi
M = Maun
Mo = Mongu
Mw = Mwinilunga
N = Nova Lisboa
T = Tsabong
U = Upington

Kalahari Group Sediments
— — Border of Botswana
- - - - % variation

Tropic of Capricorn

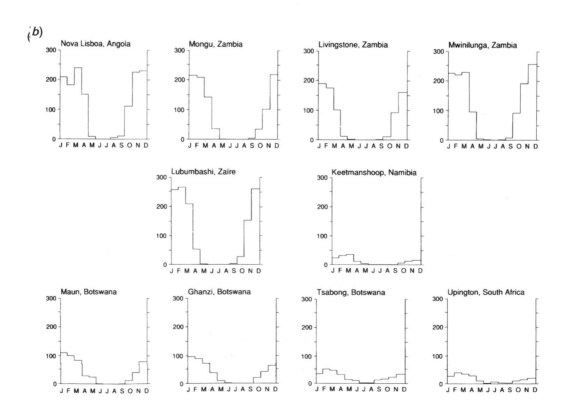

(b)

Nova Lisboa, Angola

Mongu, Zambia

Livingstone, Zambia

Mwinilunga, Zambia

Lubumbashi, Zaïre

Keetmanshoop, Namibia

Maun, Botswana

Ghanzi, Botswana

Tsabong, Botswana

Upington, South Africa

Table 4.1. *Mean daily temperature, sunshine hours and annual evaporation data for selected locations in and near the Kalahari*

| | Daily mean temperatures (°C) | | | | | | Daily mean sunshine (h) | Annual potential evapotranspiration (mm) |
| | Overall mean | Wet season | | Dry season | | Temperature extremes (°C) | | |
		Max.	Min.	Max.	Min.	Max.	Min.		
Upington (Oct.–Apr.)	21.9	32.0	16.7	22.8	5.8	43.0	−8.0	10.3	3805
Tsabong (Oct.–Apr.)	19.5	32.1	15.2	24.4	3.6	42.0	−11.0	10.6	n.a.
Keetmanshoop (Nov.–Apr.)	21.0	32.3	16.0	23.8	8.0	42.0	−4.0	10.7	3903
Ghanzi (Oct.–Apr.)	20.5	31.7	14.5	26.2	6.4	42.0	−7.0	9.8	3305
Maun (Oct.–May)	22.0	35.5	18.6	28.0	8.5	43.0	−6.0	9.1	3058
Mongu (Oct.–Apr.)	22.5	29.6	18.1	28.4	12.0	39.0	−2.0	8.0	2313
Mwinilunga (Oct.–Apr.)	20.0	30.7	15.7	27.4	8.8	35.0	−1.0	6.8	n.a.

Months in brackets refer to normal start and finish months of the wet season.

effectiveness of night-time re-radiation from the ground during the generally cloudless dry season. It can also be noted that there is not a straightforward relationship between latitude and maximum temperatures, due to the modifying effects of altitudinal variations and, during the wet summer season, cloud cover. Thus, Keetmanshoop in Namibia, at latitude 26° 36′ S and an elevation of 1066 m asl, achieves higher mean and extreme maximum wet season temperatures and has a much greater mean number of daily sunshine hours than Mwinilunga, which, being in northwestern Zambia (latitude 11° 45′ S) is much closer to the equator, and is also at a greater altitude (1361 m asl).

Diurnal temperature regimes in the Kalahari are frequently large during the dry season, generally increasing with distance from the equator. This results in less effective solar radiation receipt and longer night-time hours during which re-radiation can occur. In fact, in the Southern Kalahari the average annual incidence of days with a minimum temperature below 0°C is 10, with at least one occurrence of ground frost (Schulze, 1972).

Variations in relative atmospheric humidity occur in conjunction with seasonal and diurnal temperature regimes, the air mass affecting a particular area and distance from a source of humid air. Consequently, humidity essentially decreases from east to west and north to south in the Mega Kalahari and is lowest during the dry season months when subsiding anticyclonic conditions affect much of the region. Thus, although warmer air can hold more moisture than cooler air, so that for a given amount of moisture in the atmosphere relative humidity falls as temperature rises, the dry season humidity of the air over the

Figure 4.3 (opposite). (a) Mean annual rainfall (mm) over southern Africa, and annual percentage variability over Botswana. (b) Mean monthly rainfall (mm) for selected places in the Kalahari.

Kalahari Desert is commonly as low as 30 per cent (Schulze, 1972), due to the lack of penetration into the area by moist air streams at this time.

4.2.3 Evapotranspiration

It is not just low rainfall which contributes to desert conditions, but also high potential evapotranspiration rates. The low relative humidity and warm to hot temperatures of the Kalahari mean that average annual potential evapotranspiration rates are high, with rates measured using evaporation pans increasing towards the southwest where values exceed 4000 mm (Schulze, 1972, 1984). Values of the Budyko–Lattau dryness ratio (mean annual net radiation over the product of mean annual rainfall and latent heat of evapotranspiration: Hare, 1977) show that in the southwestern Kalahari the potential exists to evaporate between four and ten times the annual rainfall (Tyson, 1986).

4.2.4 Climatic variability and drought in the Kalahari

Given that much of the Kalahari is arid or semi-arid, the most potentially significant elements of climatic variability are changes in rainfall patterns through time. Long-term changes, at the scale of thousands, possibly hundreds, of years, can be determined from proxy data sources and are discussed in Chapter 7. The relatively short duration of meteorological records in the Kalahari (in Botswana, the longest continuous record is for Mochudi, where recordings began in 1909), make it more difficult to determine whether cycles or trends in rainfall really exist; it can also be difficult, for example, to determine whether some of the symptoms of desertification are due to changes in climatic parameters or human activities.

Some of the early Europeans visiting the Kalahari in the mid-nineteenth century made observations to suggest that the Kalahari was, at least in parts, a relatively well watered place (section 9.1.3). The datum which these observations set led subsequent investigators to suggest that the Kalahari (Wilson, 1865; Wallis, 1935) or southern Africa as a whole (Barber, 1910; Schwartz, 1919; Kokot, 1948) was experiencing a progressive increase in dryness. As Tyson (1986) notes, disputing such ideas was difficult because of the lack of meteorological data, but the concept of cyclicity rather than directionality in rainfall trends was proposed over 100 years ago (Hutchins, 1888; Tripp, 1888). Subsequent statistical analyses of the available meteorological data by Tyson, Dyer and Mametse (1975) have demonstrated that there are no southern Africa-wide nor sub-regional trends for progressive drying (or its converse) over the last 100 years or so.

Closer investigation of the meteorological records has indicated the occurrence of a statistically significant 18-year cycle of rainfall fluctuations over the summer rainfall zone of southern Africa (Tyson et al., 1975; Tyson and Dyer, 1975; Tyson, 1979, 1986). The extent and exact periodicity varies slightly within the area as a whole, but the overall trend is identifiable for all locations (Figure 4.4), while proxy data mentioned by Tyson (1979) indicate that the trend can be traced back at least to the 1840s.

The oscillation consists of alternating wet and drought decades. However, within this, wetter and drier periods occur at a range of time scales from days to years (Tyson, 1986), and regardless of the time period involved, the synoptic causes remain essentially the same: dry spells result from a build-up of high pressure over the land and wet spells from the converse. Tyson (1986) has also placed the development of the appropriate circulation patterns for precipitation

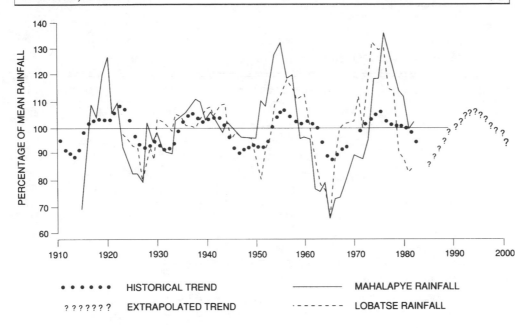

Figure 4.4. *The 18-year rainfall oscillation over the summer rainfall zone of southern Africa (Tyson, 1986). The two curves for stations in eastern Botswana are from Bhalotra (1985b).*

fluctuation on scales ranging from decades to the Quaternary in the Southern Oscillation. This describes variations in the position and intensity of the zonal Walker Circulation. During the high phase of the oscillation there is a rising limb of the Walker Circulation over central southern Africa, resulting in low pressures over the Kalahari, and increased summer rainfall. The low phase experiences a displacement of the rising limb to the Indian Ocean, and its replacement over land by the more stable falling limb, associated with dry conditions.

4.3 Vegetation and soils

The low agricultural potential of the Kalahari has resulted in an absence of detailed soil and vegetation studies. Frequently, mapping at the reconnaissance level has been directed towards non-typical regions which display the most value in human terms. Thus, most of the Middle Kalahari, and the Kalahari of Namibia and South Africa, have had a soil mapping programme of at least 1:250 000 scale for some years, while the Kalahari Desert is being mapped currently, and for the first time, at a scale of 1:1 million. It is not surprising that De Queiroz (1989: p. 11) has recently noted:

> There is little known about the soils of the Kalahari. In the absence of reliable knowledge, it is commonly assumed that the soils of that ecosystem are relatively homogeneous and sandy.

However, for the bulk of the Kalahari region the sedimentological and climatic setting mean that, at best, soils evolved from Kalahari Sand can be described as skeletal or 'weakly developed if at all' (Dregne, 1968). Soil profile development is virtually absent, though variations in texture with depth have been observed in some locations (Buckley, Gubb and Wasson, 1987a). In many areas of the Kalahari it is perhaps inappropriate to think in terms of soil, as surface materials are, above all, sediments which have very low contents of organic material (0.20–0.55 per cent: Wellington, 1955; Bergström and Skarpe, 1985) and nutrients. In such circumstances vegetation communities tend to be hardy, adaptive, and low

Table 4.2. *Levels of essential soil nutrients in the Southern Kalahari, and comparison with data from Central Australia*

Nutrient	Southern Kalahari			Central Australia		
	Mean	SE	n	Mean	SE	n
Phosphorus	4.3	1.0	64	13	1.1	220
Potassium	41	3.6	64	154	9.4	107
Nitrogen						
Dune: ridge crest	47	6	8	46	1	107
upper slope	69	5	14	56	6	32
mid-slope	82	7	14	73	5	25
lower slope	84	7	12	115	8	19
Inter-dune	101	8	16	117	16	37

Data after Buckley *et al.* (1987*a*, *b*).
Mean and standard error values expressed as parts per million.

in species diversity. The relatively well developed vegetation cover over much of the Kalahari is surprising given these prevailing soil characteristics.

Within the Kalahari region, geomorphological factors have resulted in variations in topography and drainage, and the development of different ecosystems. Examples of these geomorphologically related ecosystems include inter-dune hollows, pan and lake basins, drainage lines, riverine and swamp habitats, and areas of rock exposure. These frequently extreme and sometimes unique environments have their own soil and vegetation characteristics, which vary considerably from those of the Kalahari sandveld. Some, such as the Okavango Delta, have been subject to more comprehensive study (e.g. Heemstra, 1976; Astle, 1977; Snowy Mountains Engineering Corporation, 1989).

4.3.1 Soil characteristics

Trapnell and Clothier's (1957) study from western Zambia and that of Sims (1981) in Botswana demonstrated the close relationship between geology and soils in the Kalahari (Figure 4.5). Over most of the central and Southern Kalahari soils are developed on Kalahari Sand, and are classified as arenosols in the FAO–UNESCO classification used in southern Africa. They have poor profile development, are moderately acidic, and notable, at least throughout the Kalahari (Figure 4.5). Over most of the central and Southern Kalahari soils are developed on Kalahari Sand, and are classified as arenosols in the FAO–Skarpe and Bergström, 1986). Buckley *et al.* (1987*a*, *b*) note that even when compared with the sandy soils of central Australia (Table 4.2), which are widely regarded as particularly infertile, the dunefield soils of the Southern Kalahari are at least three times as defficient in phosporus and potassium, and two to ten times in extractable calcium. Nitrogen levels are comparable, both in absolute terms and the way in which they vary in a catenary fashion across dune ridges, with the highest levels found in interdune areas. In areas where the sand cover is thin, as around pan margins and along drainage lines, the presence of underlying calcrete frequently leads to calcic or petrocalcic horizons (Siderius, 1972).

Leistner (1967) noted the tendency for finer-textured soils to develop in depressions of all types, particularly in pans and fossil valleys, where calcium,

Figure 4.5 (*opposite*). *A descriptive soil map of the Botswana Kalahari. After Sims (1981).*

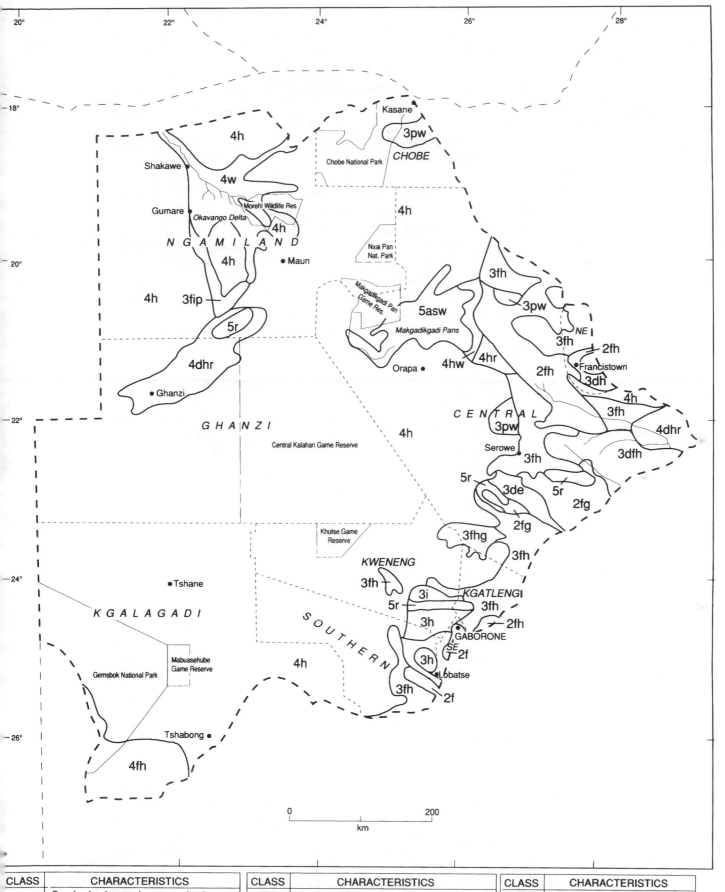

CLASS	CHARACTERISTICS
2l	Sandy clay loams, loams or clay loams with moderate soil fertility
2fg	As 2l with slight surface crusting
2fh	Sandy loams with moderate soil fertility
3de	Shallow soil with moderate erosion
3dfh	Shallow sandy soils with low fertility
3dh	Shallow sandy soils
3fh	Sandy (loamy sand) with low fertility

CLASS	CHARACTERISTICS
3fhg	As 3fh with surface crusting
3fip	Clayey soils hard or very hard when dry with low fertility
3h	Sandy (loamy sand or coarse sandy loams)
3i	Loamy or clayey soils with hard or very hard consistence when dry
3pw	Clayey soils, somewhat poorly drained
4dhr	Very shallow sandy soils with common surface stones or rock outcrops

CLASS	CHARACTERISTICS
4fh	Sand with very low fertility
4h	Sand
4hw	Sand with poor drainage
4w	Soils with poor drainage
5asw	Alkaline and saline soils with very poor drainage
5r	Rocky and / or very stony

magnesium, potassium and phosphorus levels are significantly higher (DHV Consulting Engineers, 1980). Soils in all types of hollows, including inter-dune ridges, usually fall in a spectrum between calcic luvisols and gleysols, dependent on drainage potential.

Where the sand cover is thin and the protrusion of pre-Kalahari bedrock occurs, soil characteristics are modified accordingly. Thus, the Ghanzi Ridge tends towards calcisols, while outcrops of sandstones and granitoid rocks give rise to arenosols which are sometimes indistinguishable from Kalahari Sand. Along the eastern Kalahari margin, at the Kalahari Sand-Precambrian boundary, ferralic arenosols and acrisols with pisolithic horizons are very common. A similar transition to more oxidised soils is apparent in the Northern Kalahari.

In the Mpandamatenga–Kazungula area of northeast Botswana and northwest Zimbabwe, large depressions of probable lacustrine origin contain vertisols derived from the weathering of underlying Karoo basalts (Sweet, 1971). These soils have high fertility but are extremely intractable; a large proportion of the soil moisture is held at tensions above 15 bars, making it unavailable to most plants. The resulting ecosystem, rich in wildlife, is probably unique in the Kalahari (Arup-Atkins International, 1988).

The Okavango Delta (section 5.3.1) has a topographic catena related to annual and long-term variations in flood patterns, the whole developed from the organic and arenaceous sediment load of the river, and modified by the precipitation and dissolution of sodium, calcium and silica (Snowy Mountains Engineering Corporation, 1989). Thus, perennial channels are dominated by histosols and arenosols, while islands and ridges have calcic arenosols. The intervening topographic mosaic of seasonal channels, flood plains and riverine fringes have calcic, arenic and gleyic luvisols, arenosols and fluvisols, dependent on minute changes in level and hydrology.

Lacustrine soil associations are found in the Ngami (section 5.3.2), Mababe (section 5.3.3) and Makgadkgadi (section 5.3.4) Basins, as well as parts of the eastern Caprivi/Chobe area (section 5.3.5). Again, catenas are developed from arenosols on peripheral ridges and basin margins to poorly developed gleysols and vertisols in the basin sumps. Soils in the Makgadikgadi reflect the role of that basin as a terminal sump for surface and ground water, and are frequently calcareous, sodic or saline. Arenosols occur in sand ridges, sand dunes and on higher ground. Pans at higher elevations in this complex contain gleysols, while lower pans, including Sua and Ntwetwe, are dominated by gleyic solonchak soils, with frequent silcrete duricrust development.

4.3.2 Vegetation characteristics

Much of the vegetation of the Kalahari is commonly described as savanna, such that it can appear to be '. . . no desert, and perhaps the most striking impression was that this was no Sahara but an endless panorama of grass and bush . . . giving a picture sometimes of pleasant parkland of good sized camelthorn trees' (Gaitskell, 1954). Early writings dealing with Kalahari vegetation were purely descriptive and limited in spatial extent (e.g. Lugard, 1909; Miller, 1939; Pole Evans, 1948). Subsequent studies have included regional (Weare and Yalala, 1971; Van Rensburg, 1971 – both for the Botswana Kalahari) and subregional (e.g. Trapnell and Clothier, 1957; Rushworth, 1970; Blair Rains and Yalala, 1972; Timberlake, 1980) surveys which have generally been physiognomic rather than floristic in approach. Ecological studies have also appeared (e.g. Boughey, 1963; Werger, 1978).

The southwestern area, particularly within the bounds of the South African part of the Kalahari Gemsbok National Park, has probably received the greatest depth of ecological investigation (e.g. Leistner, 1959; Leistner and Werger, 1973; Bothma and De Graff, 1973; Werger, 1978; Eloff, 1984; Van Rooyen *et al.*, 1984). Much of the recent interest in the Kalahari savanna communities has concerned their value for, and the effects of, livestock grazing, particularly within Botswana (e.g. Skarpe, 1983; Bergström and Skarpe, 1985; Skarpe and Bergström, 1986).

The term 'savanna' is itself used and defined in a number of different ways, but it is perhaps best used to describe the vegetation communities found in the tropics which are a transition between the closed tropical forests of humid areas and the open grasslands of arid areas (Bourliere and Hadley, 1970). In fact, the Mega Kalahari embraces all three of these groups, with areas within Zaïre and northern Angola having a cover of humid tropical forest and parts of the driest, extreme southwestern Kalahari possessing desert grassland. A notable paradox of the Kalahari savanna zone is the way in which a relatively limited number of plant species dominates, yet, sub-regionally, vegetation appears to differ markedly (Weare, 1971). This is because the various woody and herbaceous plants and grasses tend to combine in different proportions to create communities which often differ more in relative than absolute species composition.

4.4 Vegetation communities

The diversity and biomass of Kalahari vegetation generally increases in a northeasterly direction in line with rising mean annual precipitation values (Figure 4.6). Superimposed upon this general framework are community variations related to inter-regional differences in soil types and geomorphological effects, with the ephemeral and relict river valleys, lake and pan depressions, and intra-dunefield topography being particularly important in the case of the latter. A further important group of factors affecting vegetation characteristics, at a range of spatial and temporal scales, relates to the impact of fire and other disturbances.

4.4.1 Savanna communities

At the regional scale, Kalahari vegetation communities are dominated by savanna complexes as far north as the Caprivi Strip of Namibia and even into western Zambia (Trapnell and Clothier, 1957; Edmonds, 1976) and southern Zaïre (De Dapper, 1981b, 1988). There has long been debate as to whether savanna communities are a natural climatic climax or whether they are human-induced through the effects of burning (see section 4.5). Suffice it to say here that humans have probably affected the vegetation for thousands of years, that some form of savanna is likely to be natural in the drier areas, especially in the Kalahari Desert, and that there is evidence that savanna communities in wetter areas (western Zimbabwe, western Zambia, Angola and southern Zaïre) are at least in part due to human and large herbivore (e.g. Cumming, 1982) activities disrupting the climax deciduous forest communities.

The vegetation communities of the Kalahari Desert (Weare and Yalala, 1971; Figure 4.7) correspond with the arid savannas defined by Huntley (1982). These occur in areas receiving on average less than 650 mm of rainfall, experience distinctly seasonal growth regimes and consist largely of grassland, dominated by annuals, interrupted by trees and shrubs. The grasses are dominated by *Aristida*, *Eragrostis* and *Stipagrostis* species, with trees and shrubs frequently, but not

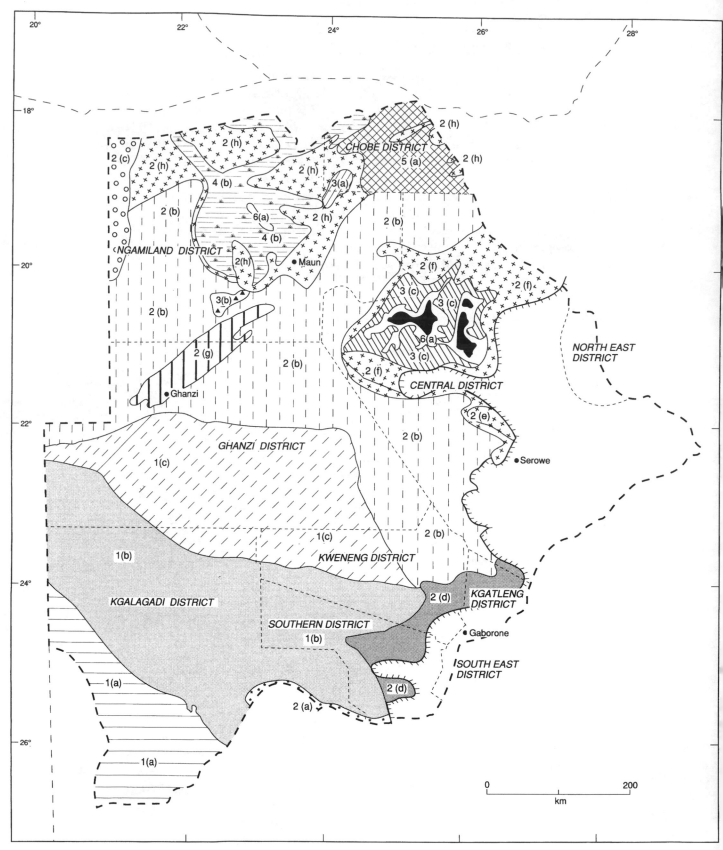

1 SHRUB SAVANNA
   a. Arid Shrub Savanna
   b. Southern Kalahari Bush Savanna
   c. Central Kalahari Bush Savanna

2 TREE SAVANNA
   a. Molopo Thornveld
   b. Northern Kalahari Tree and Bush Savanna
   c. Northwest Tree Savanna
   d. Arid Sweet Bushveld
   e. Mixed Mopane Bushveld
   f. Mixed Mopane Tree and Bush Savanna
   g. Ghanzi Bush Savanna
   h. Ngamiland Tree Savanna

3 GRASS SAVANNA
   a. Mababe Grassland
   b. Lake Ngami Savanna Grassland
   c. Delta Grassland
   d. Fringe Pan Grassland

4 AQUATIC GRASSLAND
   a. Vlei Grassland
   b. Swamp Grassland

5 DRY DECIDUOUS FOREST
   a. Chobe Forest

6 RIPARIAN FOREST
   a. Okavango Fringe Forest

Figure 4.7. Examples of Kalahari vegetation communities (numbers in brackets refer to classification in Figure 4.6).
(a) Typical ungrazed tree and bush savanna (2b), containing mature Acacia trees, up to 6 m high. Northeastern Botswana. (b) Sparse bush savanna (1c) near Dutlwe, southeast Kalahari, which has been subjected to heavy grazing pressure.

Figure 4.6 (opposite). Vegetation communities of the Botswana Kalahari. After Weare and Yalala (1971), but reference numbers modified. Note that Weare and Yalala's use of 'Northern' and 'Southern' Kalahari does not correspond with that adopted in this volume.

(c)

(d)

(c) A stand of mature Mopane woodland (part 2f), on clay-rich soils
near the Nata River in western Zimbabwe. Characteristically there is little
ground cover or understorey. (d) Mopane scrub (foreground) with mixed
woodland containing Terminalia sericia (part 2b) in the distance. Hwange

National Park. (e) Dry deciduous forest (5a) in northwestern Zimbabwe containing Teak trees (Baikaea plurijuga). (f) The Shepherd's or Witgat tree, Boscia albitrunca, found in valleys and on deep dune sand in the southwestern Kalahari (part 1a). Photograph by J. du P. Bothma.

(g) *Pure grassland on imperfectly drained vertisol clay surfaces (3d), in this case the Kazuma Depression on the Botswana–Zimbabwe border. The forest in the background occupies the linear dunes which confine the depression.* (h) *Stands of Hyphaene palm on sand islands within the area of older alluvium near Parakarungu, Lake Liambezi. The palm forms a distinct subunit of the aquatic grasslands throughout the Okavango and Linyanti–Chobe deltas. Other components of the swamp vegetation (4b) can be seen in Figure 5.3b.*

exclusively of *Acacia, Commiphora, Colophospermum* and *Terminalia* genera (Figure 4.7a).

Despite its floristic inadequacy (e.g. Skarpe, 1983), Weare and Yalala's (1971) scheme does give a good indication of the major characteristics of the Kalahari savanna communities. They divided the savannas of the main sandveld into two major types according to whether the larger plant elements are dominated by shrubs or trees. The transformation from shrub to tree dominance is, of course, gradual, but it approximates to the area where mean annual rainfall is 350–400 mm (Figure 4.3).

The shrub savannas therefore occur in the drier southwestern Kalahari. The arid shrub savanna, sometimes called 'shrubland' (Wild and Fernandez, 1967) which occurs in the driest areas of all and extends into South Africa and Namibia, is found in conjunction with the Southern Kalahari dunefield, which has an influence on the local scale distribution of plants (see below). The ground cover afforded by the clumped grasses, which include species of limited spatial distribution, such as *Scmidtia kalahariensis*, is often less than 35 per cent (Thomas, 1988a), with appreciable areas that are totally bare (Grove, 1969; Van Rooyen and Verster, 1983). Beyond the dunefield, plant densities and diversity increase in a northeasterly direction, first into the Southern Kalahari, and then Central Kalahari, bush savannas (Figures 4.6 and 4.7b), which are sometimes known as 'thornveld'. Trees and shrubs may occur in dense thickets, often dominated by stunted *Acacia erioloba* (also termed *A. giraffae*, or Camelthorn), *Acacia mellifera* and *Boscia albitrunca* (the Shepherd's or Witgat tree), interspersed with broad, grassy plains.

The remaining areas of the Kalahari sandveld are almost exclusively covered by Weare and Yalala's (1971) Northern Kalahari tree and bush savanna (Figure 4.7a; note that their use of northern, central and southern Kalahari does not correspond with that of Passarge (1904) or this volume). The vegetation of this vast area consists mostly of grassy plains dotted with low shrubs, often of *Acacia*, *Grewia* and *Croton* species (Weare, 1971), interspersed with belts of trees which often occur on sandy ridges. As well as fully grown examples of the species which form the shrubs, other important species include *Burkea africana*, *Lonchocarpus nelsii* and *Combretum zeyheri*, but perhaps most common of all is *Terminalia sericea*, the presence of which is sometimes taken to indicate deep sands.

A number of notable variations occur within the area dominated by tree savanna. These may be due to the geological effects, for example the Ghanzi Ridge supports a modified form of bush (Blair Rains, 1969), which could, however, be due to adverse grazing pressures. Also significant is the appearance of *Colophospermum mopane* (the Mopane or Butterfly tree) which is a significant, in parts dominant, component of the tree savanna north of about 22° S (Figure 4.6). As well as occurring in mixed woody thickets, especially when in shrub form, this tree often occurs in dense, pure stands, to the virtual total exclusion of other species (Figure 4.7c), possibly where soils have a high clay content (Wellington, 1955). The cause of the notable differentiation between Mopane scrub (Figure 4.7d) and Mopane tree savannas, with a lack of transition from the former to the latter type over time, are unknown, although changes in herbivore grazing patterns, and the susceptibility of this species to frost, have been put forward as causes. Likewise the rapid expansion of Mopane scrub into grassland areas since the mid-nineteenth century (section 8.3.3), most noticeable in the Mababe Depression, has been explained by a reduction in the herbivore population.

4.4.2 Woodland and forest communities

North and east of the Kalahari Desert, where rainfall amounts are generally higher, the Kalahari Sand often supports woodland, especially where deep rooting is expediated by deep, soft sands. In some areas, woodland is virtually continuous, while in others it is zoned and forms a component of what Huntley (1982) has termed moist savanna. In northeastern Botswana and neighbouring western areas of Zimbabwe and Zambia, dry deciduous woodland occurs both as continuous stands and as belts associated with dune ridges (Boughey, 1963; Flint and Bond, 1968; Weare, 1971; Thomas, 1984*a*). This is often dominated by *Baikaea plurijuga* (the Rhodesian Teak), with mature examples attaining heights of 20 m (Figure 4.7*e*), *Pterocarpus angolensis* and *Burkea africana*. These can form a closed canopy woodland, with an understorey of species such as *Baphia obovata* and *Ochna pulchra*, or a more dispersed woodland savanna which includes abundant *Terminalia sericea* and *Combretum hereroense*. Degradation of the woodland has occurred in many areas. In southern Zaïre, this has led to replacement of the woodland, known as *Miombo*, by *Dilungu* grassland (De Dapper, 1988); in western Zambia, continuous canopy teak forest has widely been replaced by two-storey woodland (Williams, 1986).

4.4.3 Intra-dunefield variations

The dunefields of the Kalahari, which are discussed in detail in Chapter 6, support notable variations in vegetation communities, not only at the inter-dunefield scale between the different systems, which is a reflection of their position within the southwest–northeast climatic and vegetation gradient, but within the dune systems themselves (see Figure 6.2). The southern dunefield (see section 6.2.1) occurs within the arid shrub savanna zone, but the linear dune ridges tend to support lower plant densities, with clumped *Eragrostis* and *Aristida* grasses, than the interdune areas. These possess more typical shrub savanna (Huntley, 1982) with sub-regional variations (Van der Meulen and Van Gils, 1983) and individual tall trees, commonly *Acacia* spp. (Grove, 1969). The higher dunes and dune ridge intersections also appear to be favourable sites for tree growth (Bothma and De Graff, 1973; Buckley, 1981; Eriksson *et al.*, 1989; Figure 4.7*f*), probably because of the water retention capabilities of the deep dune sands and the relative stability of the dune intersections.

The eastern and northern dunefields, which as their names imply occur in northeastern and northern Botswana and neighbouring territories (see sections 6.3.1 and 6.3.2), are stable dune systems which in both cases possess higher biomass vegetation communities on the dune ridges than in the inter-dune areas. North and west of the Okavango Delta the northern dunefield generally supports open bush and tree savanna on the dune ridges, dominated by *Burkea africana* and *Pterocarpus angolensis* and grassland with more sparse shrubs in the inter-dune areas (Grove, 1969). In the eastern dunefield, the soft, sandy ridges generally support the dry deciduous woodland appropriate to their regional setting and described above. By contrast, interdune areas are generally grassed, perhaps due to their more compacted and finer sediments, and frequent seasonal waterlogging (Thomas, 1985), though in some cases they support shrub or Mopane savanna. The intermediate ridge flanks and areas where vegetation has been disturbed often possess mixed scrub savanna woodland dominated by *Terminalia sericea* (Flint and Bond, 1968; Thomas, 1984*a*).

4.4.4 Riverine, lacustrine and swamp communities

The Middle Kalahari landforms associated with permanent or seasonal standing water, either today or in the past, have distinct vegetation communities. The Okavango Delta has been identified as one of the botanically richest areas in southern Africa, with 1061 species recorded at a species/area ratio of 0.0545, a level of diversity surpassed only by the Fynbos region of the southern Cape Province (Snowy Mountains Engineering Corporation, 1989). These communities range from savanna woodland to riverine forest to aquatic grassland to perennial swamp, following the catena from sand island to permanent water.

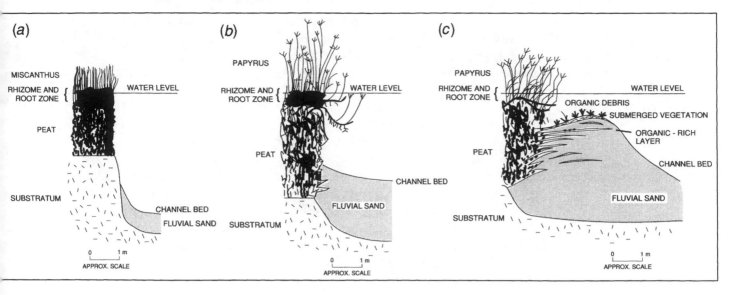

Figure 4.8. Schematic sections of channel margin vegetation in the Okavango Delta, after McCarthy et al., 1988b. The long, flexible papyrus rhizomes are susceptible to breakage in high water velocities which means that M. junceus tends to dominate the outer banks of bends where flow is rapid (a) and C. papyrus the inner banks (b). Sand bars may develop here, becoming colonised by submerged species such as Otellia ulvifolia and Potamogeton thunbergii (c).

Within this topographic continuum there are also variations due at least in part to the frequency and degree of seasonal flooding (Biggs, 1976).

The aquatic communities of the channels, lagoons (*madiba*) and swamps are composed of communities of emergent, floating and submerged species. Emergents occur as marginal swamp grassland along channel margins, in which *Cyperus papyrus*, *Phragmites australis* and *Miscanthus junceus* are important components (McCarthy *et al.*, 1988*b*). These are usually rooted in submerged channel peat deposits and, due to differences in the growth habits of the three species, occupy different positions within sinuous channel systems (Figure 4.8). Comparable channel margin communities are associated with the perennial Chobe River in the Caprivi Strip, although the floating fern *Salvinia molesta* (Kariba weed) is also present in this river. Although *Salvinia* has invaded parts of the Okavango, since 1986 it has been subject to an intensive and successful control programme. Adjacent to these perennial swamps are floodplain systems containing communities of seasonal swamp, dominated by *Schoenoplectus corymbosus* and *Cyperus articulatus*, and flooded grassland, containing some 77 species (Snowy Mountains Engineering Corporation, 1989).

Riparian woodlands occur in belts up to 200 m wide on islands and along channels. They contain a variety of deciduous trees, including *Combretum hereroense* and *imberbe*, *Garcinia livingstoni*, *Ficus sycomorus* and species of *Acacia*, mostly *A. nigrescens*. These riparian woodlands are also found beyond the Okavango on the Boteti and Thamalakane Rivers, and along the Chobe. In the northern Mega Kalahari, Malaisse (1975) noted that strips of riverine forest protrude into the *dilungu* grassland along the valleys of permanent rivers.

Sand islands within the area of older alluvium to the east of the Okavango Delta and south of the Linyanti River can be readily identified from the air because they support significant stands of palm trees. The Makgadikgadi, Ngami and Mababe depressions all support distinct grassland communities (Weare and Yalala, 1971), although bush encroachment by *A. tortillis*, and, as already noted in the Mababe, by *Mopane*, has been considerable. Smith (Snowy Mountains Engineering

Corporation, 1989) suggests that old waggon routes through the Ngami basin can be identified easily from the 1951 aerial photography by the alignment of *Acacia* trees, which have grown from seed pods propagated in cattle dung. At the periphery of the lake basins tall *A. erioloba* dominates deep sand ridges, while the Baobab (*Adansonia digitata*) is a striking tree on rock outcrops, escarpments and shorelines. Towards the Makgadikgadi the vegetation becomes increasingly halophytic; an obvious component of the landscape in the western Makgadik-gadi, also found in parts of the Okavango, is the palm *Hyphaene petersiania*.

The smaller pan depressions of the Central and Southern Kalahari display variations in the degree of surface vegetation (Boocock and Van Straten, 1962), which ranges from total bareness to a well developed grass cover of *Sporobolus*, *Odyssea* and *Cynodon* species (Figure 4.7g). Differences probably reflect variations in the periodicity of inundation by standing water, the depth of the groundwater table, surface sediment salinity and the efficacy of deflation processes.

The dry river valleys support a denser and sometimes more mature vegetation cover than surrounding areas. In the southwestern Kalahari, Weare and Yalala (1971) specifically identified the Molopo thornveld community (Figure 4.7) as an important variant within the shrub savanna. It generally comprises the same species as neighbouring areas, but within a kilometre or so of the valley bottom examples of *Ziziphus mucronata* and Acacia trees, in particular *A. erioloba* attain greater heights.

4.5 Disturbances affecting Kalahari vegetation and soils

The regional classification of Kalahari vegetation communities (Figure 4.6) does not make apparent the small-scale spatial variations in plant communities which occur and persist at a variety of time scales as a result of a number of disturbance factors which range from fire to the feeding habits of wild and domestic animals.

4.5.1 Fire

The role of fire in the evolution of savanna vegetation communities has long been debated, with some arguments supporting the notion of savannas being fire climaxes and others that they would exist as climatic climaxes regardless of fire. Whatever the merits of these different views, it is likely that fire, caused both by lightning strikes, and by deliberate and accidental human actions, has influenced Kalahari vegetation communities for a considerable period. Deliberate burning has been carried out both as an element of hunting strategies (probably since the Stone Age) and as a grazing management procedure in order to stimulate and improve the quality of grass and shrub growth. In general, while some nutrients are lost into the atmosphere during burns, fires release the nutrients stored in dead organic material, returning them to the soil in a form which might promote the growth of new shoots. Skarpe (1980) found that the best shrub species for browse food were the best survivors of fires and that grass growth improved in the years immediately following a burn. Fires are still deliberately employed in this latter context in the Kalahari, with aerial imagery of the Kalahari clearly showing the spatial impact which it can have on the vegetation (Figure 4.9). A study by Field (1978) estimated that 10 per cent of the rangeland in Botswana was burnt deliberately in 1975, though this activity has since been outlawed in its uncontrolled form by the Botswana government, on the disputed grounds that it has a detrimental impact (Skarpe, 1980) on plant communities.

The occurrence and impact of fires are likely to vary both from place to place in

Figure 4.9. An aerial photograph of part of the Kalahari west of the Ngami Basin showing the impact of burn scars on the landscape.

the Kalahari and from year to year. Generally, but paradoxically, it is likely that fires are hotter and therefore more destructive when rainfall amounts are higher. This is because plant growth and therefore the build up of fuel material is greater in wetter areas. As rainfall in the Kalahari is seasonal regardless of total annual amounts, with fires most likely to occur towards the end of the dry season when the vegetation is driest, higher rainfall amounts do not have a supressive influence through a wetting effect. Consequently, fires probably increase in significance across the Kalahari rainfall gradient in a northeasterly direction (Arntzen and Veenendaal, 1986), but their occurrence may also have a temporal dimension with a greater prevalence, even in the driest southwestern Kalahari (Buckley *et al.*, 1987*a*) during or immediately following wetter years or periods.

Within the Kalahari savanna zone, fires not only cause a reduction in the presence of fire-sensitive species, but they are also generally size-selective. In the western central Kalahari, Skarpe (1980) noted the greater resistance of *Boscia albitrunca* compared to *Acacia* species and the tendency for lower tree canopies (at about 2 m above the ground surface) to be destroyed and taller individuals to survive. The shrub layer also tended to survive, resprouting from ground level in the post-fire period, unlike the lower trees and larger shrubs which were killed below ground as well as above.

Repeated burning therefore might lead to a propensity for larger individual trees to survive together with regeneration of the bush scrub layer. The destruction of smaller trees and the survival of larger ones is not always the case, however. In the deciduous woodland area of western Zimbabwe, Boughey (1963) and Flint and Bond (1968) note the particular susceptibility of *Baikiaea plurijuga* to burning and subsequent replacement by dense *Terminalea sericia* scrub. In the Southern Kalahari, the exclusion of fire as part of a management strategy may well have contributed to the destruction of large individual examples of *Acacia erioloba* in the Nossop Valley during a single fire reported by Buckey *et al.*, (1987*a*), due to the considerable growth of tree base vegetation in intervening fire-free years.

4.5.2 Animal activities

The Kalahari has supported, and in some areas still supports, large populations of wild herbivores. Populations of domestic herbivores, notably cattle, have probably existed on the fringes of the Kalahari for 2000 years in conjunction with pastoralist cultures and today appear to be growing not only in absolute numbers but also in terms of the area of the Kalahari which they affect. At the local scale, particularly around water sources, and possibly more extensively, herbivore populations are therefore a further influence upon the dynamics of Kalahari vegetation communities, through their grazing and browsing activities.

Several authors have pointed out that herbivores have had a profound influence upon the structure of African savanna communities. According to Cumming (1982), however, it is only a limited number of species, which include elephant, buffalo, wildebeest and zebra, that cause changes in community structure through their feeding habits, because others have evolved over a long time period during which their food plants also evolved efficient defence systems. For example, many *Acacia* species do not grow thorns when canopy height exceeds 5 m, which is about the upper height of giraffe browsing (Foster and Dagg, 1972).

The significance of grazing and browsing damage is likely to have been relatively well distributed under totally natural conditions because the removal of food plants by over-use in one area would have caused the animals to move elsewhere in the search for food, allowing plants to recover. Under conditions of restricted movement and confinement, however – which occur increasingly today due to land use competition and apply both to domestic livestock and to wildlife restricted to the limits of National Parks – impacts are likely to be more significant and possibly more long-lasting (Thomas 1988*e*; section 9.4.3).

Of the wild herbivores, elephants are particularly important in this respect, as they have the ability to browse both delicately and selectively or in a destructive way which results in the death of whole trees. Their impact is exemplified from

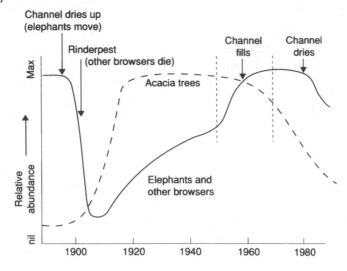

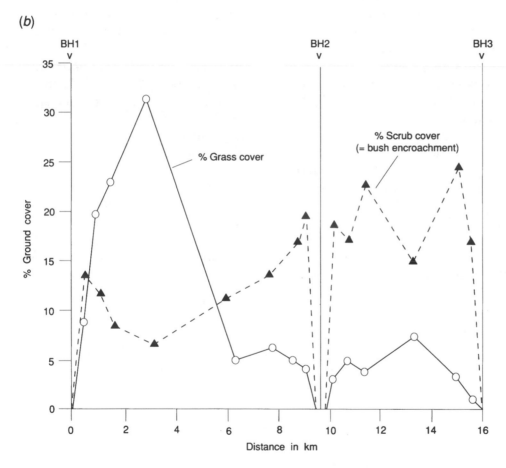

Figure 4.10. (a) Acacia dynamics in Chobe National Park, Botswana during the twentieth century. Acacia abundance appears to oscillate in an inverse relationship with the size of elephant and other browser populations in the area. These animals are attracted to the area by the all-year-round availability of water in the Chobe River; periods when water is available elsewhere or when disease affects game populations cause a reduction in browser numbers in the area and consequently a recovery in the Acacia tree population. After Lewin (1986). (b) Simplified vegetation patterns around three boreholes in Kweneng District, eastern Kalahari. Permanent use of the boreholes for water by cattle results in the creation of a bare 'sacrifice zone' in the immediate vicinity of the bores due to trampling, and the reduction or eradication of the grass cover in a zone several kilometres wide due to grazing pressures. The removal of grasses commonly allows rainfall to percolate to greater depths in the sand, allowing shrub species to grow in importance, causing the process known as 'bush encroachment'. After Martens (1971).

Sibungwe National Park, Zimbabwe, located on the eastern extremity of the Kalahari Sand, where Guy (1976) calculated that a single elephant would destroy 1500 trees annually. If this figure is representative, it is likely to be significant that the Sibungwe elephant population quadrupled to 11 000 and the range area available was reduced by 70 per cent between 1955 and 1980 (Cumming, 1981), with Anderson and Walker (1974) showing that one-third of the area had suffered bush scrub encroachment at the expense of trees.

It is difficult to assess the significance of such changes in terms of vegetation patterns within the Kalahari Desert, but damage and destruction of Kalahari woodlands by herbivores have been noted in other areas too (Flint and Bond, 1968; Lewin, 1986). In the Chobe area, cyclical fluctuations in the abundance of Acacia trees during historical times have been linked to fluctuations in the size of the elephant population induced by disease and hydrological changes (Figure 4.10a). Although it can be argued that these changes are not detrimental and are part of the natural dynamics of savanna vegetation communities (Lewin, 1986), the increasing provision of water at permanent pumped boreholes, supplying both wildlife in the Kalahari's sizeable National Parks (e.g. Mitchell, 1961; Cumming, 1981) and cattle communities in the drier Southern Kalahari, results in all-year-round grazing and browsing pressures and a reduction in the opportunities for vegetation communities to recover (see Chapter 9). In the Kalahari savanna zones, many authors (e.g. Martens, 1971; Seitshiro, 1978; Arntzen and Veenendaal, 1986; Thomas, 1988e) have noted that the year-round use of these borehole water sources by cattle can result in the superimposition of local-scale radial vegetation patterns upon the landscape (Figure 4.10b), characterised by the eradication of palatable species and the encroachment of bush scrub (e.g. Tolsma, Ernst and Verwey, 1987). However, given the complexity and range of factors which are likely to interact and influence the dynamics of Kalahari vegetation, in particular the climatic gradient, rainfall cycles, fire and animals, and the lack of longer-term data concerning vegetation dynamics, it cannot be automatically assumed that these changes represent permanent modifications to the nature of Kalahari plant communities.

Part Two

GEOMORPHOLOGY AND ENVIRONMENTAL CHANGE

5 The geomorphology of the Kalahari: rivers and lakes

THE LANDSCAPES of the Kalahari have been formed by a range of geomorphological processes operating against a background of tectonic activity, climatic change and, increasingly, human use of the environment. The dominant processes are those involving the activity of water and wind on the land surface, and the movement of water, with associated chemical activity, within the groundwater zone.

5.1 Kalahari landscape overview

The observations of early European travellers, who viewed, from ground level, a homogeneous landscape of gently undulating sand plateaux, interspersed with occasional valleys and pans which were, inevitably, dry in their hour of need (e.g. Selous, 1893), encouraged the conception of the Kalahari as a desert or 'thirstland'. Areas within the Kalahari where water is more freely available, such as the Okavango–Makgadikgadi system, became the foci of travel routes, and were described from the mid-nineteenth century onwards (e.g. Livingstone, 1858a; see section 1.3.2). By the 1930s the greater part of physiography of the region had been mapped at a reconnaissance level, and basic descriptions of the landforms had become available (e.g. the 1935 1:250000 map of the Bechuanaland Protectorate, reproduced by the Botswana Society, 1984). Since then, advances have been made in two directions; first, in the resolution of our observations of the landscape, and second, in our understanding of its origins and evolution.

On the first point, our knowledge of the landforms of the Kalahari has been increased immeasurably by the adoption of aerial photography and satellite imagery to provide an overall view (Mallick et al., 1981; Williams, 1986), an epoch ushered in by oblique photographs taken of the Zambezi and Okavango by Captain Meredith and Lieutenant Tasker from two DH9 biplanes on Alex du Toit's expedition of 1925. Now we are able to see, in toto, massive landforms that can be only fragmentarily observed on the ground due to their great size, and, in some cases, the limited relative relief which they possess. These landforms are related to, and provide evidence of, climatic and environmental changes that have taken place in the Quaternary. The resolution of dating methods is not yet sufficient to explore the complete Quaternary history, but there is sufficient evidence, presented in Chapter 7, to identify climatic episodes, wetter, and possibly drier, than present within the past 40000 years.

The implications of these landforms were explored in Grove's seminal paper of 1969; since then, considerable research has been carried out on various components of the landscape. Arid phases have been considered to be represented by extensive fields of linear dunes, which have been described for the Middle and Southern Kalahari by Lancaster (1981, 1986a) and for northern areas by Thomas (1984a), though debate exists on the correlation of dunefields, the role of wind directional changes and their significance as indicators of aridity (Lancaster, 1988; Thomas, 1988b).

Episodes of more substantial precipitation have been inferred from the presence of a massive fossil palaeolake system, termed Lake Palaeo-Makgadikgadi, which links the Makgadikgadi, Ngami and Mababe Basins to the Okavango and adjacent rivers (Cooke, 1976, 1980, 1984; Cooke and Verstappen, 1984; Shaw, 1985a, 1986, 1988a), and which exhibits a high degree of control by tectonic factors. Studies have also been made of other climate-related landforms, such as pans (Lancaster, 1978a, 1978b, 1979), ephemeral rivers (Heine, 1982) and caves (Cooke, 1975a, 1984; Cooke and Verhagen, 1977).

The water balance, too, is implicated in the evolution of landforms. Water in the sub-surface zone is continually affected by evaporation and recharge to produce one of the most comprehensive duricrust suites in the world, the characteristics of which have been noted in section 3.5. In turn, some of these duricrusts are associated with surface landforms, such as pans and valleys, suggesting that ground water plays an active part in the formation of these landforms.

Despite these advances, some landforms and some areas of the Kalahari remain poorly understood, particularly in the Northern Kalahari where field access is limited. This chapter examines the fluvial and lacustrine landforms of the Kalahari in the light of present knowledge, while aeolian and rock landforms are discussed in Chapter 6. Pans are included in the latter because although their development is now in some cases being considered in terms of groundwater activity, deflation has often been cited as a major control on their formation. The palaeoenvironmental implications of landforms form the subject of Chapter 7.

5.1.1 General drainage characteristics

Drainage within the greater Kalahari area generally reflects the precipitation gradient from north to south and the structural development of the subcontinent, and has encompassed a number of drainage changes since the division of Gondwanaland, which have been described in section 3.2.2. The sand plateau of the Northern Kalahari has exoreic drainage, forming the watershed between upper Congo tributaries in the north and the Okavango and Zambezi systems to the south. These rivers are perennial, although they exhibit strongly seasonal regimes. In the southern basin, however, surface drainage is strongly endoreic or absent altogether, with the single link to the Atlantic, provided by the Molopo and its tributaries, rarely carrying water for more than a few kilometres at a time. The Middle Kalahari is a zone of contrasts, with a central, highly saline sink for both surface and ground water provided by the Makgadikgadi Basin, adjacent to the considerable freshwater resources provided by the Okavango and Zambezi within the fault zone. Throughout the Middle and Southern Kalahari there is a close link between surface and sub-surface flows, the latter contributing greatly to weathering, duricrust type and development, and inevitably, to the formation of landforms.

5.2 The perennial rivers of the Northern Kalahari

By virtue of its higher precipitation levels, the Northern Kalahari possesses all the

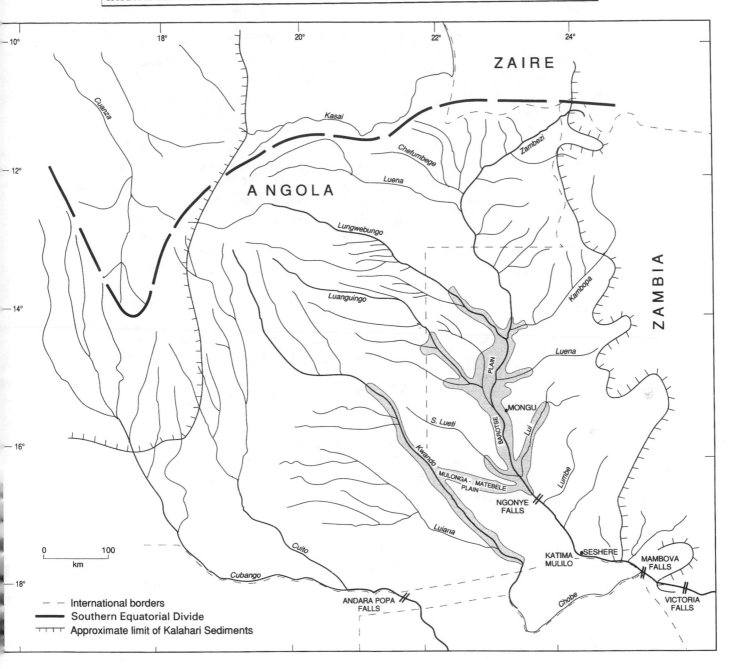

Figure 5.1. The drainage of the Northern Kalahari.

perennial rivers of the Kalahari. Though the perennial Orange River in part marks the southern limit of the Kalahari Desert, neither it, nor its palaeoenvironmentally significant tributary, the Vaal (Helgren, 1979) impinges directly on the desert and it will not therefore be considered in this volume. The rivers of the Northern Kalahari can be divided into the perennial river systems which terminate or skirt the Middle Kalahari, the Cubango/Okavango, Kwando/Chobe, and Zambezi, and those which are tributaries of the Congo system, notably the Kasai (Figure 5.1). The two systems are separated by the Southern Equatorial Divide watershed such that the former flow in an essentially southerly direction in the Northern Kalahari and the latter to the north, although the Kasai flows eastwards in its upper reaches. The Congo tributaries, flowing away from the Kalahari proper, will not be discussed here, while the southward flowing rivers are considered below.

5.2.1 Rivers south of the Southern Equatorial Divide

The sources of the Cubango/Okavango, Kwando/Chobe and Zambezi all lie close to the Southern Equatorial Divide watershed, where mean annual rainfall is in the region of 1300–1400 mm, in the Angolan Highlands or south of the section of the watershed located on Kalahari sediments and known as the Kalahari Col (see section 1.2.2). As befits a river of its size, there are several headwaters which could equally lay claim to being the true source of the Zambezi, including the Chefumbage and Luena (Figure 5.1), but several authorities (Gibbons, 1904; Steel, 1917; Wellington, 1955) consider that the source (in Wellington's words) is 'a bog in dense bush on the southern slopes of the south equatorial watershed, about 30 miles north of Mwinilunga.' Of greater hydrological significance are the respective sizes of the headwater basins, which, taking the region north of the Middle Kalahari, are 300 400 km² for the Zambezi, 129 500 km² for the Kwando/Chobe and 121 700 km² for the Cubango/Okavango (derived from data in Wellington, 1955).

Whereas the Kwando, which becomes the Linyanti or Chobe in Botswana, has its source within the Mega Kalahari, the Cubango, which becomes the Okavango, rises on the crystalline Benguela Plateau of eastern Angola, west of the section of the watershed known as the Kalahari Col. From its source it flows about 300 km SSW before passing into the Northern Kalahari, where it remains for the rest of its course before reaching the Okavango Delta. The Zambezi takes a short detour from the bog described by Wellington, around a Karoo sandstone hill, before entering the Northern Kalahari, though other headwater streams do rise within the Kalahari itself. It can essentially be regarded as a river of the Mega Kalahari throughout the part of its course known as the Upper Zambezi, which extends to the confluence with the Chobe at the Mambova Falls, and perhaps along the stretch of the Middle Zambezi as far as the Victoria Falls.

5.2.2 The Cubango/Okavango

Compared with its terminal delta, the Cubango River, which becomes the Okavango at its confluence with its major tributary, the Cuito, has itself received relatively little detailed investigation, though it has been considered to be in many ways the most remarkable river in southern Africa (Wellington, 1955). Although the annual hydrograph of the Okavango has a seasonal regime reflecting rainfall patterns in the catchment, with the flood peak reaching the distal end of the Delta in May or June (e.g. Shaw, 1986), the high rainfall and run-off of the upper reaches of the catchment mean that high flow levels can occur even in the driest months, with, for example, Wellington (1955) reporting a flow of 85 cumecs 225 km downstream from the source in July 1953.

On its course through Angola the channel of the Cubango is for the most part gently incised into the Kalahari Sand, but in places greater and more obvious incision of up to 30 m has been achieved where the sub-Kalahari geology has been exposed and more upstanding lithologies breached (Schönfelden, 1935). After marking the border between Angola and Namibia for nearly 400 km on an almost due east course, along which it is joined by the Cuito, the river enters a 40 km section of rapids which crosses Damara Sequence quartzites. The southern end of this section of the river, which is in a valley that is over 3 km wide and cut into the Kalahari plateau by about 15 m, is marked by a quartzite ridge that ponds back the flow to form a channel 300 m wide. The ridge is eventually topped to form a series of rapids known as the Popa Falls, with a height of about 5 m. Thereafter, the river turns southeastwards to enter the 'Panhandle' of the Okavango Delta.

5.2.3 The Upper Zambezi and Kwando

Among several streams in the Zambezi headwaters in Angola and Zambia there are three major ones (Figure 5.1): the Zambezi itself, whose early stages pass over a series of rapids, and the Angolan Luena and Chefumbege, which both rise on the Kalahari Sand. South of the confluence of the Angolan Luena and Zambezi, the river enters a substantial alluvial floodplain, referred to as the Barotse Plain by Wellington (1955) and the Bulozi Plain by Williams (1986). This is flooded annually between January and June and extends for some 180 km, attaining a maximum width in excess of 30 km to the south of Mongu, before narrowing and terminating north of the Ngonye Falls (Williams, 1986). The Upper Zambezi is joined by some notable tributaries between the Angolan Luena confluence and the falls. The major east bank tributaries in this section are the Kabompo, Zambian Luena and Lui, and from the west, the Lungwebungo (Lunge–Bungo in Angola), Luanguingo (Luanginga) and Southern Lueti. The lower sections of all these tributaries are alluviated, being extensions of the Barotse Plain, while the trellis-like Angolan headwater tributary systems of the Lunge-Bungo and Luanginga are notable in that they appear to represent the superimposition of drainage on an ancient linear dune landscape (Thomas, 1984a, b; see section 6.3.2).

The Ngonye Falls mark the beginning of a section of the river, 130 km in length and extending to Katima Molilo, along which the Kalahari Sand has been partially breached and a series of rapids marks the passage of the Zambezi across outcrops of lower Kalahari (Barotse Formation) sandstones and quartzites (Money, 1972). Unlike Money, Williams (1986) considers that the passage of the river across these outcrops is insufficient alone as an explanation of the change in character of the Zambezi valley from a broad alluvial plain to a narrower, constricted feature. Instead, he advocates tectonic uplift, associated with the downwarping in the Middle Kalahari, as the cause of the channel incision. At Katima Molilo, the river enters the swamp zone of the Middle Kalahari where, at the Mambova Falls, it is joined by the Chobe (Figure 5.3a) before heading eastwards towards the Victoria Falls and leaving the Kalahari.

The southeasterly course of the Zambezi between the Angolan Luena confluence and Katima Molilo is paralleled to the west by the Kwando, which becomes the Linyanti and ultimately, the Chobe. Though significantly narrower, the flood plain of the Kwando is twice as long as the Barotse Plain, as the channel lacks a section of rapids where significant incision through the Kalahari Sand has occurred. Its course below the confluence with the Luiana does, however, have a steep gradient, representing a fall of about 75 m in 100 km (Seiner, 1909). A notable feature of the relationship between the Zambezi and Kwando is the Mulonga-Matebele Plain, which is a broad belt of alluvium connecting the two river valleys at latitude 16° S (Figure 5.1). The plain does not possess any notable fluvial features today (Williams, 1986), although it is subject to flooding in the wet season and possesses numerous small pans (Thomas, 1984b). Its origin has been the subject of considerable speculation, (e.g. Verboom and Brunt, 1970; Verboom, 1974; Westerhof, 1976) and although Williams (1986) discounts the possibility of it being an earlier course of either the Kwando or Zambezi, its location upstream of the zone of tectonic uplift suggests that it may have played a role as an outlet for ponded seasonal floodwaters. Williams (1986) suggests that its orientation indicates an origin postdating the formation of the linear dunes of the northern dunefield (section 6.3.2), which is supported by the fact that the dunes flanking the plain appear to have been truncated and in some cases destroyed by the action of water (Thomas, 1984b).

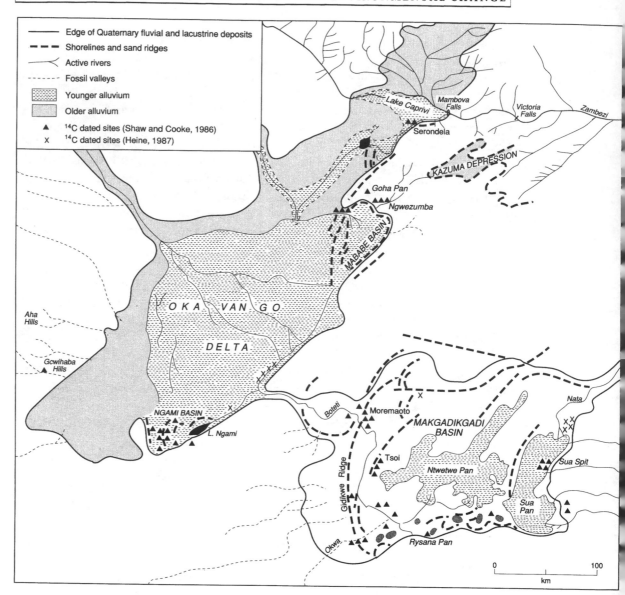

Figure 5.2. The drainage of the Middle Kalahari. After Shaw and Cooke (1986).

5.3 The fluvio-lacustrine landscapes of the Middle Kalahari

At 18° S, the three major sub-parallel rivers draining from the north, the Okavango, Kwando/Chobe and Zambezi, reach the tectonically active zones of the Gwembe Trough and Okavango Graben, where extensive alluviation is offset by active movement along a series of SW–NE trending faults. Tectonic adjustment and climatic change has led, in the past, to the formation of Lake Palaeo-Makgadikgadi, encompassing the Okavango Delta, parts of the Chobe–Zambezi confluence, and the Ngami, Mababe and Makgadikgadi Basins (Figure 5.2).

The complexity of this regional network has long been recognised (Livingstone, 1858a, b) and described (Passarge, 1904; Du Toit, 1926; Wellington, 1955; Grove, 1969). Its sheer size, however, has delayed comprehension of its function. The area covered by lacustrine landforms in northern Botswana amounts to 60 000 km², equivalent to that of the present Lake Victoria in East Africa (Cooke, 1980). To this may be added the fluvio-lacustrine landforms of western Ngamiland, the Caprivi Strip and, possibly, the alluvial corridor to the

Kafue in Zambia to reach an area of approximately 120 000 km², making Lake Palaeo-Makgadikgadi second in size in Africa to Lake Chad at its Quaternary maximum (Rognon, 1980). The following sections describe the component parts of Lake Palaeo-Makgadikgadi, and provide an assessment of its configuration. The age and implications of the system are examined in Chapter 7.

5.3.1 The Okavango Delta

The Okavango River enters the Middle Kalahari at Mohembo through the Panhandle, a narrow swamp controlled by NW–SE faults, before extending into a delta-shaped system of swamps and anastamosing distributary channels (Figure 5.3b), essentially an alluvial fan, covering some 6000–13 000 km² (UNDP/FAO, 1977), depending on the prevailing flood and precipitation characteristics (Figures 5.4). The term 'delta' is not as precise as 'delta-fan' which has been adopted by McCarthy et al., (1988a), but remains in common usage and has been retained here. The gradient below the Panhandle is extremely low; the entire delta slopes from about 1000 m asl at Mohembo to 931 m asl at the Boteti divergence (Cooke, 1976).

The delta has five 'limbs' of distributaries, the most easterly of which, the Magwegqana or Selinda Spillway, occasionally carries water to the adjacent Linyanti River, and hence towards the Zambezi. The Maunachira and Thaoge systems drain towards the Mababe and Ngami Basins respectively, while the central distributaries, the Mboroga and Jao, are terminated by the distal Kunyere and Thamalakane Faults, which again distribute water towards the Ngami and Mababe Basins via the Kunyere River and the bidirectional Thamalakane River, with an additional offtake via the Boteti River towards the Makgadikgadi Basin.

The mean budget of the delta amounts to about 15.5 km³ annually (15.5 × 10⁹ m³ yr⁻¹), of which two-thirds is from inflow initiating from the Angolan Highlands, with the remainder from precipitation over the delta itself. Whereas the precipitation occurs mainly between October and March, the flood peak passes Mohembo in April, and usually reaches the distal end of the delta in July. There are thus two distinct inputs to the regime. The distribution of the water takes the form of a permanent swamp, which occupies the upper delta and the main distributaries, and seasonal swamp resulting from the annual flood, which is concentrated in the lower delta.

The total amount of water entering the delta varies from year to year, but some 96 per cent is lost to evapotranspiration or ground water, while the remainder leaves as outflow from the distributaries. As the swamps occupy only 30–60 per cent of the 22 000 km² of alluvial sediments comprising the delta area, it follows that the delta is a dynamic system in both space and time (Hutton and Dincer, 1979), with a discrete and unusual set of geomorphological processes. The amount and route of flow within the delta changes on a seasonal basis in response to variations in the hydrological and climatic regimes, vegetation blockage, siltation, human- or animal- (particularly hippopotamus) induced diversion and to minor tectonic events. The nature of this flow has been summarised by Wilson and Dincer (1977: p. 33):

> The conical shape (of the Delta) favours the spreading of flood waters, but the nature of the spreading is in some ways analogous to water being spilled on a flat table top, the area of spill being nearly proportional to the amount spilled.

Studies by McCarthy et al., (1987, 1988a) have shown that switching of flows within the anastamosing channel system occurs on a cycle of approximately 100 years following aggradation of the channel bed. The channel is gradually raised above, but confined by, the surrounding swamp by up to 4 m of accumulated peat, which eventually causes the channel to become moribund. After abandon-

Figure 5.3. (a) The Chobe River has a broad and deep, meandering channel which occupies the southeastern sector of the Middle Kalahari swamp zone. In the background is the Chobe Escarpment. (b) A typical channel (foreground), fringing swamp (melapo, middleground) and island sequence within the Okavango Delta.

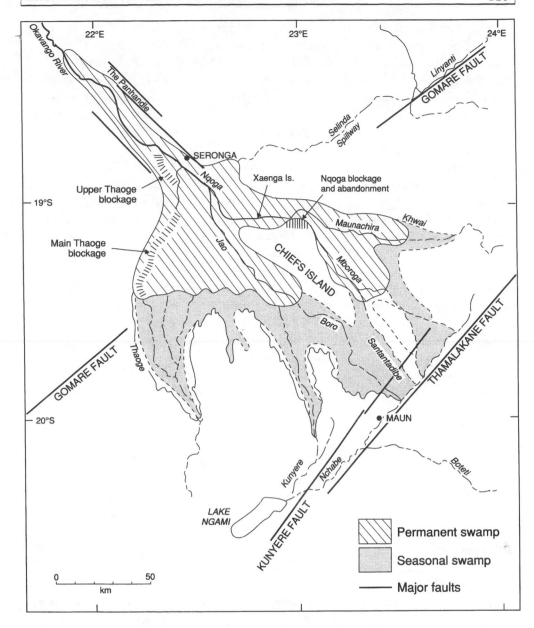

Figure 5.4. The main features of the Okavango Delta. After Hutchins et al. (1976a, b) and McCarthy et al (1988a).

ment the channel peat becomes desiccated and destroyed by fire, leaving a sand ridge some 30–40 cm above the area of the surrounding swamp, prior to reflooding.

Inflow and outflow records kept for Mohembo, at the head of the Panhandle, and Maun, at the distal end of the delta, since 1930 have provided some data on gross changes in the delta regime, particularly with reference to shifts between the lower distributaries and regime changes following the Maun earthquake of 1953 (UNDP/FAO, 1977; Snowy Mountains Engineering Corporation, 1987: vol. 2). Changes on the historical timescale, with particular reference to *Papyrus* blockage and partial abandonment of the Thaoge system since the nineteenth century, have been noted by Wilson (1973) and Wilson and Dincer (1977), while Shaw (1983, 1984) has identified major fluctuations in the hydrological regime since 1849.

Patterns of sedimentation established within the Okavango Delta are interesting in the context of rates and environments of deposition of the Kalahari Group sediments. Present sedimentation can be considered in terms of solute, flotation and suspended/bedload components (Shaw, 1989a). The latter has been estimated at 0.6 million tons (400 000 m³ yr⁻¹) (UNDP/FAO, 1977) with a mean grain size of 0.2–0.4 mm in the upper delta, grading down to 0.2 mm in the distal sector. The sampling programme adopted to date has been far from comprehensive; other estimates include 2 million tons yr⁻¹ (Snowy Mountains Engineering Corporation, 1986) and, based on measurements at one site in a low flow period, of 110 000 tons yr⁻¹ (McCarty, Stanistreet and Cairncross, 1990). These figures also suggest that deposition within the delta is not even, with the bulk of the sediment being deposited in the lower Panhandle, with increased gradients and flow velocities in the actual delta. Such fluvial disequilibrium could be a permanent feature, arising from tectonic movements, channel avulsion or long-term variations in the hydrological regime.

The flotation component, composed largely of *Phragmites* species, *Cyperus papyrus* and *Vossia cuspidata* detritus, has long been considered important in the blockage of channels (Wilson, 1973; McCarthy *et al.* 1986a), but it is difficult to estimate in terms of volume as it travels only short distances. McCarthy *et al.* (1990) estimate that some 30 000 tons enter the Panhandle annually, a small fraction of the total biomass produced within the delta itself. Its deposition in peat layers has provided an analogue for the deposition of Permian coal beds elsewhere in southern Africa (Cairncross *et al.*, 1988).

The role of solutes is also becoming increasingly apparent. ¹⁸O studies of the Okavango (Dincer, Hutton and Khupe, 1978) have indicated increasing evaporation downstream, while groundwater analysis (Hutchins *et al.*, 1976b) shows a complex pattern of low Total Dissolved Solids (TDS) (<1 g l⁻¹) calcium-dominated water near the channels, and older water (TDS of 1–5 g l⁻¹), rich in chlorides and sulphates, in parts of the lower delta. Salt concentrations are altogether surprisingly low, given the prevailing evaporation rates in the delta and the formation of massive salt deposits in the adjacent Makgadikgadi. Recent studies have indicated that one reason for this may be the precipitation of salts, particularly calcite, trona and thermonatrate, on the shorelines of islands within the swamp complex (McCarthy, McIver and Cairncross, 1986b).

Summerfield (1982) has also indicated that silica concentrations in Okavango water are high (mean 55 ppm for 24 samples) and increase downstream. Although silcrete is not found at the ground surface anywhere within the delta, recent investigations in the vicinity of the Thamalakane and upper Boteti Rivers (Snowy Mountains Engineering Corporation, 1987: vol. 2) indicated laminated crypto-crystalline silcrete forming to the limits of exploratory drilling at 20 m. Contemporary formation of silcrete is thought to be rare, and has implications for the interpretation of the Kalahari Group sediments.

5.3.2 The Ngami Basin

Flow from the distal sector of the Okavango Delta reaches the Thamalakane fault line, and flows southwest along the Thamalakane and Nchabe (or Lake) Rivers, losing a considerable proportion of the flow at the Boteti bifurcation. The Nchabe, in turn, meets the Kunyere River, flowing along the parallel fault line, at the village of Toteng, and they pass as a single channel into the Ngami Basin (Figure 5.5a).

This fault-controlled sedimentary basin contains 400–650 m of waterlain sands called the Okavango Series by Hutchins *et al.* (1976a, b) and is subject to high

Figure 5.5 (opposite). (a) The geomorphology of the Ngami Basin. (b) The geomorphology of the Mababe Basin. Both after Shaw (1985a).

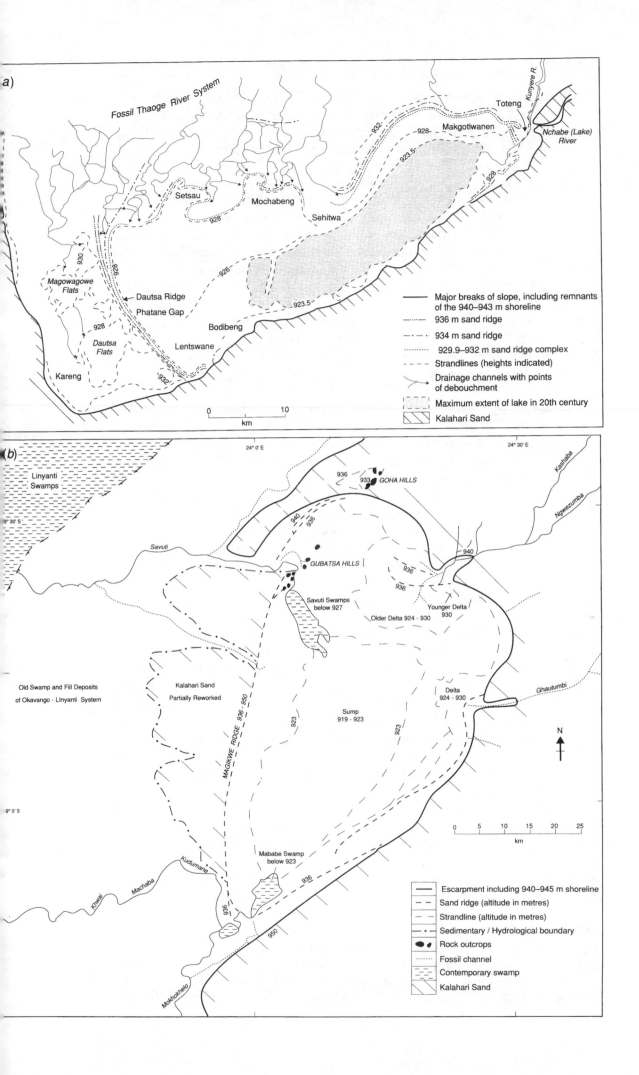

a)

Fossil Thaoge River System

Toteng
Kunyere R.
Makgotlwanen
Nchabe (Lake) River
Setsau
Mochabeng
Sehitwa
932
928
923.5
928
926
Magowagowe Flats
930
926
Dautsa Ridge
Phatane Gap
Bodibeng
928
Lentswane
932
Dautsa Flats
Kareng

0   10
km

Major breaks of slope, including remnants of the 940–943 m shoreline
936 m sand ridge
934 m sand ridge
929.9–932 m sand ridge complex
Strandlines (heights indicated)
Drainage channels with points of debouchment
Maximum extent of lake in 20th century
Kalahari Sand

(b)

24° 0' E
24° 30' E

Linyanti Swamps

8° 30' S

936
933 GOHA HILLS
Kashaba
Ngwezumba

940
936

Savuti

GUBATSA HILLS
936
940
936

Savuti Swamps below 927
Older Delta 924 - 930
Younger Delta 930

Old Swamp and Fill Deposits of Okavango - Linyanti System

Kalahari Sand Partially Reworked

Delta 924 - 930
Ghautumbi

N

MAGIKWE RIDGE 936 - 950
923
Sump 919 - 923
923

9° 0' S

0   5   10   15   20   25
km

Khwai
Machaba
Kudumane
Mababe Swamp below 923
936
936
950
Mokhokhelo

Escarpment including 940–945 m shoreline
Sand ridge (altitude in metres)
Strandline (altitude in metres)
Sedimentary / Hydrological boundary
Rock outcrops
Fossil channel
Contemporary swamp
Kalahari Sand

levels of seismic activity indicating normal faulting (Scholz, 1975). The southeastern side of the basin forms a low escarpment of Ghanzi Group limestones and sandstones with a veneer of Kalahari Sand at a minimum altitude of 950 m asl; the basin itself is underlain by Karoo sediments.

The basin possesses a series of concentric beach ridges and shorelines enclosing an area of 1800 km², far greater than the area of the contemporary ephemeral Lake Ngami, which reached a maximum of 250 km² in 1925 (Du Toit, 1926). On the northern side this fossil lake is defined by a series of ridges separating the lake flats from the now abandoned anastamosing channel system of the Thaoge River, which, prior to its final desiccation, had some 31 identifiable distributaries at the lake littoral. The basin is also bisected by a 25 km long complex of sand ridges, known collectively as the Dautsa Ridge, separating a higher area of flats and pans in the west from the contemporary lake. The Dautsa Ridge itself is bisected by the Phatane Gap, where a fossil channel passes through the ridge complex as a series of *en echelon* hollows separated by parallel ridges. To the west of the Dautsa Flats an older ridge system separates the basin from an area of old alluvium aligned to the fossil channel system of the Groot Laagte, which rises on the Namibian watershed.

The shorelines (Table 5.1) occur in a range of altitudes equivalent to area:volume thresholds in the lower Okavango Delta, which would stabilise the Ngami Lake at a given altitude. With the exception of the >940 m level, represented by a ridge along the south and west sides of the basin coincident with beds of round quartz gravels in a calcrete matrix, or a change in the colour of the sand overburden, all of the ridges are sharply defined, and are composed of uniform rounded to subrounded medium (mean grain size [M_0] between 2.3 and 2.8 Φ) sand, white to yellow in colour, that is typical of the Okavango bedload at the present time. The dominant source of the sediment appears to have been the now defunct Thaoge system; the alignment of the basin and the spits and zetaform embayments within it reflect a dominant E/ENE wind direction at the time of formation.

The Dautsa Ridge reaches a maximum altitude of 943 m along a line of sand dunes on top of the central ridge, which consistently equates to the 936 m level. In its present form it is probably of a different generation from the 945 m ridges noted in the Makgadikgadi and Mababe Basins; 945 m shoreline features on the southeastern side of Ngami show a poor state of preservation. The central ridge itself bifurcates in the northern section to form parallel, outward-facing beaches enclosing a lagoon feature 5 km long and up to 100 m wide, now occupied by a series of wet season pans, suggesting that it originated as a low and intermittent barrier beach, interrupted by the Phatane Gap. The 936 m ridge has been partially eroded south of Phatane, and the 934 m ridge has grown at an oblique angle to it, suggesting a depositional hiatus.

The 936 m ridge appears again at Magotlwanen, in the northeast of the basin, while the shorelines which terminate the Thaoge inflows are of the 930–932 m series. The shoreline of the contemporary maximum Lake Ngami lie at 923.5 m, some distance into the basin.

The interpretation of the sedimentary sequence found within the Ngami Basin, particularly in the NE sector, suggests some extensive and relatively recent changes in the nature of the lake (Snowy Mountains Engineering Corporation, 1987: vol. 2). Well sorted, medium grain sands comprise the bed material to a depth of at least 25 m in this area, although clay lenses also occur. Near the surface, however, the bed-sand is covered with a 3 cm layer of fine aeolian sand, in turn covered with up to a metre of diatomaceous earth, rich in *Surirella*, *Navicula* and *Nitzschia* species, suggesting still and slightly alkaline conditions at the time

Table 5.1. *Altitudes of landforms in the Ngami and Mababe Basins*

Altitude (in metres above sea level)	Lacustrine features	Fluvial features
Ngami Basin		
940–945	Shoreline	
936	Dautsa + Magotlwanen Ridges	Maximum altitude of Thamalakane thalweg
934	Dautsa + Magotlwanen Ridges	Terrace on upper Thamalakane
930–932	Three ridges in series	Flood plain–Boteti Junction (932) Thalweg–Boteti Junction (930) Thaoge Inflow (930)
928	Strandline	Flood plain–Kunyere/Nhabe Confluence
926	Strandline	Thalweg–Kunyere/Nhabe confluence
923	Strandline – max.contemporary lake	
919	Sump	
Mababe Basin		
940–945	Magikwe Ridge + shorelines	
936	Ridge + beach	Ngwezumba and Gautumbi Bars Ngwezumba Estuary Savuti Overflow
930	Strandline	Ngwezumba Delta Mababe Bar and Terrace Savuti Terrace 1
929		Savuti Terrace 2
927		Savuti Terrace 3
926		Savuti Marsh Tsatsarra Delta
923	Strandline	Mababe Marsh
919	Sump	

of deposition (S. Metcalfe, pers. comm.). This, in turn, is overlain by reddish-white, baked, laminated clays, consistent with the firing of diatomaceous earths and clay layers, possibly by the burning of *Phragmites* beds. These are covered by a surface layer of silts and clays. The whole sequence suggests a cycle of desiccation and still-water sedimentation.

5.3.3 The Mababe Basin

Following the Thamalakane River northeast from Maun, the thalweg reaches a maximum altitude of 936 m asl close to the village of Shorobe, then decreases towards the Mababe Depression, suggesting the possibility of bidirectional flow under conditions of increased water availability. The Thamalakane (called the Mokhokhelo in its lower reaches) is joined by the Khwai (also called Mochaba or Kudumane) and enters the Mababe Depression at its southern extremity. Water from the Okavango can also enter the basin by a more indirect route via the Savuti channel (Figure 5.6), an offshoot of the Linyanti system, which has a tenuous connection to the Okavango via the Magwegqana (Selinda) Spillway. The history of flows in the Savuti shows it to be highly erratic in its regime, probably because

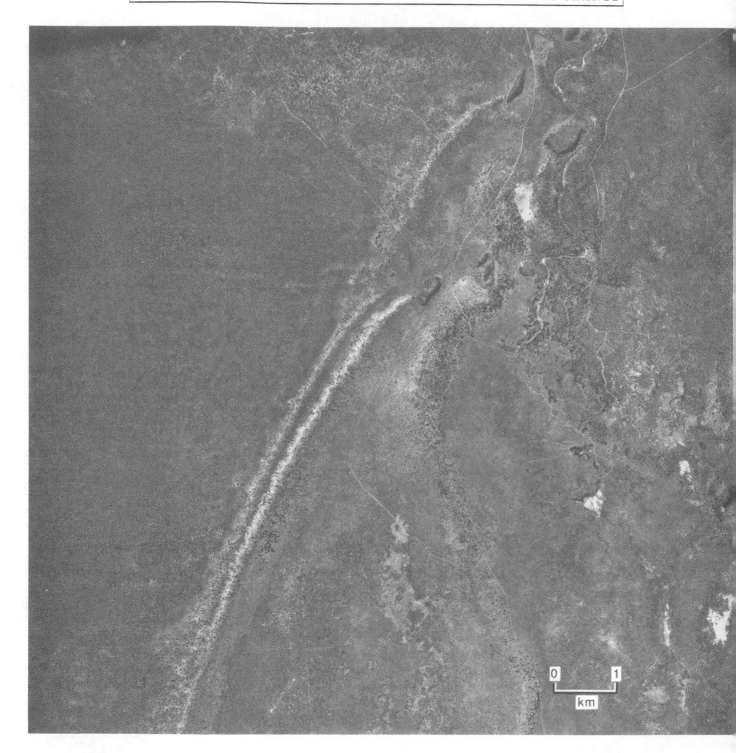

Figure 5.6. Aerial photograph showing part of the Mababe Depression, near Savuti. The curved feature running north–south is the Magikwe Ridge, which at this location bifurcates to form tombolo-like features. This is also one of the few locations where the ridge is breached, in this case by the Savuti Channel, which runs north–south in the northeastern corner of the picture. The features in the same area which are in shadow on their eastern sides are the Gubatsa Hills. (See also Figure 6.10a).

of adjustments along the Linyanti and Chobe Faults (Shaw, 1984). Previous courses are indicated by the fossil Tsatsarra channel and the network extending northeast along the foot of the Kalahari Sand escarpment which extends towards the Chobe.

On the eastern side streams drain from the Kalahari–Zimbabwe watershed, notably the Ngwezumba, a sand river which originates in the Kazuma Depression. Both the Ngwezumba and the Gautumbi have well developed delta features, although only the Ngwezumba carries water, and that only in years of exceptional rainfall. During the last hundred years the Mababe has received little inflow and has not contained a standing lake in human memory, although oral tradition supports the existence of 'Lake Mababe' in the eighteenth century (Tlou, 1972: p. 152). The Mababe Marsh dried out in the earlier years of the twentieth century (Campbell and Child, 1971), while the Savuti Marsh received inflow up to the 1880s (Stigand, 1923), and then again from 1958 to 1983.

The Mababe Depression (Figure 5.5b) itself is a heart-shaped basin of some 90 km × 50 km at its maximum dimensions, with the greatest width coincident with the dominant E/ENE fetch. It can be viewed as a twin of the Ngami Basin, with the eastern side comprising an escarpment at approximately 950 m asl formed by downthrow along the Thamalakane Fault; sediment thickness varies from 600 m along the fault to a few tens of metres in the Savuti area, where outcrops of Kgwebe Formation porphyries form the inselbergs of the Gubatsa Hills. The western side of the basin is defined by the massive compound Magikwe Ridge, which runs for 75 km from Mababe to the Goha Hills, interrupted only by the gaps at Tsatsarra and Savuti. The ridge is not as sharply defined as the Dautsa Ridge, and for the most part consists of two parallel ridges at 940–945 m, with summit dunes reaching more than 950 m asl. These ridges are best seen south of Savuti (Figure 5.6), where they bifurcate to form tombola features, as illustrated by Grove (1969: p. 202). A 936 m ridge is also apparent on the eastern face of the Magikwe Ridge, particularly between Savuti and Goha.

The altitudes of fluvial and lacustrine landforms are shown in Table 5.1. As with Ngami, the landform suite is dominated by the 940–945 and 936 m levels. The 936 m level is represented not only by ridges, but by cobble and quartz pebble beaches at three leeward sites in the Gubatsa Hills. On the eastern side of the basin the shorelines are less pronounced, probably because of the sheltered location. However, the Ngwezumba has an abandoned channel at 940 m, a sheltered estuary at 935–936 m, and two 936 m bars at the estuary mouth, declining into a suite of deltaic sediments down to 924 m. These bars clearly overlie earlier channel alignments, and indicate that the Ngwezumba was an active river at the 936 m lake level, which was achieved on at least two occasions. Similar, but less well developed features, occur at the mouth of the Gautumbi.

Lower strandlines indicate stages in the final desiccation of Mababe. These include fossil channel alignments of the Savuti and Tsatsarra, a series of three paired terraces on the Savuti between 930 and 927 m, and bars in the Mababe River system at 936 and 930 m asl. An interesting 936 m level, including a cobble beach, is also apparent around the Goha Pan, adjacent to the Goha Hills to the north of Mababe. Although it is unlikely that the Pan ever formed part of the Mababe lake, filtration through the Magikwe Ridge to equalise the levels is a plausible explanation.

5.3.4 The Makgadikgadi Basin

Some 18 km south of Maun the Thamalakane bifurcates at an altitude of 930–932 m into the Nchabe and Boteti Rivers. This latter channel is the main inflow to the Makgadikgadi Basin, 100 km to the east (Figure 5.2). At present it carries, aided by human interference, the greater part of the Thamalakane outflow, yet this volume (averaging 0.3×10^9 m^3 yr^{-1}) is barely sufficient to reach into the Makgadikgadi most years, and the Boteti itself is clearly a misfit stream (Cooke, 1980) with a conveyance capacity at least equal to the full Okavango flow (UNDP/FAO, 1977). Other inputs to the Makgadikgadi are also small compared with the size of the feature. Inflow from the east and northeast via the Nata, Semowane, Mosetse, Lepashe and Mosupe Rivers is sufficient in good years to inundate part of Sua Pan (Cooke, H.J., 1979a), while other drainage lines, notably the Okwa and Nungu systems in the southwest and north respectively, do not function at all under the current climatic regime. It is probable, however, that the basin acts as a terminal sink for ground water in the Middle Kalahari.

The Makgadikgadi Basin covers some 37 000 km^2; its suite of fluvial, lacustrine and aeolian features is on a far grander scale than those of the Mababe or Ngami. Coinciding with a series of SW–NE structures in the underlying Basement Complex and Karoo Sequence rocks (Baillieul, 1979; Smith, 1984), which have given pronounced orientation to pan alignments within the basin, the whole may represent a horst and graben structure (Cooke, 1980), with the scale of faulting indicated by the juxtaposition of Karoo and Basement rocks in the narrow strip between Sua and Ntwetwe Pans. The sub-Kalahari geology is further complicated by the emplacement of dyke swarms on a NW–SE alignment.

The south and east sides of the Basin are defined by major escarpments of Karoo sandstones and basalts (Stansfield, 1973), while the Gidikwe Ridge (Grey and Cooke, 1977), again a complex of parallel sand ridges at 940–945 m asl, with occasional dune summits above this level, extends for 250 km in the west and north, broken only by the Boteti and the fossil channels of the Okwa river. This separates the main basin from extensive flats below 950 m asl which extend on the western side of the ridge to the edge of the linear dune fields 50 km distant. These flats are probably lagoonal, as they terminate all of the linear drainage lines of the dune country, such as the Passarge and Deception Valleys, with the exception of the Boteti and Okwa, which Cooke (1980) believes had sufficiently stable regimes to continue flow across the area once desiccation of the flats had commenced. The flats are also covered by east–west lineations, only identifiable on satellite imagery, which have been interpreted as both lacustrine and aeolian features (section 6.3.1). The morphology and orientation of the Gidikwe Ridge suggests that, with the Magikwe and Dautsa Ridges, it has a common origin as an offshore barrier bar, its lacustrine origins confirmed by sediment characteristics and the presence of bivalve shells, in particular Ostracoda spp. (Cooke, 1980). Certainly it has acted as a barrier to the Boteti inflow on a number of occasions (Cooke and Verstappen, 1984), and three levels of ponding have been identified along the Boteti to the west of the ridge to an altitude of up to 940 m (Shaw et al., 1988). Tectonically derived tilting and derangement of the ridge seems to have taken place after the formation of the 945 m level (Cooke and Verstappen, 1984), as indicated by the divergence of the 945 m ridge from its 920 m asl neighbour. Traces of ridges which appear to be both older and, possibly, higher than the Gidikwe Ridge appear to the north of the Makgadikgadi, predating the later stages of dunefield development in this area, and interrupting the drainage of the fossil Nungu River (Mallick et al., 1981).

The basin is occupied by a series of pans or playas, notably Sua Pan, with an area of approximately 3000 km^2 (Ebert and Hitchcock, 1978) and the slightly larger Ntwetwe Pan. Sua is the lowest member of the pan sequence, with a sump level of 890 m asl. It has an almost flat surface of saline clays and salt efflorescence (section 3.6.3), overlying 50–100 m of interlayered clays and sands. The presence of highly saline ground water close to the surface gives rise to the distinctive geomorphological environment of saline playas, which is characterised by deposition and chemical evolution of sediments and salts during inundated and drying phases, followed by wind erosion and dust transport to the depth of ground water when the pan is dry, to produce so called 'Stokes surfaces' in the near-surface profile (Fryberger, Schenk and Krystinik, 1988; Shaw and Thomas, 1989). It also offers the prospect of almost unlimited resources of evaporite deposits. A number of smaller pans lie to the west and south, as well as Lake Xau (Dau), which is sporadically flooded by the Boteti River. These pans have either saline clay or grassed surfaces, the latter dominant above the 920 m level, depending on relative relief and drainage characteristics, together with dune configurations.

Within the basin a series of strandlines and ridges form prominent features (Cooke, 1979a, 1980), and have been accurately mapped over much of the western, southern and eastern sectors. In the west the 940–945 m level can again be traced on the Gidikwe Ridge, to the west of the lagoonal flats, and on the Boteti River above the Moremaoto Gorge. In the northern basin it can be identified in parts of the extension of the Gidikwe Ridge which runs east as far as the Nata River, and to the east and south occurs along the Karoo escarpment. The 936 m level is absent from the Makgadikgadi, except in the upper Boteti, but a 920 m shoreline circumscribes the basin, being distinct throughout except in the vicinity of Tsau on the Boteti River, where it has been destroyed by subsequent sedimentation. It occurs as a calcrete-capped inflexion to the east of the Gidikwe Ridge, as a series of gravel-capped spits, bars and ridges at the base of the escarpment, and as distinctive bevelled surfaces, covered with rounded gravels, on the islands of the lower pans, particularly Kubu and Kokonje Islands in Sua. In parts parallel ridges have formed, indicating a hiatus in deposition. The lowest complete shoreline is that of the 912 m level, visible mainly in the south, particularly around Mopipi, and east of the basin, where it forms the 'root' of the 40 km long Sua Spit.

Below 912 m the lake which occupied the basin evidently fragmented into a series of smaller water bodies coincident with present pans. Cooke (1980) has identified beaches at 908 and 904 m in Sua, perhaps coincident with a 905 m level tentatively identified by Ebert and Hitchcock (1978), while Ntwetwe Pan has an extensive flat at 910 m, much dissected in the past by the meandering course of the lower Boteti. This surface, like the present surface of Sua Pan, has extensive spreads of silcrete gravels, which Grey (1976) has ascribed to polyphase diagenesis of Karoo sandstones and basalts, but which could equally be due to precipitation within existing sediments under alkaline groundwater conditions (section 3.5.4).

A series of fossil deltas has been imposed on to the pan and ridge landscape by the Boteti River, related to the desiccation of the Makgadikgadi Basin. Three of these lie to the west of the Gidikwe Ridge at Tatamoge, Makalamabedi and Moremaoto (Cooke and Verstappen, 1984; Shaw, Cooke and Thomas, 1988). These are evidenced not only by deltaic sediments and shorelines, but by the presence of fossil meander loops indicating points of reduced stream velocity. East of the Ridge major deltas occur at Tsoi and to the north of Rakops, the former cutting across the 920 m shoreline. These appear to decrease in age downstream, as the lower deltas have cut extensively into earlier deposits. Downstream of Rakops minor delta features relate to the contemporary regime, for example, where the Boteti enters Lake Xau. Here, fluvial sedimentation is very thin (Breyer, 1982) and corresponds to the low sediment loads encountered in contemporary flow. Similar features, now abandoned, occur where the lower

Boteti enters and leaves Ntwetwe Pan, and reaches its end in the western part of Sua Pan.

5.3.5 Lake Caprivi: the Chobe–Zambezi confluence

The Gwembe Trough is a tectonically active zone characterised by extensive alluviation, in which both the Zambezi and Kwando Rivers are confined by fault-controlled escarpments which delimit the trough. The rivers themselves are controlled by a series of NW–SE and SW–NE faults, which give a strongly rectilinear drainage pattern. This is particularly noticeable on the Kwando, which pursues a dog-leg course along the Linyanti, Liambezi and Chobe Faults, with swamps along the first of these faults, and an ephemeral lake along the second. The Chobe has its confluence with the Zambezi at the Mambova Falls at 926 m asl (Figure 5.2), a massive Karoo basalt ridge along a N–S fault, which impedes the flow of the rivers, and, during the annual flood peak on the Zambezi, causes backflooding of the Chobe for some 20 km upstream. Below the Mambova Falls the rivers again diverge briefly around Impalera Island, then flow as the Zambezi down a steeper section of the trough to Victoria Falls some 100 km distant.

Apparent on satellite imagery is the presence of two alluvial land systems between the escarpments. The lower alluvium occupies the eastern Caprivi Strip and the area between the Linyanti River, Lake Liambezi and the Chobe Escarpment. This contains the seasonal swamp and floodplain area of the two rivers, has identifiable channel systems, mainly superimposed meander loops, and is vegetated with floodplain grassland and *Papyrus* swamp. Northwards, the land rises to a belt of older alluvium similar to the older deposits of the Okavango. This is covered with *Colophospermum mopane* woodland, with seasonal swamps in depressions, pans and channel remnants. This, in turn, gives way to linear dunefields on a line through Katima Mulilo on the Zambezi. The zone of younger alluvium contains a number of lacustrine features, suggesting that it is not entirely formed by floodplain deposition (Shaw and Thomas, 1988). To the south of Lake Liambezi a number of N–S oriented sand ridges up to 20 km in length occur. These ridges exhibit the orientation, profile and grain size characteristics as the Dautsa, Gidikwe, and Magikwe Ridges, and are likely to have a similar origin as offshore bars during a period of lacustrine inundation. The summits of the ridges are remarkably uniform in height; the two largest, now the sites of the villages of Satau and Parakarungu, are at 936 m, while the other ridges in the complex lie between 932 and 936 m. An area of lagoonal sedimentation is represented by diatomaceous earths in the lee of the Parakarungu Ridge, indicating deposition under agitated conditions in clear water of moderate pH (S. Metcalfe, pers. comm.).

Further evidence for a lacustrine episode is found along the Chobe Escarpment between Kachekabwe and the Mambova Falls. For the most part the Escarpment is formed of Karoo sandstones and basalts, covered with a layer of colluvially worked Kalahari Sand, which would effectively disguise any water-related features that may occur on it. At three points, however, there are alluvial terraces above the present flood plain. The largest of these is a gently sloping terrace between 934 and 938 m asl at Serondela, composed of silty sand overlying 3 m of calcrete derived from alluvium. This contains concentrations of freshwater gastropod shells, mostly *Lymnaeae* and *Bellamya*, a still-water genus, first reported by Bradshaw (1881; see section 3.6.2).

The morphology of the Mambova Falls suggests that they would provide a considerable impediment to large volumes of water. At present the channels between the chain of islands of Karoo basalt pass between steep cliffs up to 5 m in

Table 5.2. *Altitudes of lacustrine features in Lake Palaeo-Makgadikgadi*

Altitude (in metres above sea level)	Ngami	Mababe	Makgadikgadi	Caprivi
940–945	Shoreline	Ridge	Ridge	
936	Multiple ridges	Multiple ridges		Ridges
934	Ridge			Terrace
930–932	Three ridges	Strandline		
926–928	Strandlines			Mambova Falls
923	Strandline	Strandline		
920			Ridges + shorelines	
919	Sump	Sump		
912			Ridges + shorelines	
890			Sump	

height, and the summit of the outcrop on the Botswana side, Commissioner's Kop, lies at 935 m, with similar levels on Impalera Island (Namibia) and the Zambian bank.

It has been proposed (Shaw and Thomas, 1988) that the extent of the younger alluvium corresponds to the extent of a lake of some 2000 km^2 termed Lake Caprivi, which covered the eastern Caprivi Strip westwards to a line with Katima Mulilo and the Mababe Depression. Heights of lacustrine features within this lake are at an altitude of about 936 m asl, as are the altitudes of the falls on the Zambezi at Katima Mulilo, and the overflow channel from the Savuti (Mababe) at the foot of the Chobe Escarpment. In turn, Lake Caprivi would correspond to the 936 m level encountered in both Mababe and Ngami.

At present no evidence has been discovered for the existence of levels above 936 m at the Zambezi–Chobe confluence. Although the belt of higher alluvium, mentioned earlier, lying to the west of Lake Caprivi almost certainly exceeds this altitude, there are no altitudinal data available to test its geomorphological relationships.

5.3.6 The configuration of the Middle Kalahari palaeo-lakes

The wealth of data now available on the geomorphology of the Okavango and its associated rivers and basins indicates considerable and far-reaching changes to the environment in the past, particularly within the period of the Quaternary. It is evident that periods of extensive lakes were interspersed with periods of aeolian activity in the basins. The geomorphological evidence indicates two major palaeo-lake stages (Table 5.2; Figure 5.7); the higher at the 940–945 m level is termed 'Lake Palaeo-Makgadikgadi' (Grey and Cooke, 1977), and represents the fossil lake at its greatest extent, covering all three basins and the lower part of the Okavango Delta, with the assumption that the Okavango, and adjacent delta systems, such as the Groot Laagte, were active at the same time. The water is likely to have occupied much of the upper Zambezi trough and extended NE towards the Kafue River, though evidence here is scarce. Estimates of its size vary from 60 000 km^2 (Cooke, 1980) to 80 000 km^2 (Mallick *et al.*, 1981) though these do not take into account the Zambezi sector.

A second stage can be seen by linking the 936 m landforms in the Ngami and Mababe Basins along the Thamalakane Fault, with extensions to Lake Caprivi.

This can conveniently be termed 'Lake Thamalakane' (Shaw, 1988b), and would have been controlled by a number of hydrological thresholds at the 936 m level, including the thalweg of the Thamalakane River, the offtake of the Savuti, and ultimately, the Mambova Falls. Lake Thamalakane probably had an area of about 7000 km² in the delta area, but would also have overflowed to the Makgadikgadi Basin, supplying either the 920 m or the 912 m level. Below this level, Lake Thamalakane would have fragmented to its constituent basins, with their own responses to climate and hydrology.

It is obvious that the complexity of the evidence suggests implications for the tectonic, climatic and hydrological history of the region, and failure to understand these complexities has led, in the past, to serious misconceptions about the Kalahari environment. These themes are explored in Chapter 7.

5.3.7 The Etosha Basin

The fault-controlled Etosha Basin has many visual similarities to the Makgadikgadi. Infilled with up to 450 m of sands and gravels over a limestone base, the dominant landforms are a series of pans, of which Etosha Pan, covering about 6000 km², is the largest (Figure 5.8). These are separated by a series of calcrete scarps between 1 and 20 m in height. The floor of Etosha Pan is composed of saline clays, which may be inundated during the wet season from inflow through broad channels, termed *oshanas* or *omurambas*, which distribute local precipitation. The water table is close to the surface, and springs are common, particularly in the southeast of the pan. The ground water is relatively fresh in comparison with the Makgadikgadi, and there is less silcrete development of the type associated with saline environments.

Wellington (1938, 1939) has suggested that Etosha once received inflow from the Angolan Highlands via the Cunene River, which distributed the water into the basin via a series of shallow, anastamosing *oshanas*, similar in function to the present-day Okavango. The capture of the Cunene by a coast-wise stream from the direction of the Ruacana Falls, possibly during the Pliocene, deprived the Etosha lake of its inflow; the Cunene bed now lies 13 m below the level of the Etaka Oshana, previously the principal distributary.

Apart from the presence of stromatolites at Insel, suggesting a lake level 8 m above the present floor, and which lie beyond the range of carbon dating (section 7.4.3) there is no evidence for lacustrine episodes of the type noted in the

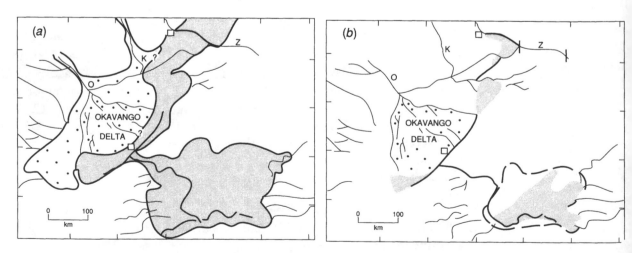

Figure 5.7. Configuration of (a) the Lake Palaeo-Makgadikgadi Stage and (b) the Lake Thamalakane Stage. After Shaw (1988a).

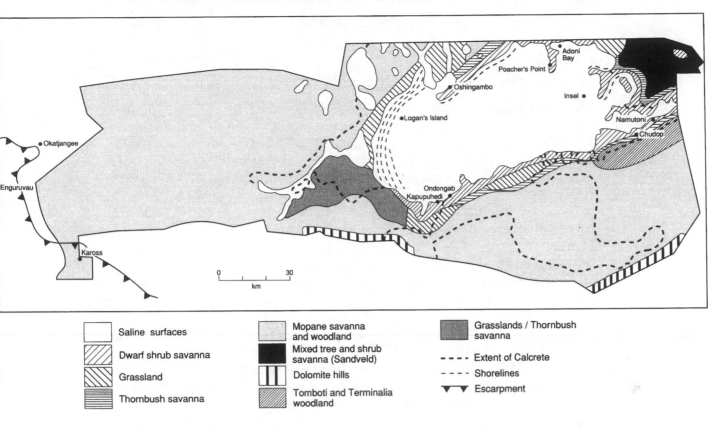

Figure 5.8. *The Etosha Basin, showing major vegetation communities and geomorphological features, within the area of the National Park. Data from several sources including Rust and Weinke (1976).*

Okavango–Makgadikgadi system. Rust (1984, 1985) has described the geomorphology of the pans within the framework of episodes of morphodynamic stability and activity within a semi-arid climate. Stability is associated with calcrete development, while aeolian activity, scarp retreat and basin sedimentation occur during active phases. The evolution of the pan is thus largely dependent on the fusion of smaller pans by scarp retreat, and the cycle of inundation and evaporation common to enclosed basins (Shaw and Thomas, 1989).

5.4 The ephemeral drainage of the Southern Kalahari

The Southern Kalahari is dominated by a broad, featureless interfluve at 1000–1100 m asl, called the *Bakalahari Schwelle* by Passarge (1904). This separates the Middle Kalahari, with its endoreic drainage, from the Molopo River and its tributaries, which drain to the Orange River, and hence to the Atlantic. The Bakalahari Schwelle has little coherent channel development, drainage takes the form of small, seasonally inundated depressions known as pans (section 6.4) which, although present throughout much of semi-arid southern Africa, attain great densities in this area, with a peak of one pan per 22 km² around Kukong (Lancaster, 1978a). Such channels that are developed form the headwater sections of dry valleys, called *mokgacha* (plural: *mekgacha*) in Botswana and *laagte* in South Africa, which trend towards the Makgadikgadi or the Molopo River network. Research over the past decade has indicated that some pans and mekgacha have a common origin, although mekgacha have received little scientific interest beyond their function in exposing the sub-Kalahari geology (e.g. Ludkte, 1986; Aldiss, 1987a).

5.4.1 The mekgacha of the endoreic Kalahari

Dry valleys draining towards the Makgadikgadi Basin and Okavango are found in an arc from the Aha Hills area on the Namibian watershed, through the Bakalahari Schwelle to the periphery of the Kalahare sediments along the Kalahari–Zimbabwe rise in eastern Botswana (Figure 5.9). The largest of these features is undoubtedly the Okwa–Mmone system, which drains from both Namibia in the west and the vicinity of Molepolole and Kanye in the southeast as far as the Gidikwe Ridge in the Makgadikgadi Basin, a potential catchment of 90 000 km². Other important mekgacha networks drain towards the western fringe of the Okavango (e.g. the Groot Laagte, Ngamasere, Xaudum and Gcwihabedum networks), and to the western and southern fringes of the Makgadikgadi (e.g. the Deception, Passarge and Letlhakane Valleys). Surface flow within the mekgacha is currently uncommon, but may occur over a short distance in relation to intense precipitation, as happened in the Letlhakane in 1969 (Mazor et al., 1977).

The mekgacha have a distinct morphology, comprising a set of valley forms from source to debouchment. The initial catchment stage is usually a broad, shallow, clay-floored valley with little evidence of surface flow, except where hills and rock outcrops are adjacent, as in the Aha and Gcwihaba Hills. This initial section can be developed in a variety of parent material: the upper Letlhakeng tributaries are developed in Kweneng (Karoo) sandstone (Shaw and De Vries, 1988), granite gneiss in the Xaudum (Wright, 1978), and Kalahari Sand in the Serorome. This initial section varies in length from a few hundred metres up to 80 km, as in the case of the Mochaweng mokgacha in Letlhakeng.

The initial valley form inevitably gives way in areas of greater relief to an incised section of rectilinear, flat-bottomed valleys, with steep sides and abrupt valley heads, frequently with abandoned spring lines at the nick point, or along the valley sides. These sections are associated with the presence of duricrusts, particularly silcrete, calcrete and a bewildering range of materials in the intermediate calsilcrete and silcalcrete categories. So frequent are the changes in duricrust composition over short distances that, in a detailed analysis of duricrust composition in part of the Letlhakeng system, Gwosdz and Modisi (1983) could conclude that the only certitude is an antipathy in the relationship between CaCO₃ and silica in these materials. However, Shaw and De Vries (1988) noted an association between calcrete, low hydraulic gradients and fine-grained parent rock, while silcretes tended to form from arenaceous sediments. Ferricrete and fersilcretes have also been recorded.

The duricrust is derived from the alteration of Kalahari Group lithologies or the sub-Kalahari bedrock; studies of the Letlhakeng system indicate duricrusting of basal Kalahari pebble beds to produce a calcrete conglomerate, and the alteration of Karoo sandstones, siltstones and mudstones (Shaw and De Vries, 1988), while the four distinct calcretes of the Gcwihaba Valley are associated with dolomite (Cooke, 1975a). The duricrust, in turn, is resistant to erosion, and may form low bluffs or vertical cliffs up to 10 m in height which have been weathered, possibly beneath a veneer of Kalahari Sand, to produce pseudo-karstic features such as pavements, karren, tafoni and even caves (Figure 3.6b).

There is a consensus that the valley side duricrusts are 'old' (Wright 1978; Mallick et al., 1981), certainly older than the overlying sands from which, as already noted, they are separated by a pronounced unconformity. Some investigators consider them to be of Pliocene age (Netterberg, 1969a; Goudie,

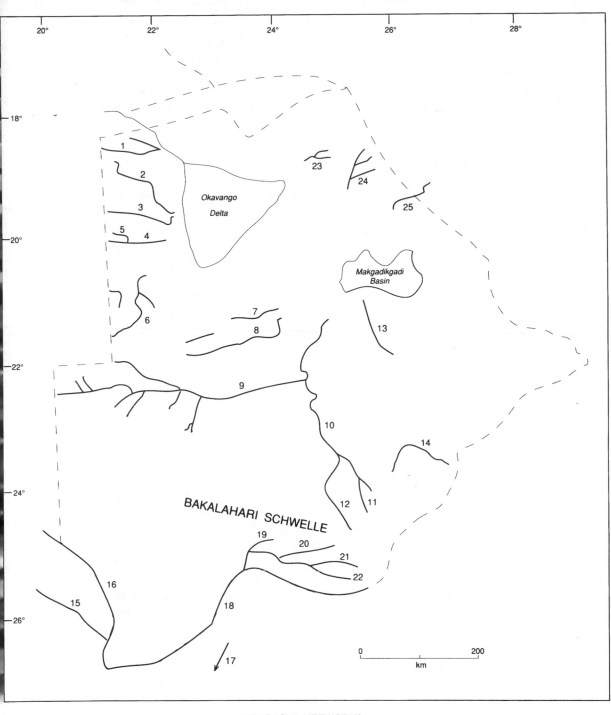

MEKGACHA NETWORKS

1	NGAMASERE	10	MMONE	18	MOLOPO
2	XAUDUM	11	LETLHAKENG	19	MABUASEHUBE
3	QANGWADUM	12	NALEDI	20	UKWI
4	EISEB	13	LETLHAKANE	21	MOSELEBE
5	GCWIHABEDUM	14	SEROROME	22	SEKHUTANE
6	GROOT LAAGTE	15	AUOB	23	GHAUTUMBI
7	PASSARGE	16	NOSSOP	24	NUNGA
8	DECEPTION	17	KURUMAN	25	LEMEMBA
9	OKWA				

Figure 5.9. Mekgacha networks in the Kalahari.

1973), though it is probable that some formed earlier, even since the deposition of the basal Kalahari lithologies. Boocock and Van Straten (1962) suggest that the mekgacha have incised into the duricrusted African surface of Dixey (1956a). In some cases, as in the Xaudum (Wright, 1978), duricrust formation appears to have preceded tectonic adjustment. A 'younger' calcrete has also been identified forming from unconsolidated sediments in valley bottoms to a depth of 10 m, in relation to the contemporary watertable (Foster et al., 1982).

Moving away from the higher ground at the periphery of the Kalahari basin, the gorge sections of the mekgacha decrease in relief and merge into 'dambo' or 'vlei' type valleys (Mackel, 1974; Acres et al., 1985) which are essentially shallow, linear depressions without obvious channels. These are invariably confined to the Kalahari beds; drilling in the Ngamiland and central Kalahari mekgacha has indicated sands and duricrusts to depths of 50 m or more (Jack, 1980). However, limited rock outcrops have been reported; for example, granite in the middle Okwa (Aldiss, 1987b) and marble in the Groot Laagte (Ludkte, 1986). The lower valleys have low gradients, a tendency to alignment by linear dunes, seasonal flooding in valley floor pans and depressions, and may be partially or wholly obscured by windblown sand. Some of the valleys terminate in delta sediments within the Okavango or major internal basins, others simply dissipate into the sandveld.

Conventional wisdom suggests that mekgacha were formed by fluvial erosion during periods of wetter climate, though precise mechanisms for this erosion are not cited. There is certainly evidence for increased flow in the past: the large delta of the Okwa beyond the Gidikwe Ridge (Cooke and Verstappen, 1984) is one example, while the 4600 km² area of deltaic sediments laid down by the Groot Laagte network adjacent and perpendicular to the older Okavango alluvium is even more impressive. Further evidence for the past presence of water, probably in the form of swamps, comes from shell beds and lignite deposits to a depth of 55 m against a fault in the bed of the Xaudum drainage (Jack, 1980), and freshwater molluscs in the lower Okwa have been dated to c. 15 000 BP (section 7.4.2). Jack (1980) goes as far as ranking the mekgacha by the apparent age of fluvial activity, with the Ngamiland valleys younger than those of the central Kalahari, a hypothesis that requires further investigation.

Fluvial erosion, however, does not explain the morphology of valleys in their gorge sections, in particular the low bifurcation ratios of tributaries, the increase in size of the gorge sections over relatively short distances, or the lack of channels in the valley floors. This particular morphology was recognised by Peel (1941) in Libya, and attributed to spring sapping and groundwater flow, an idea that has been revived in recent years with new interest in groundwater landforms (e.g. Higgins, 1984), and the possibility of analogues in extraterrestrial landscapes (Baker, 1980). A number of factors suggest that the headwater sections, at least, are formed by deep weathering along preferential groundwater flow paths against a background of long-term uplift of the Kalahari rim. In the first instance, a relationship between valley alignment and faults has been established in a number of locations, including the Xaudum (Wright, 1978), the Naledi (Mallick et al., 1981) and Deception Valley (Coates et al., 1979). This relationship reaches an extreme in the case of the Serorome Valley, which is oriented west and north before being diverted by the Zoetfontein Fault eastwards to join the Limpopo network, thus becoming exoreic.

In the Letlhakeng system the faults appear to be in the Precambrian basement, with alteration of both the overlying Karoo sandstone and decomposition of the sub-Karoo dolerite to a depth of 125 m in a test borehole (Von Hoyer, Keller and

Rehder, 1985), with similar weathering depths recorded in the Naledi mokgacha (Buckley, 1984).

Secondly, there has been a long association between mokgacha and groundwater availability (e.g. Chapman, 1886), which has led in recent years to the establishment of wellfields for mining enterprises in mekgacha, as in the Letlhakane and Naledi Valleys. It has been suggested that mekgacha alignments are a principal method of recharge in the Kalahari Group strata as a whole (Mazor et al., 1977; Verhagen et al., 1978; section 3.7). The presence of fossil spring lines and spring tufas, together with historical evidence for abundant springs, as in Letlhakeng during the nineteenth century (Campbell and Child, 1971), suggests that both valley and duricrust evolution have taken place against a background of gradually lowering watertable, although human interference has accelerated the process.

5.4.2 The Molopo system

The Molopo network represents a transitional stage between permanently dry mekgacha and the seasonal rivers typical of the semi-arid hardveld. All four of the major valleys, the Molopo, Kuruman, Auob and Nossop, together with smaller networks, such as the Moselebe, rise on bedrock, and it is this which contributes to their unusual hydrological regimes, which vary between the rivers.

The Kuruman has the most reliable flow (Figure 5.10a), and at its upstream end is a perennial river, fed by a series of dolomite springs, including the famous Eye of Kuruman, which has yielded a constant 750 m³ per hour since at least 1820. In most years the water rapidly percolates into the stream bed a few kilometres downstream, but it may flow throughout the length of the river in high rainfall years, in a channel of 'sand river' type. The Molopo flows sporadically in its upper reaches, but the water does not proceed beyond Water's End, between Werda and Tshabong. Thereafter, the Molopo has typical mekgacha form, as do its tributaries in the Moselebe network. The Auob and Nossop have partially dune-obscured, broad, flat beds, which are typical of flash floods in arid environments (Figure 5.10b).

This intermittent fluvial activity has left terraces and fluvial sediments, especially within the Molopo (Rogers, 1936) and the Kuruman. Some of these features have been dated to the late Quaternary (Heine, 1982), while high terraces, beyond the confines of the present valley, are clearly much older. Flooding occurs sporadically in response to short-term precipitation events, particularly during years of higher rainfall. Floods have been recorded in the Nossop in 1806, 1963 and January 1987, in the Kuruman in 1891–92, 1894, 1896, 1915, 1917, 1918, 1920, 1974–77 and 1988–89 and in all four rivers of the Molopo network in 1934 (Clement, 1967; Verhagen, 1983). For the most part these floods occur in response to rainfall over the upper catchment and are short-lived, with the water being rapidly absorbed into the river bed. They may, however, be of considerable magnitude: the 1934 Nossop flood was estimated by the newspapers of the time as being 450 feet wide, travelling at 6 mph (Clement, 1967), while the Nossop flood of 1987 threatened damage to farming infrastructure.

The size of the Molopo Valley, up to 40 m deep in places, led many early European travellers to surmise that the channel was the relict of a former great river draining the Transvaal. Wellington (1929, 1955) suggests that capture of its headwaters by the Vaal River has accounted for its demise, while Smit (1977), Bruno (1985) and Rathbone and Gould (1982) have considered it to be a deeply incised pre-Kalahari valley.

Again, a large range of calcretes and silcretes form the valley sides. Boocock and Van Straten (1962) noted a propensity to nodular silcrete, calc-conglomerate and calc-sandstone in the lower Molopo, with calcrete in the upper Molopo and silcrete in the Moselebe tributary. Rogers (1936) comments on the presence of calcrete conglomerates, up to 6 m thick, above the bed of the river. Silcretes, accompanied by the typical rectilinear valley form, are also found in the Auob and Nossop in the Gemsbok Park.

Figure 5.10. (a) The Kuruman River in flood in 1989 within the linear dune system of the southwestern Kalahari, some 80 km from its source. (b) The Nossop valley, 30 km south of Twee Rivieren. Calcrete and silcrete cliffs are exposed in the valley sides, but their full height is obscured by sand blown into the valley from neighbouring linear dunes.

6 The geomorphology of the Kalahari: aeolian, pan and rock landforms

6.1 Aeolian landforms

GIVEN THE PROPENSITY for the wind to transport available sand in arid environments, it is not surprising that the Mega Kalahari, which represents the largest continuous sand sea on earth, possesses a sizeable body of aeolian landforms, primarily sand dunes. Lieutenant Hodson (1912: p. 21) regarded the dunes of the southwestern Kalahari to be '. . . by far the most dreary and depressing part of the desert', but for others they have proved to be a major source of interest. Unlike that early policeman, we now have the benefit of aerial photography, satellite imagery and even photographs taken on space shuttle missions which together have greatly assisted in the identification and description of the Kalahari dune systems. But the explanation of their development, distribution and climatic and environmental significance relies upon a combination of detailed fieldwork and recourse to models and theories of desert dune development and their relationship to a range of climatic parameters.

To this end, the dunes found in the Kalahari Desert and in adjoining parts of Zimbabwe, Zambia and Angola, and even in areas of Zaïre, are now well described, but accounting for their development in terms of modern and ancient environmental conditions is still fraught with difficulty. This is a consequence of several factors, including the nature of the predominant dune types found in the Kalahari and the general question of the climatic and atmospheric conditions conducive to their development, whether some of these dunes are currently geomorphologically active or relict features, and, in the case of the latter, the considerable problem of dating the time of development. However, in considering these specific issues, it is necessary to outline the types of dunes found in the Kalahari and their distribution.

6.1.1 Dune types of the Kalahari

Sand dunes are only one of a number of categories of aeolian depositional landforms (Thomas, 1989), ranging in scale from the sand seas of regional extent (including the Kalahari) to the small aeolian ripples which can develop upon any small, exposed patch of sand. Dunes can themselves be further categorised according to the wind environment in which they are found and their relationship (orientation) to formative winds, the nature of dune movement and the morphology of the dune. Of the 11 basic dune types included in the classification scheme of Thomas (1989), six are found in the Kalahari: parabolic dunes, blowouts, barchan dunes, transverse ridges, linear ridges and seif dunes (Table 6.1 and Figure 6.1).

Table 6.1. *A summary of the dune types present in the Kalahari*

Dune type	Location
Linear dunes (general)	All three Kalahari dunefields are dominated by linear dunes (est. 85% of the southern dunefield) These may be divided into linear ridges and seif dunes:
Linear ridges	Low degraded forms dominate the eastern and northern dunefields. Partially vegetated forms dominate the southern dunefield
Seif dunes	Linear forms with sharp sinuous crests: present in the drier parts of the southern dunefield and where the vegetation of linear ridge crests has been destroyed
Transverse dunes (general)	Within and beyond the western margin of the Makgadikgadi Depression:
Barchan dunes	On the floor of Ntwetwe Pan
Transverse ridges } Barchanoid ridges }	West of the Gidikwe Ridge
Parabolic dunes	Southern dunefield and Bakalahari Schwelle:
Lunette dunes	On downwind margins of pan depressions
Nested parabolics	Isolated patches in southern dunefield, where vegetation cover disturbed
Blowouts	Widespread location, especially in the southern dunefield, where local vegetation disturbance has occurred. In some places, may evolve into parabolic and nested parabolic forms

Because the Kalahari contains dunes of different ages, some of which have been subject to degradation by non-aeolian geomorphological agents, they are not always the pristine examples found in text books. Furthermore, the occurrence of most Kalahari dunes in dunefields (sometimes called ergs) means that dunes may interact with each other giving rise to compound and complex forms (McKee, 1979), the former when dunes of the same type coalesce and merge, the latter when dunes of different basic types are superimposed. Nevertheless, it is useful to discuss briefly the basic types of dune which are present in the Kalahari.

Linear ridges and seif dunes

Following detailed field investigations in Israel and Namibia (Tsoar, 1978; Livingstone, 1986) and the general abandonment of an unsubstantiated theory of linear dune development related to large-scale vortices of atmospheric instability, linear dunes, sometimes also called longitudinal dunes, are now widely believed to develop in regimes possessing two dominant directions of sand-moving winds which in theory may be up to 180° apart, or in wide unimodal regimes (see Thomas (1989) for a more detailed explanation of dune development). The net result of this is that linear dunes extend downwind approximately parallel to the resultant sand-moving direction, with sand passing along their lengths rather than the dunes themselves migrating in any significant way.

The nature of linear dunes is somewhat complicated by the fact that dune form may vary in response to a number of other environmental variables. In the case of linear dunes the most important of these is probably vegetation. Plants may colonise linear dunes because of their relative stability and the propensity for

Dune type		Number of slip faces	Major control on form	Formative wind regime	Nature of movement
Blowout		0	Disrupted vegetation cover	Various	May extend down wind
Parabolic dune		1	"	Transverse Unimodal	Slow, nose migration
TRANSVERSE DUNES					
Barchan dune		1	Wind regime and sand supply	"	Forward migratory
Barchanoid ridge	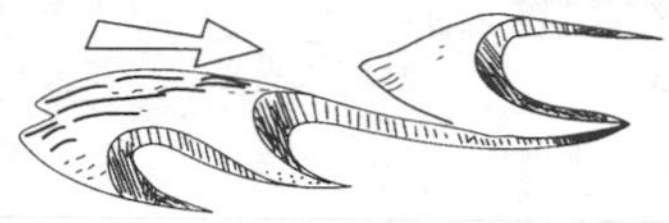	1	"	"	"
Transverse ridge		1		More directional variability than for barchans	"
LINEAR DUNES					
Linear ridge		1 – 2	"	Biomodal / wide unimodal	Extending
Seif dune	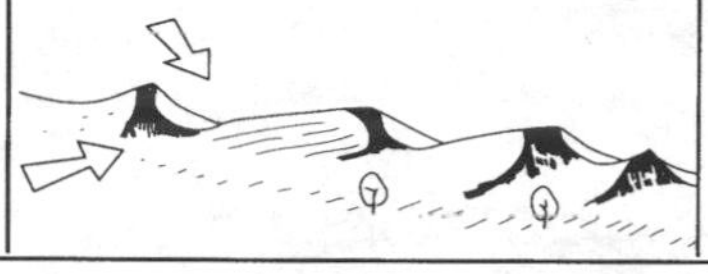	2	"	Biomodal	"

Figure 6.1. Basic dune types found in the Kalahari, showing their principal characteristics. Differences between the three categories of transverse dunes shown here are not always clear but in the Kalahari a distinction can be made between the barchan-like forms in Ntwetwe Pan and the transverse ridge features to the west of the Makgadikgadi Depression (see Figure 6.4). After Thomas (1989).

aeolian sand to be an important sediment for water retention in arid environments (e.g. Tsoar and Møller, 1986). Generally, linear dunes can be divided into two major categories: linear ridges, which have a relatively subdued form, a small, single slip face and a partial vegetation cover (Figure 6.2a) that can restrict sand movement and cause the dune to be orientated parallel to the strongest sand moving direction; and seif dunes which are mostly devoid of vegetation and possess an active, sharp, sinuous crest that forms in response to sand movement from more than one direction. Both types are found in the Kalahari (Thomas, 1988c). The status of an individual linear dune or dunefield may change in response to climatic and other factors that may cause changes in the vegetation cover.

Figure 6.2 (a) The sparsely vegetated crest of a linear dune ridge in the southwest Kalahari. The photograph, looking northwest, was taken 10 km north of Twee Rivieren. Note the denser vegetation in the interdune straats. (b) One of the nested parabolic dune patches in the southwest Kalahari. Several patches occur within the area of linear ridges between the Nossop and Auob valleys; this one lies 55 km north of Twee Rivieren. Photograph by J. du P. Bothma. (c) Southwest Kalahari linear dunes displaying active 'seif' crests, near Aroab, Namibia. (d) Vegetated relict dune ridges in western Zimbabwe. The photograph looks south from the crest of one ridge to the adjacent ridge.

Parabolic dunes and blowouts

Parabolic dunes (Figure 6.2b) are often related to blowouts in that they can develop through the aeolian activity ensuing from the disturbance of a vegetated or partially vegetated sand surface. Under such conditions the term parabolic dune is given when a more clear-cut dune form develops; sand is removed from the disturbed area and transported downwind, becoming trapped on the surrounding vegetated surface, with the nose of the dune pointing downwind. This may migrate downwind with the arrival of further sand, while the arms remain anchored by vegetation.

Blowouts, which may or may not mark incipient parabolic formation, are often small features, but they are becoming increasingly common in the Kalahari as disturbance of the natural vegetation cover occurs in conjunction with human activities. However, given the partial ground cover and clumped nature of the natural vegetation in much of the Kalahari, especially the drier areas, aeolian processes can naturally move sand to accumulate as small mounds or 'coppice dunes' among grasses and shrubs (Thomas, 1988c).

Parabolic dunes which develop in conjunction with pan depressions are perhaps more important as a component of the Kalahari landscape than those described above. These dunes, often termed 'lunettes' (Hills, 1940), represent the pan-margin accumulation of sand and sand-sized clay pellets deflated from the ephemerally or permanently dry depression surfaces, and have been intensively studied in the southwestern Kalahari (see section 6.4.2).

Transverse and barchanoid forms

As the name implies, transverse dunes or ridges form obliquely to the formative wind direction. The term 'barchan' is applied to a type of transverse dune which

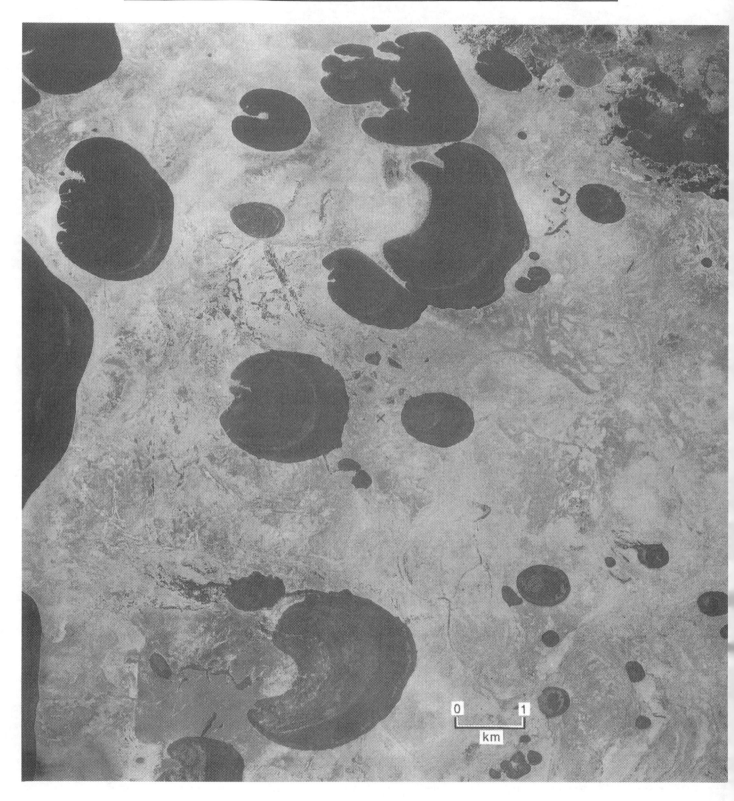

(a)

Figure 6.3. (a) Aerial photograph showing barchan or barchanoid forms in the northwestern part of Ntwetwe Pan. These sand features have relative relief which is usually less than 5 m and grassed surfaces (Cooke, 1980). In some cases contour-parallel vegetation banding can be distinguished, marking the height of past water levels within the pan (Grove, 1969). These features are widely thought to be the remnants of barchan dunes, though their development as subaqueous forms cannot be totally excluded.

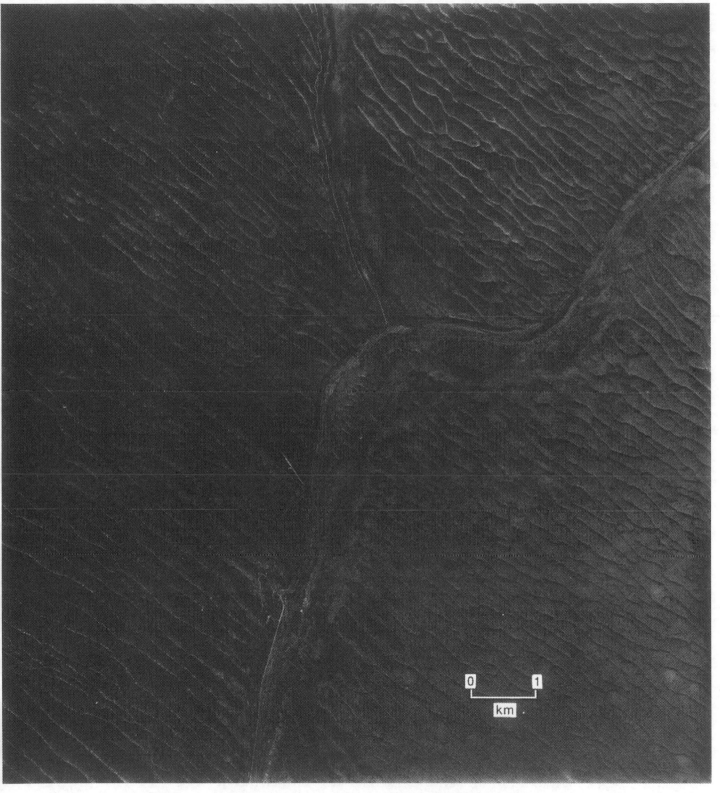

(b)

(b) *The confluence of the Molopo and Nossop (the north–south tributary) valleys occurs within the southwestern Kalahari linear dunefield. To the south of the confluence, the eastern side of the Molopo valley, small, closely spaced linear dunes can be identified, developed from the sediment within the valley rather than directly from the Kalahari Sand of the main dunes (Thomas and Martin, 1987).*

possesses a markedly curved form, with the 'horns' of the dune pointing downwind. Features possessing a morphology between these two are referred to as barchanoid dunes or ridges, though in practice making distinctions between barchan, barchanoid and transverse types is not always easy. These types of dune are relatively uncommon in the Kalahari, but they have been identified by Grove (1969), Cooke (1980) and Mallick et al. (1981) within the Makgadikgadi Basin (Figure 6.3a) and beyond its western margins. The presence of barchan or barchanoid dune forms within the depression would seem appropriate on geomorphological grounds, as these migrating dunes develop where sand supply is limited (Wasson and Hyde, 1983) and often upon hard surfaces. With a prevailing easterly wind direction, the supply of sand would be limited within the depression, and the surface of the depression would present a hard mud or saline layer over which migrating forms could readily pass. However, the identification of these features as dunes has not been firmly verified, with Besler (1983) preferring Cooke's (1980) alternative explanation of development as subaqueous bedforms.

6.1.2 Dune occurrence in the Kalahari

In a study of the relative abundance of different desert dune types, Fryberger and Goudie (1981) noted that about 85 per cent of the 100 000 km² or so of the southwestern Kalahari Desert was represented by a linear dunefield, the remaining 15 per cent being classified as sand sheets. In turn, linear dunes represent 99 per cent of all dune forms present. In fact, throughout the Mega Kalahari, linear forms are clearly the most common dune type, but their characteristics vary according to their regional setting: in the southwest where aridity is greatest they may possess only a sparse vegetation cover and display significant sand movement, whereas in wetter areas such as western Zimbabwe and southern Angola they may support a significant woodland cover and display no evidence of aeolian activity, suggesting that their development was a feature of some antiquity.

Other dune types are linked to specific environmental conditions. Lunettes are found on pan margins, but only within the Southern Kalahari. They are notably absent from the small pan depressions of the eastern Middle Kalahari and western Zimbabwe, and from the large pans of western Zambia (Goudie and Thomas, 1985). Transverse features are restricted, as discussed above, to the vicinity of the Makgadikgadi depression. Parabolic dunes (excluding pan margin lunettes) are also of minor significance in the context of the Kalahari as a whole, but a recent study has shown that groups of nested parobolic dunes attain local importance in the Southern Kalahari (Eriksson et al., 1989).

The total pattern of the Kalahari linear dunes has been described as a 'wheelround' not unlike the pattern of dunes found in the arid core of Australia (e.g. Goudie, 1970). Examination of the morphology, orientation and ecology of the Kalahari dunes has led to their division into a number of component dunefields. This has been carried out with regard to the common linear dunes, although other forms have also been considered within this framework. For our purposes, the division into southern, eastern and northern dunefields (Thomas, 1984a; Figure 6.4) is employed in the following detailed consideration of the Kalahari dunes.

6.2 The southern dunefield

The southern dunefield occurs between latitudes 23° S and 28° 20' S and longitudes 18° E and 22° 30' E in the driest, southwestern, part of the Kalahari, an

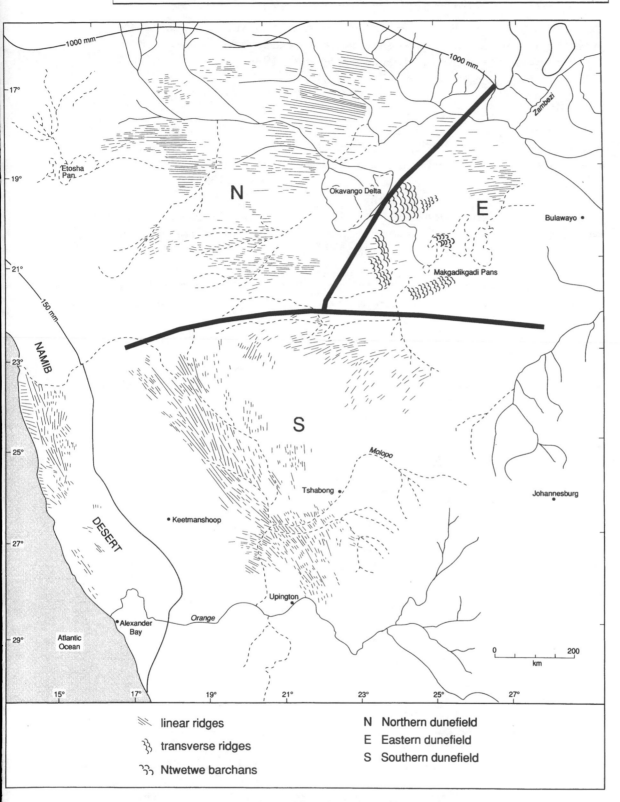

Figure 6.4. The three major dunefields of the Kalahari, dominated by linear dune forms.

area receiving on average between approximately 200 and 300 mm of rainfall per annum. Linear dunes dominate, both as simple and compound forms, and as linear ridges and seif dunes (Thomas, 1988c), but lunettes also occur on some pan margins and from aerial photographs appear to act or to have acted as sediment 'feeders' for linear dunes which continue downwind from these pan margin features (Grove, 1969; Figure 6.5). Grove (1969) also mapped an area of transverse ridges near the Nossop River, but this has not been confirmed by subsequent investigations. Small patches of nested parabolic dunes have, however, been identified within the linear dunefield (Eriksson *et al.*, 1989).

6.2.1 Linear dunes

Following an early description by Lewis (1936) and discussions of the forms and patterns by Goudie (1969, 1970) and Grove (1969), the linear dunes of this area have received more recent detailed and quantified investigations of morphology, dunefield morphometry and sedimentology (Thomas 1986*b*, 1988*c*; Lancaster, 1986*a*, 1988; Thomas and Martin, 1987).

The overall pattern of dunes (Figure 6.4) consists of parallel to subparallel ridges orientated from northwest to southeast, but this masks considerable variability within the system. Towards the confluence of the Nossop and Molopo the overall orientation becomes less meridional than further to the north, and south of this point dunes also feed into the system from a westerly direction. The southern margin of the system is marked by the Orange River, and as Lancaster

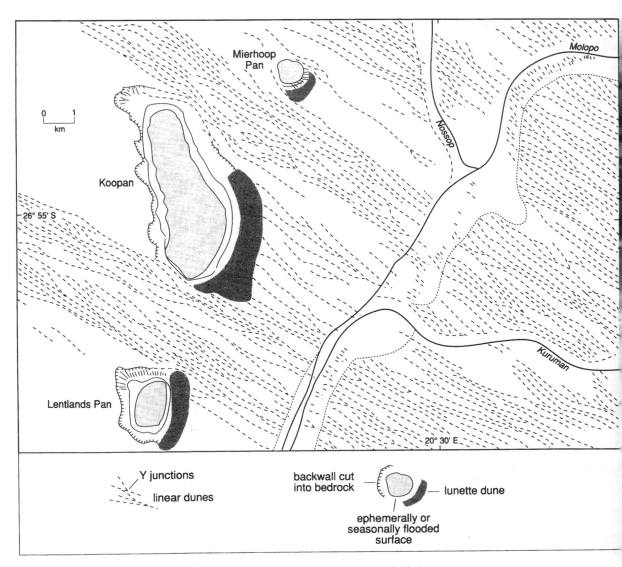

Figure 6.5. Detailed map of part of the southern Kalahari dunefield showing dune orientation and spacing, Y junctions, pans and pan margin lunette dunes. The major lunette dune on the margin of Koopan, which is itself cut into the rock floor of a major duneless corridor, clearly acts as a source of sediment for the linear dunes which extend downwind from it.

(1988) notes, the dunes become less distinct north and east of the Nossop and Molopo valleys; their northern limit appears to be at about 23° S. However, the 'feint furrow patterns' of Grove (1969) and Mallick et al.'s (1981: p. 21) 'well preserved elongate lingoid dunes' and isolated elongate sand 'humps' in central Botswana may represent a southwest–northeast curvature of the same system.

Just as dune orientation within the system varies, so does the form of the dunes and dunefield morphometry. Partially vegetated linear ridges dominate, but ranging in height from about 5 to 25 m, up to 250 m wide and 2 km apart (Thomas, 1988b). A widely noted characteristic is that dune cross-sectional asymmetry prevails, with the steepest flank on the west or southwestern side. This may possess a small slip face at the angle of repose for dry sand (about 30–35°), but Lancaster (1988) reports an overall mean slope value of 13.4° for a sample of 22 dunes, compared to 8.4° for the northeast or east flank.

The crests of the linear ridges are generally not sharp but rounded (Figure 6.2a), often supporting sparse, clumped grasses (e.g. Van Rooyen and Verster, 1983) that act as foci for local sand accumulation, giving rise to a hummocky, undulating appearance (Thomas, 1988c). Dune flanks are better vegetated, but still with appreciable areas of bare sand, while the interdune areas (called 'straats') have been described as open to sparse low savanna scrub woodland (Huntley, 1982).

6.2.2 Sedimentary characteristics

Compared to other dunefields, the southern Kalahari dune system is composed of relatively coarse sand. A study by Lancaster (1986a), based on 22 sample points within the dunefield, has suggested that the particle size characteristics of the Kalahari Sand which forms the dunes varies at two scales within the dunefield: across dune profiles (Table 6.2) in response to the expected sorting processes which are now understood to be a normal facet of aeolian transportational processes (Thomas, 1989); and within the dunefield in the regional direction of sand transport.

The dunes have developed in response to regional wind regimes, thus aeolian sand transport has essentially been in a south to southeasterly direction. Lancaster (1986a) notes that the dune sands become finer and better sorted from northern and western areas towards the southeast. Therefore, the finest sands ought to occur in the vicinity of Upington. This is also a reflection of linear dunes as sand passing forms rather than migrating dunes per se.

It is, however, important to recognise that this picture is complicated by sediment contributions being made to the system from a number of other sources. These include sand transported into the southwestern part of the system from the west of the lower Molopo valley (Fryberger and Ahlbrandt, 1979; Figure 11.2 in Thomas, 1989); finer sand derived from the floors of the dry river valleys within the system (Figure 6.3b), which is generally paler and greyer than the reddish-yellow sand which composes most of the dunefield (Thomas and Martin, 1987; Lancaster, 1987); and, thirdly, sand which is contributed to the system from the floors and fringing lunettes of pan depressions. It has also been noted that the 'source bordering' dunes which develop from such 'intra-ergal' sand sources differ somewhat in morphology and spacing from the main dunes of the system (Lancaster, 1986b, 1988; Thomas and Martin, 1987).

6.2.3 Linear dune patterns

Lancaster (1986a, 1988) has suggested that most of the linear dunes in the southern dunefield are simple forms, i.e. individual forms which do not interfere

Table 6.2. *Averaged grain-size characteristics across linear dunes, southern dunefield*

Twenty-two dunes sampled; values in phi units,* calculated using Folk and Ward's (1957) method.

Location on dune		Grain-size characteristics			
		Mean	Standard deviation	Skewness	Kurtosis
Crest	$\bar{x}$	2.16	0.49	0.14	0.52
	SD	0.20	0.12	0.09	0.03
NE dune flank	$\bar{x}$	2.21	0.62	0.05	0.52
	SD	0.19	0.16	0.12	0.02
SW dune flank	$\bar{x}$	2.26	0.59	0.07	0.53
	SD	0.18	0.15	0.09	0.04
Interdune	$\bar{x}$	2.12	0.90	0.02	0.52
	SD	0.32	0.28	0.17	0.177

Data from Lancaster (1986a).
* A log scale, whereby 0 phi = 1.0 mm, 1 phi = 0.5 mm, 2 phi = 0.25 mm, etc.

or interact with other adjacent dunes. However, this masks the fact that in parts of the dunefield, and locally throughout, the pattern is more complex. (Goudie, 1969, 1970; Thomas, 1986a; Lancaster, 1986b; Thomas and Martin, 1987; Eriksson et al., 1989). Compound forms occur where adjacent ridges interact, and complex forms where the linear dunes extend from pan lunettes or are superimposed by parabolic dune patches. In the context of the whole dunefield, Lancaster's (1987) estimate of 33 to 50 per cent of its area being occupied by simple ridges seems appropriate.

In the northern part of the dunefield, equatorward of about 26° S, simple single ridges appear to dominate the system (e.g. Breed and Grow, 1979). They are widely spaced (Figure 6.6), commonly over 500 m apart and 10–20 m high (Lancaster, 1988). These are the most vegetated dunes in the system. To the south, the pattern increases in complexity; dune spacing tends to decrease and many ridges merge at 'Y' junctions, taking on a dendritic appearance in some areas (Goudie, 1969), with a number of pattern variations being apparent (Thomas, 1986a).

'Y' junctions are an important component of this and a number of other linear dune systems. In most cases (e.g. '999 cases in 1000' in the northwestern Simpson Desert, Australia – Mabbutt and Wooding, 1983) the junctions are open in an upwind direction. These 'normal' junctions represent the merging of adjacent dune ridges, probably under the influence of deflecting side winds. In the central and southern areas of this dunefield, ridges are very close together in comparison with other linear dunefields (Thomas, 1988b), rarely exceeding 300 m apart, which increases the probability of ridge deflection resulting in merging.

Analyses in the Kalahari (Thomas, 1986a) have shown that in the central area of the southern dunefield, over 15 per cent of the junctions are open downward. This 'reverse' type of junction, uncommon in other linear dunefields, may be a consequence of a number of factors, though it is currently not possible to determine their exact cause. According to Breed et al. (1979), wind regimes in the area are currently more complex than those normally to be expected in areas of linear dune development, which may have led to greater dune deflection. Alternatively, it may result from the superimposition of two different patterns of linear dunes.

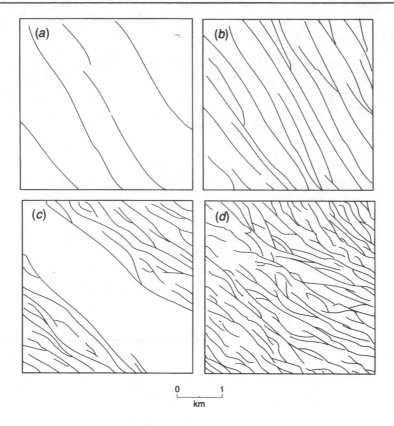

(a) Northern area – widely spaced straight ridges
(b) Uniform dendritic pattern
(c) Broad duneless straats separating coalescing dendritic dunes
(d) Complex reticulate pattern

Figure 6.6. Variations in the linear dune patterns in the southern Kalahari.
After Thomas (1986a) and Lancaster (1988).

6.2.4 Linear dune morphometry

The height and spacing of linear dunes is determined by the availability of sediment for dune development and the characteristics of the wind regime. In the Southern Kalahari (Thomas, 1988b), as elsewhere (e.g. Wasson and Hyde, 1983), the further apart ridges are, the higher they become. This can be explained by the fact that, if sand supply is approximately constant, the greater the number of ridges the less the amount of sand available per ridge. However, this relationship does not generally hold for individual ridges, but for 'blocks' of dunes (Figure 6.7), which may partly be due to the existence of dune junctions and the effects of dune merging and division.

Where merging at junctions results in a reduction in the number of dune ridges, new dunes tend to begin a short distance downwind, maintaining the mean spacing within a particular part of the system (Thomas, 1986a, 1988b). Even in the central part of the dunefield, where the pattern of dunes is rather complicated and varied, morphometric analyses using a number of statistical parameters (Thomas, 1986a) have demonstrated that the dunefield appears to have attained a degree of steady-state equilibrium between the wind, sediment supply and dune patterns and forms.

6.2.5 Seif and parabolic dunes

Although partially vegetated linear ridges dominate this dunefield, in some areas dune development locally reflects the presence of an even sparser plant cover. At some localities, overgrazing has significantly reduced the vegetation cover on both dunes and in interdune straats. On the ridges, this has led to enhanced aeolian activity, especially in the crestal zone, allowing sand transport to occur in response to all potential sand-moving winds. This has resulted in the enlargement of slip faces and the development of sinuous, active crests (Figure 6.2c); in other words, the linear ridges change to seif dunes (Thomas, 1988b). In interdune areas, small coppice dunes can develop as sand becomes trapped around the remaining shrubs and grass clumps.

Where the sand-moving wind regime is more unimodal, localised devegetation has resulted in the development of parabolic dunes, often in clusters or 'nests' (Figure 6.2b). This has been observed to occur between the Auob and Nossop valleys, in relation to larger trees within the dunefield, which tend to grow at the junctions between higher linear ridges, where their deep root systems exploit the moisture retained within the sand (Eriksson *et al.*, 1989). The trees support a well utilised herbivore micro-ecosystem, which leads to vegetation pressure and the development of blowouts (Eloff, 1984).

6.3 The eastern and northern dunefields

The eastern and northern dunefields, though also dominated by linear duneforms, differ markedly from the southern field in that, occurring in wetter areas of the Mega Kalahari, they support considerably more vegetation, including significant woodland communities. This has been widely regarded as a clear indication that these are relict dune systems, the presence of which is an indication of greater past aridity in the Kalahari (see below and Chapter 7). The northern and eastern dunefields, respectively called northern dunes Group A and Group B by Lancaster (1981), are differentiated largely on the grounds of orientation and latitude, with the northern system centred upon northern Botswana, southern Angola and western Zambia and the eastern system best developed in western Zimbabwe.

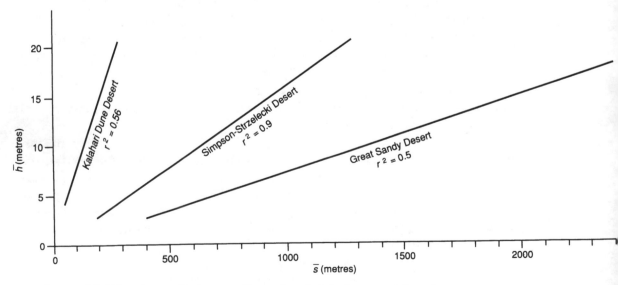

Figure 6.7. The relationship between linear dune height (measured from the interdune area to the dune crest) and spacing (measured as the crest-to-crest spacing), for blocks of dunes in the southwestern Kalahari and in other dunefields. After Thomas (1988b).

6.3.1 The eastern dunefield

The major part of the eastern dunefield extends from east of the Gwayi River in western Zimbabwe to an area to the north and west of the Makgadikgadi depression in Botswana, including the 'Gidikwe lineations' of Grey (1976) (Figure 6.4). This dunefield consists almost exclusively of linear ridges, but the transverse features found in the western part of the depression (Grey, 1976; Cooke, 1980) can also be considered as part of this dunefield.

Flint and Bond (1968) used both aerial photographs and ground survey to explain the origin of the 'sand ridges' in Hwange National Park in western Zimbabwe. The first warden of the park, Ted Davison, has described how during the 1930s he mapped the area from the ground and was puzzled by the profusion of parallel dry water courses on the sand (Davison, 1977). Topographic map makers were also confused and one edition of the 1:250 000 map sheet of the area is covered by parallel blue lines as if the region comprised a vast river network with the divides consisting of low, sandy interfluves. These low, grassed areas are now known to be the interdune straats, but the considerable development of wooded vegetation on the sand ridges (Figure 6.2d) gives them a markedly un-dune-like appearance.

It is this distinction between dune ridges and interdune vegetation communities which makes the dune system apparent in aerial imagery (Flint and Bond, 1968; Thomas, 1984a). Gaps in the mapped pattern of the dunes may not be real, and in many locations they represent areas where the vegetation zonation has been destroyed by fire, elephant browsing, and human activities. The main dune ridges of the eastern dunefield form a gentle arc which is least distinct in the east, extending from east of the Gwayi River, where the orientation of tributary streams may also preserve the pattern of former linear dunes (Thomas, 1983/4), across Hwange National Park before terminating at longitude 25° E in Botswana. The orientation of the arc changes from 95° to 275° in the east, 90° to 270° in the centre and 80°–265° at the Zimbabwe–Botswana border (Thomas, 1984a).

The ridges are generally considerably broader and more widely spaced than those in the southern dunefield. Ridge widths have been recorded as ranging from 500 m to 2500 m (Thomas, 1984a), with a wavelength of 1500–2500 m (Lancaster, 1981). They are also generally low, with Flint and Bond (1968) citing relative relief ranging from only 2 to 4 m. However, some ridges are clearly much higher than this (see Figure 6.2d), and Thomas (1984a) surveyed ridges at the Botswana–Zimbabwe border with a mean height of 22.5 m.

The overall impression is nonetheless one of a very degraded linear dunefield (Flint and Bond, 1968; Lancaster, 1981; Thomas, 1984a, b). The low height and great spacing of the ridges has led to their being described as the residual stumps of formerly greater dunes, perhaps comparable with the linear dunes of the Namib Sand Sea, which have been degraded by processes such as sheet wash (Flint and Bond, 1968; Thomas, 1986b). Analysis of the ridge sediments (Flint and Bond, 1968; Thomas, 1984b, 1985) has identified the presence of much higher proportions (in excess of 5 per cent) of silt and clay than are normally found in active dune sands, which has also been interpreted as evidence of ridge degradation.

While dominated by the linear ridges described above, there are three additional landform units which can be considered to be components of the eastern dunefield. Grove (1969) and Cooke (1984) both described an area of degraded barchan dunes on the floor of Ntwetwe Pan (Figure 6.3a). Second, to the west of the Makgadikgadi Basin, the Gidikwe lineations are clearly visible in aerial photography and satellite imagery and were mapped by Grove (1969) as an area of longitudinal dunes with a wavelength of 200 m. Subsequent field investigations (Massey, 1974; Grey, 1976) have revealed that they have neither topographic nor vegetational representation on the ground. Features without topographic representation formed by aeolian processes are called 'wind streaks' (e.g. Greeley and Iversen, 1985) and result from the erosion and deposition of sediments with different grain-size characteristics. Grey (1976) has, however, suggested that the Gidikwe lineations may have a subaqueous origin. Finally, to the west of the lineations, and also west of the northern end of the Gidikwe Ridge, are areas of transverse and barchanoid dune ridges, up to 20 m high and 2 km apart (Grey, 1976; Mallick et al., 1981).

6.3.2 The northern dunefield

The northern dunefield appears to consist almost exclusively of linear ridges with subdued and rounded crests, termed alab dunes by Grove (1969), which are now known to extend from the west bank of the Zambezi River in Zambia (Williams, 1986) to the Etosha Pan in the west. The system is particularly well represented on the northwestern side of the Okavango Delta (e.g. Grove, 1969; Lancaster, 1980), where some ridges have an unbroken length in excess of 200 km. The ridges again support a mixed savanna woodland vegetation which aids their distinction from the grass- and scrub-covered interdune straats. Like the eastern dunefield, this system also forms an arc, but the 'peak' of the northern arc occurs along longitude 21° E rather than 27° E (Thomas, 1984a; Figure 6.4).

To the northwest of the Okavango Delta the ridges are up to 25 m high and 1000–2500 m apart (Lancaster, 1981), and where topographic obstacles such as the Tsodilo Hills are encountered, the sand has piled up on the eastern sides (Grove, 1969), creating dune-free shadow zones to the west which are up to 50 km long (Lancaster, 1981). The ridges are abruptly terminated at the northwestern edge of the Okavango Delta, either because of degradation and truncation by the waters of the delta itself, or possibly because of earth movements associated with the faultline which marks this edge of the delta (Jones, M.T., 1962; Jones, C.R., 1982). The ridges to the northeast of the Okavango Panhandle have been subjected to flooding of the interdune areas and considerable degradation (Grove, 1969; Mallick et al., 1981). In western Zambia, where ridges have a wavelength of about 300 m (Williams, 1986), and at the eastern end of the Caprivi Strip of Namibia, parts of the dunefield also appear to have experienced flooding, and the deposition of alluvium in interdune areas.

In Angola, north of latitude 17° S and west of longitude 21° 30', the ridges become more broken, though the former pattern of linear ridges can be inferred from the orientation and trellis-pattern of the tributary network of the Lungé–Bungo River (see Thomas, 1984a) as far north as 13° S. The true northern limit of the dunefield is unknown, but De Dapper (1979a, b, 1981a, 1985) has inferred the presence of linear and transverse dune relics from the micro-relief features of the Kalahari Sand-covered Shaba plateau region, near Kolwezi in Zaïre, latitude 11° S.

6.3.3 The question of dune activity in the Kalahari

The vegetated dunefields have frequently featured in palaeoenvironmental deductions for the Kalahari, with the orientation and position of the different linear dune systems inferred to indicate shifts in the position of the southern African anticyclone, around which the predominantly easterly (and south to southwesterly in the southwestern Kalahari) dune-forming winds blew (e.g. Lancaster, 1980, 1981; Thomas, 1984a; see section 7.4.6 for discussion of this topic). One of the major assumptions which has been made in this context is that the presence of vegetation on most types of desert sand dunes is an indication that the features are no longer active and therefore that they are of relict or fossil status. The question of the significance and status of desert dune vegetation has recently received attention (Ash and Wasson, 1983; Tsoar and Møller, 1986; Thomas and Tsoar, 1990), which has some bearing on the interpretation of the Kalahari dune systems (Thomas, 1988c).

While vegetation can stabilise dunes, and therefore can indicate their relict status in terms of aeolian activity, it can also act as a focus for sand accumulation and dune growth, as well as contributing to the morphological development of dunes (Thomas and Tsoar, 1990). To some extent, which of these roles dune vegetation plays depends on the type of dune concerned. In the Kalahari, all dune types, as well as dunes in all three dunefields, possess a vegetation cover, albeit differing in community structure and density from place to place, so that the question of whether dunes are stabilised, fossil features requires careful consideration.

The linear dunes in the eastern and northern dunefields, together with other dune types where they are present, are generally very well vegetated, though Flint and Bond (1968) estimated that ground cover on the linear ridges in western Zimbabwe was only 50 per cent. The vegetation cover takes on the form of woodland which varies from open savanna communities to the north and west of the Okavango Delta (Grove, 1969), to denser mixed woodland in which mature teak trees (up to 20 m high) or secondary scrub savanna dominate, in western Zimbabwe (Boughey, 1963; Flint and Bond, 1968; Thomas, 1984a). The nature of the vegetation, together with other factors such as ridge degradation, the lack of internal bedding structures and modification of the original wind-derived sedimentary characteristics confirm the fact that these two Kalahari dunefields are relict features, dating from a time or times when climate and other environmental factors were conducive to aeolian activity.

The linear dunes of the southern dunefield are in many ways analogous (Thomas, 1988c) with those investigated in the Simpson Desert, Australia, and the Sinai-Negev in Egypt and Israel (Ash and Wasson, 1983; Tsoar and Møller, 1986). In these studies vegetation on linear dunes was not necessarily found to be an indicator of the relict status of the dunes, but rather, at ground cover densities of up to 35 per cent, to allow significant sand movement to occur (Ash and Wasson, 1983), and also to contribute to the morphological development of the dunes (Tsoar and Møller, 1986; see section 6.1.1).

In the southern dunefield, the form of the seif dunes and the morphology of the linear ridges, which show sand accretion around grass clumps, the presence of small slip faces and widespread rippled surfaces (Thomas, 1988c), together with the systematic sedimentary variations across dune profiles (Lancaster, 1986a), all suggest that the dunes are not relict features *per se*, but that they are still responsive to aeolian processes. This has considerable implications for the incorporation of this dunefield in palaeoenvironmental studies, though this is not to say that the nature of the dunefield has not changed, for example from dominance by seif dunes to today's linear ridges, in response to changes in precipitation amounts and vegetation communities during the Quaternary period.

6.4 Pans

Pans are small, closed basins containing ephemeral lakes, characteristic of arid and semi-arid regions of low relief. They occur throughout the countries of southern Africa, with the possible exceptions of Lesotho and Malawi (Shaw,

1988b), and have been identified as an important component of the southern African landscape in areas with an annual rainfall of less than 500 mm yr^{-1}, particularly where the surficial materials are either shales or unconsolidated sands (Goudie and Thomas, 1985). Pans are found throughout the Kalahari, as far north as Zambia, where mean annual rainfall exceeds 1000 mm yr^{-1}. They may also occur in the Northern Kalahari, but have not been described in the scientific literature.

6.4.1 The distribution and morphology of pans

The term pan encompasses a variety of basin forms and sizes even within the Mega Kalahari, ranging from small depressions a few metres in diameter to major structural basins such as the Makgadikgadi (see section 5.3.4). A common and distinctive type occurs in the Southern Kalahari and on the Bakalahari Schwelle, particularly between the settlements of Tshane, Kokong and Mabuasehube (Lancaster, 1978a), around the Nossop–Molopo confluence and in the vincinity of the Nossop at Aminuis (Lancaster, 1986b). They are characterised by a sub-circular, sub-elliptical or kidney shape, and the presence of one or more crescentic dunes, called 'lunettes' (Hills, 1940) (section 6.1), anchored to the southern or southwestern sides of the basin. These pans commonly lack surface inflows, though short, poorly developed channels feed into some depressions, supplying run-off to the pans during major rainfall events. Pan sizes tend to range from 1 to 16 km², with depths up to 20 m. The pans also appear to have a preferred orientation parallel to the prevailing winds (Lancaster, 1978a; Figure 6.8).

Lancaster (1978a) observes that the majority of pans of this type occur within the Kalahari Sand, although where the sand cover is thin, as at Khakaea and Mabuasehube, Karoo and Precambrian basement rocks may be exposed. The pans are also associated with the presence of calcrete, which tends to form a rim around the pan periphery, and may extend for several hundred metres beyond the pan depression, usually along the courses of ephemeral inflows. Some of the higher level calcretes may contain algal mats and stromatolites, as in the Aminuis area (Lancaster, 1986b), suggesting groundwater discharge. Silcretes sometimes occur on the pan surface, or within the calcrete surround, as a result of calcium replacement or by direct precipitation (Summerfield, 1982), while ferricretes are also present in the eastern Kalahari, where they may form a second and distinct rim above the calcrete zone, as at Lekwatsi Pan northeast of Mochudi.

In most of these pans the surface is a chemically active alkaline environment composed of laminated calcareous clays, sands and possibly salt deposits (section 3.6.3), which have been formed by the processes of shallow-water lacustrine deposition, interrupted by periods of groundwater-controlled aeolian deflation. Boocock and Van Straten (1962) classified the Kalahari pans as grassed, ungrassed (clay) and saline, although surface characteristics are likely to result from a delicate balance in the groundwater-deflation relationship rather than variation in formative processes themselves (Butterworth, 1982). Bruno (1985) notes that grassed pans tend to have sandy soils and a gentle gradient, together with a lower groundwater table, and suggests a reduction in geochemical activity and partial burial by aeolian sands in this pan type.

Pans of less mature aspect are encountered throughout the Kalahari region in topographic hollows, where drainage impedence and the tendency for fines to migrate downslope have created conditions for the ponding of water. Linear pans with limited calcrete development are common in interdune hollows in the vicinity of the Auob and Nossop Rivers in the southwest Kalahari. In western Zimbabwe Goudie and Thomas (1985) have recorded 2449 small pans at a density of about 1 pan per 6.7 km² and have noted that the pans are usually found in inter-dune hollows, with pan density at its greatest where linear dunes are best developed. None of the pans in this area has associated lunette dunes. In Zambia, to the east of the Zambezi River, another area of high pan density, the pans of the sandveld around Mongu (termed 'plains') are larger than the Southern Kalahari features, Itundu Plain reaching 24 km² in area, and may contain permanent water, as at Lake Makapaela (Williams, 1986).

Other topographic lows exhibiting pan development include the floors of mekgacha, where depressions are filled seasonally by rainfall, and between-beach ridges in the major lake basins. The Dautsa Pan, a linear feature some 10 km long and 50 m wide, between the two 936 m asl Dautsa shorelines, is a good example of this type (see section 5.3.2).

On a smaller scale the term pan also includes small waterholes extended by the trampling of animals (Weir, 1966). These circular to sub-circular features, floored by vertisolic clays, are common in the Mopane woodland belt of Hwange, Zimbabwe, and the Mababe Depression. Such depressions frequently retain water into the dry season and undergo considerable turbation and the destruction of surrounding vegetation from the activities of domestic stock and wild animals, particularly elephant, buffalo and hippopotami.

6.4.2 Pan margin (lunette) dunes

Lunette dunes (Figure 6.8) have been identified in relation to over 110 pans in the southern Kalahari by Mallick et al. (1981), where they occur on the leeward side in relation to prevailing wind direction (Lancaster, 1978b). Some pans display two fringing dunes, each with different sediment characteristics: an outer dune, up to 30 m high and composed of sand and a smaller, inner feature with a higher clay content (Lancaster, 1978b; Goudie and Thomas, 1986), and up to 15 per cent CaCO₃. Lancaster (1986b) has recorded a suite of three lunettes at Koes Pan, Namibia.

6.4.3 The origin of pans

Early studies of pans in southern Africa (Alison, 1899; Passarge, 1911) suggested that they were formed as a result of erosion by animals, which congregate to utilise both water and sodium salts. This view has persisted until recently (e.g. Flint and Bond, 1968). Weir (1969) has demonstrated a cycle of pan enlargement caused by stock or game animals, but the limited size of such features clearly does not explain the evolution of the larger pans; rather, it accounts for the small depressions, with diameters measured in tens of metres, including those that occur within the pans themselves.

Thomas (1988d) suggests that termite activity, which is intense at the periphery of small waterholes, may play a part in their formation by the creation of subsurface cavities and the accumulation of minerals such as sodium which attract animals.

Lancaster (1978a, b) followed Grove (1969) in suggesting that aeolian deflation was the dominant process in pan formation, interrupted by episodes of deposition of fine sediments during humid or lacustral phases. He cites as evidence the general absence of bedrock, the orientation of pans parallel to the prevailing winds, and the existence of lunette dunes. Here the outer sandy lunette

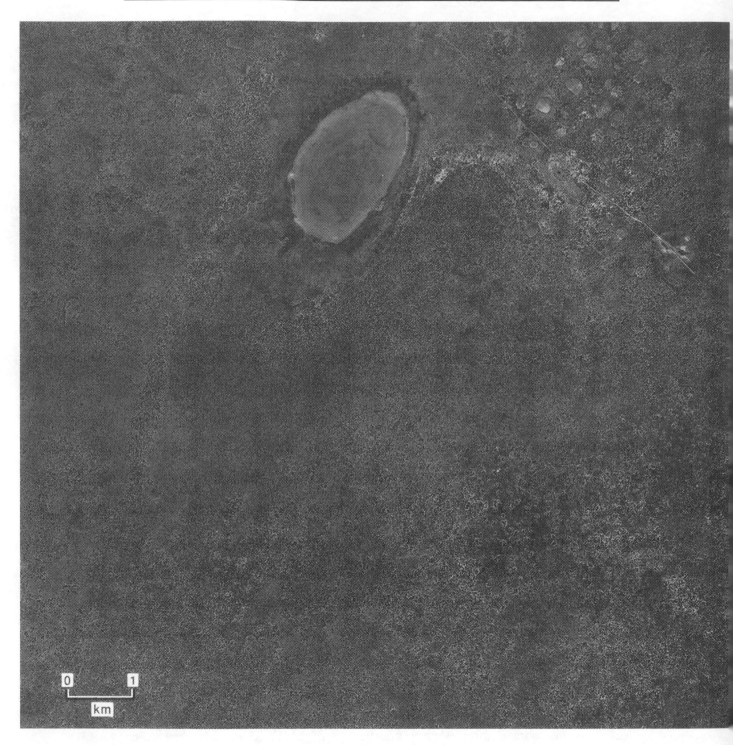

0 1
km

(a)

Figure 6.8. Air photographs of typical western Kalahari pans, (a) One Pan, (b)
Kgatlwe Pan, both in Botswana, the latter displaying shallow water inundation
of the surface. On the photographs fringing lunette dunes can be identified on the
margins of both pans due to vegetation and surface characteristics. Kgatlwe Pan

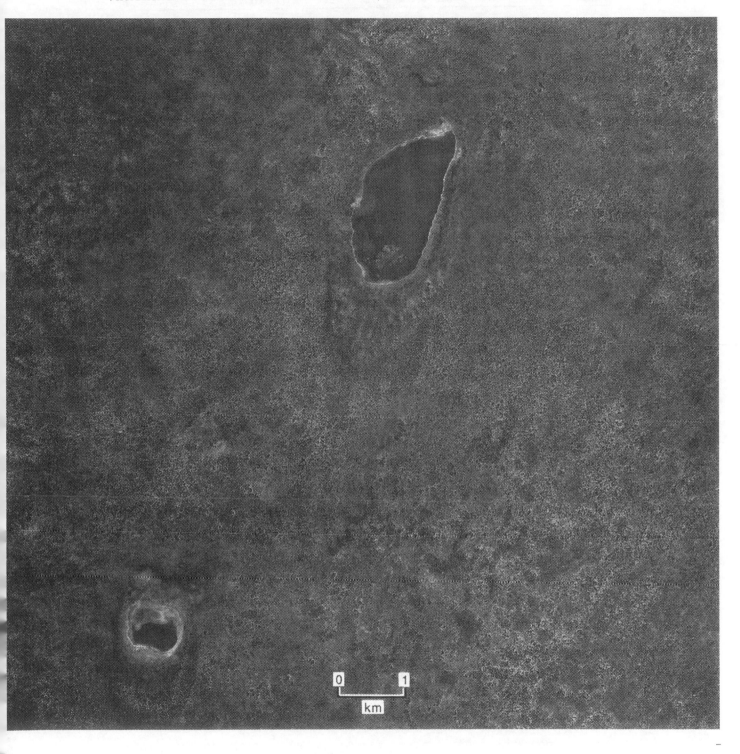

b)

possesses a lunette on the SSW side of the depression, while One Pan possesses one on the SE side, with a further possible lunette marked by the darker surface area further away from the depression.

represents sediment deflated during initial depression formation and an inner lunette, with a higher silt and clay content, results from the subsequent deflation of material from the pan surface.

The second argument does not take into account that in areas where pans occur within linear dunefields (see Figure 6.5), the orientation of the linear dunes will have an influence on pan orientation, especially as they are located in the inter-dune troughs. Goudie and Thomas (1986) have also pointed out that the mean sand, silt and clay content of the outer and inner lunettes does not always vary in the systematic way suggested by Lancaster, with some outer lunettes having a greater fines content than their inner counterparts, and in all cases the sand content of the lunette sediment being of the order of 90 per cent. This suggests that the relationship between aeolian deflation and pan and lunette formation is rather more complex than previously supposed. The presence of a pan duricrust suite, too, cannot be explained by a purely deflationary origin. Nevertheless, the surface evidence provides a strong case for aeolian activity as an agent of influencing pan morphology, particularly if changes in environmental conditions are invoked.

Investigations of pans in three dimensions in the Kalahari are still few in number, but have provided some interesting results. A number of pans have shown duricrust development at depth, as, for example, at Tshane Pan, where 10 m of calcrete overlies 35 m of silcrete, itself overlying weathered dolerite (Mallick et al., 1981). Similar results have been obtained from geophysical and geochemical studies of Mogatse Pan and Lokware Pan near Kukong, taken to represent typical clay and grass pans respectively (Butterworth, 1982; Farr et al., 1982), which have revealed weathering and duricrust formation to a depth of 30 m, together with underlying geophysical anomalies. These suggest that pans are formed by weathering along groundwater flow paths following sub-Kalahari lineations, and that pans are even capable of migration over long periods of time (Farr et al., 1982). Similar zones of groundwater convergence have been noted along pre-Kalahari drainage in the lower Molopo area (Arad, 1984) and around Tshane and Hukuntsi (Bruno, 1985). The importance of salt weathering in pan environments has been recognised recently (Cooke, 1981; Goudie and Watson, 1984), and it is probable that the salinity of the groundwater is a factor in the effectiveness of deep weathering. In Arad's (1984) study area pans were developed in locations where the Kalahari Group sediments shared a saline aquifer with the highly weatherable Dwyka Group, while Bruno (1985) suggested that freshwater aquifers are perched close to the surface of pans by calcrete layers acting as aquicludes. Changes in the zonation of fresh and saline ground waters may occur as a result of 'flushing' during periods of higher precipitation (section 3.7). Recent work by Lancaster (1986b) has placed more emphasis on the role of ground water.

The Kalahari pans thus appear to be polygenetic features. The concentration of fines in topographic depressions of varying origins will form impermeable layers, leading to cycles of lacustral deposition and aeolian activity, noticeable not only where lunettes are present but also where linear pan forms are found adjacent to dune ridges. Larger pans often have a tectonic origin, enhanced by deep weathering and ground water activity, and may thus be similar in genesis to mekgacha.

6.5 Rock landforms

Although rock outcrops are uncommon in the Kalahari core, they form striking landforms of considerable geomorphic significance. Inevitably the outcrops reflect the lithology of the sub-Kalahari geology (Chapter 2) and the rocks encountered belong to a variety of stratigraphic units from the Precambrian basement to the Karoo Sequence. The role of rock outcrops in the formation of mekgacha (section 5.4) and pans (section 6.4.2) has already been noted. Other rock outcrops form broad ridges of low relief, such as the limestones and sandstones of the Ghanzi Group, and major escarpments controlled by faults, including the Karoo basalts and sedimentary rocks of the Mosu Escarpment, south and east of Sua Pan, and the Chobe Escarpment.

At the periphery of the Kalahari there is a gradual transition from sandveld to hardveld, with an increasing presence of escarpments and hill massifs. In parts the Kalahari Basin is terminated by such escarpments, as in the Gaap Escarpment south of Kuruman, and in western Zimbabwe, where the Kalahari Sand overlies the dissected Karoo basalt plateau. To the west and east of the Kalahari Desert the transition tends to be more gradual, although substantial ranges of hills overlook the Kalahari in the vicinity of Shoshong and Serowe. In the south the decrease in thickness of the Kalahari sediments allows considerable exposure of bedrock, as in the Langeberge and Korranaberg in the Kuruman area.

6.5.1 Inselbergs

Isolated hills or inselbergs, rising upto 150 m above the surrounding plains, are landforms restricted mostly to the Middle Kalahari (Figure 6.9) where they are formed from a variety of rocks of different lithologies (Table 6.3). Where the parent rock is crystalline, the inselbergs may assume specific inselberg forms, such as boulder-strewn hills ('nubbins') and castellated block forms ('castle koppies' : Twidale, 1981).

The origin of inselbergs has generated considerable debate since they were first observed in Africa by European travellers (e.g. Passarge, 1895; Bornhardt, 1900; see Twidale, 1988, for a recent review of inselbergs in southern Africa), with forms present in the Middle Kalahari offering little evidence of mode of origin. Although structural control may be present, most inselbergs are associated with shallow depths of Kalahari sediments and regional groundwater recharge (section 3.7), and thus coincide with regional swells in the sub-Kalahari geology. As the Kalahari is predominantly a depositional environment it is reasonable to assume that they have formed from compartments of relatively resistant rock by the processes of sub-surface weathering (etchplanation) on the pre-Kalahari land surface envisaged by King (1942).

In common with most exposed inselbergs, present processes are limited to minor sub-aerial weathering and mass movement. Hills which lie within the major lake basins frequently show evidence of lacustrine processes. Thus both Kokonje and Kubu Islands in Sua Pan have summit beach gravels at the 920 m level (section 5.3.4), whilst the Gubatsa Hills in Mababe (Figure 6.10a) have cobble beaches between hills at the 936 m level.

6.5.2 Caves

Caves occur in calcareous rocks, particularly the dolomites of the Proterozoic Transvaal and Damara Sequences, exposed to the west of the Okavango and on the southern and southeastern margins of the Kalahari. The physical development of the cave systems is limited; in some cases all that remains is the cave infill. However, these sites have proven value as sources of palaeoenvironmental data for time spans ranging from the late Quaternary to the Plio-Pleistocene boundary.

The most important of these sites is Drotsky's Cave in the Gcwihaba (Kwihabe)

Table 6.3 *Location of inselbergs in the Middle Kalahari*
(Key to Figure 6.9)

Location	Geological formation	Rock type
1 Tsodilo Hills 2 Aha Hills 3 Gcwihaba Hills 4 Koanakha Hills	Damara Sequence	Dolomites and quartzites
5 Mabeleapodi Hills 6 Tsau Hills	Ghanzi Group	Quartzites, shales, sandstones, mudstones
7 Khwebe Hills 8 Ngwanalekau Hills 9 Haina Hills 10 Gubatsa Hills 11 Goha Hills	Kgwebe Formation	Acid porphyries and volcanoclastics
12 Shinamba Hills	Ghanzi Group	Quartzites, shales sandstones, mudstones
13 Kedia Hill	Karoo Sequence	Sandstone
14 Kubu Island 15 Kokonje Island	Post-Karoo intrusives	Dolerite

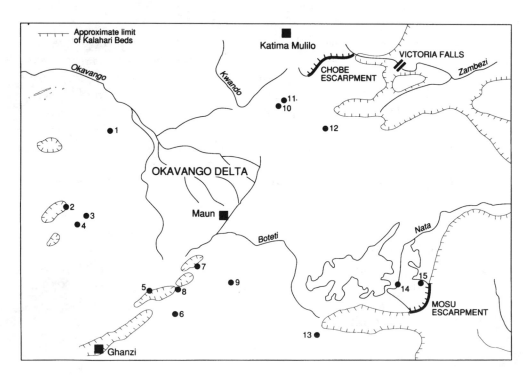

Figure 6.9. Distribution of inselbergs in the Middle Kalahari (see also Table 6.3).

Figure 6.10. (a) One of the inselbergs of the Gubatsa group in the Savuti area of the Mababe Depression. It is linked by a tombolo landform to the Magikwe Ridge, which is shown in the background (see also Figure 5.6). (b) Stalactites in Drotsky's Cave, Gcwihaba Hills. Recent tectonic activity is indicated by the fracture and displacement of the sinter. Photograph by John Cooke.

Hills, first described by Wayland (1944) and studied in detail by Cooke (1975*a*). The cave is a horizontal feature on two distinct levels running for 350 m through the largest of the hills comprising the Gcwihaba Group (Figure 6.11). It follows structural weaknesses in the rock and is predominantly solutional, although large-scale collapse is also present. The cave system contains a variety of phreatic features such as rock arches, rock blades and solutional hollows. It has been extensively faulted, as evidenced by large cracks and fault breccias (Figure 6.10*b*), some of which may be of recent origin.

Cooke's (1975*a*) study of the sinter formations within the cave, together with analysis of wind-derived sands in the cave sediments, suggests that there are at least four generations of sinter growth, representing a cycle from phreatic conditions though draining and sinter growth during vadose phases to reflooding and resolution. This cycle has occurred on at least three occasions, and may have been interrupted by arid intervals in which deposition of Kalahari Sand from the cave exterior becomes the dominant process, as at present. The radiocarbon dating of different generations of sinter, together with calcretes from the adjacent Gcwihabadum (Kwihabe) Valley have been used as the basis for late Quaternary palaeoclimatic reconstruction of the area.

Also located in the Gcwihaba and adjacent Koanakha (Nqumtsa) Hills are the remnants of ancient caves, evidenced by cave sediments and breccias now exposed sub-aerially in fissures on the hillsides. These cave sediments are fossiliferous and are thought to be of equivalent age to the Plio-Pleistocene deposits of Makapansgat in South Africa (Pickford and Mein, 1988). Two deep sink holes have also been located in the Aha Hills (Cooke and Baillieul, 1974), but contain little depositional material.

On the southern margin of the Kalahari the caves of the Gaap Escarpment (Figure 6.12) have been extensively studied from a number of disciplinary viewpoints, thus supplementing geomorphological and palaeoenvironmental studies of the escarpment itself (Butzer *et al.*, 1978). The first cave deposit studied was the infill of the Hrdlička Cave at Taung, destroyed by quarrying in the 1920s, which yielded the Australopithecine skull known as the 'Taung Child' (Young, 1926; Dart, 1926). The morphology and deposits of this cave have been painstakingly reconstructed from documentary evidence and adjacent exposures (Peabody, 1954) and have been extensively reinterpreted over recent years (Butzer, 1974).

Equus Cave lies 750 m from the site of Hrdlička Cave, and has formed within a tufa apron in the upper part of the Thabaseek Gorge. Excavation of the cave fill has yielded sediments derived from aeolian, soil wash and mechanical weathering

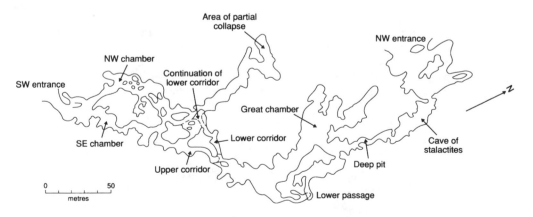

Figure 6.11. A plan of Drotsky's Cave. After Cooke (1975a).

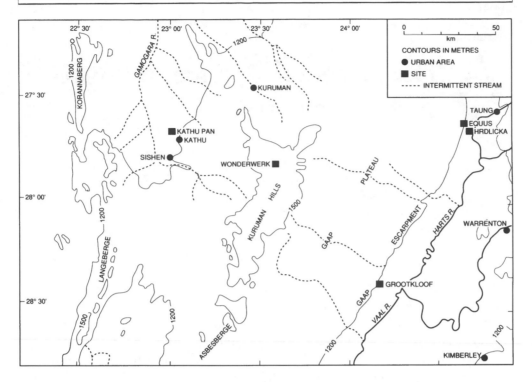

Figure 6.12. Location of cave sites in the northern Cape. After Beaumont et al. (1984).

processes (Butzer, 1974). This cave has also been the focus of recent palaeoenvironmental studies (Beaumont *et al.*, 1984). Smaller caves have also been recorded along the Gaap Escarpment, particularly in the Grootkloof area (Peabody, 1954).

Peabody also described Wonderwerk Cave in the Kuruman Hills 100 km to the west of Taung. This cave is a massive solutional cavity some 140 m long and up to 20 m wide, following the near-horizontal bedding planes in the dolomite bedrock. Recent work on this cave has been summarised by Beaumont (1979) and Butzer (1984*a*, *b*). The cave contains sediments dating back to the Middle Pleistocene (Beaumont *et al.*, 1984), including gravel beds derived from spalling or frost-fracturing, stalagmites, man-introduced organic materials, micromammalian fossils (Avery, 1981), pollen (Van Zindren Bakker, 1982*b*) and artefacts of Late Stone Age or more recent origin.

Two dolomite caves have been mentioned by King (1951) and mapped by Cooke (1975*b*) from the Lobatse area to the southeast of the Kalahari. Both caves are composed of caverns connected to the surface by collapsed sink holes, and lie at different topographic positions. Cave II is developed on two levels and has a maximum vertical depth of c38 m, with extensive formations of flowstones and sinters. Recent research has indicated the potential value of palaeoenvironmental data from this site (Shaw and Cooke, 1986).

7 The palaeoenvironmental history of the Kalahari

THE STUDY of past environmental and climatic conditions, (palaeoenvironments and palaeoclimates) has undergone rapid progress in all parts of the globe over the past three decades, particularly in the field of late Quaternary studies (the past 130 000 years) in the tropical continents. This progress has resulted from advances in the techniques available for the study, collection and dating of proxy data, the geomorphological, archaeological, pedological and biological information from which past environmental conditions can be inferred. In turn, this study has developed within a rapidly evolving methodological framework, originally drawn from studies in temperate latitudes.

These advances have been by no means uniform, for they rely on the availability of suitable sites for study, and on scientific input. Although climatic changes in the Kalahari had been suggested as early as the 1850s (Livingstone, 1858a), neither of the two conditions was fulfilled until about 15 years ago; even then, the nature of the Kalahari presented specific difficulties in this scientific field. The understanding of the palaeoenvironmental history of the Kalahari, therefore, has drawn strongly on inferences and models from other regions, sometimes to detrimental effect. The palaeoenvironmental framework that is now beginning to emerge is far from complete, but it offers a fascinating insight into the development of the science.

This chapter examines the historical development of Quaternary studies and the problems encountered in the Kalahari environment. Recent studies are reported from a variety of sites in the Southern and Middle Kalahari, and palaeoenvironmental scenarios are developed on a range of time scales.

7.1 The historical development of Quaternary studies

The roots of the study of the Quaternary, the geological era covering approximately the last 2 Ma, including the Pleistocene (about 2 Ma–10 000 yr BP) and Holocene (last 10 000 years) epochs, lies in the Northern Hemisphere in the first half of the nineteenth century, with the realisation that the northern continents had undergone episodes of extensive glaciation, accompanied by much colder climates than those presently experienced. Scientific endeavour in a range of disciplines has subsequently been directed towards an understanding of the extent, timing and duration of these glacial events, and the intervening interglacials, together with their effects on global systems, such as changes in sea level. The history of this phase has been recorded elsewhere (e.g. Charlesworth, 1957; Bowen, 1978).

An early development was the Alpine four-glacial model (Penck and Bruckner,

Table 7.1. *Correlation of African Pluvials and the Alpine Glacial Sequence*

Geological time	Alpine Glacial	African Pluvial	Wet phases
Holocene			Nakuran
			Makalian
U. Pleistocene	Würm	Gamblian	
		– Third Interpluvial –	
M. Pleistocene	Riss	Kanjeran	
		– Second Interpluvial –	
	Mindel	Kamasian	
		– First Interpluvial –	
L. Pleistocene	Günz	Kageran	

After Deacon and Lancaster (1988).

1909) which persisted throughout the early twentieth century, and was adapted by researchers in East Africa (Gregory, 1921; Wayland, 1934) to explain the observed evidence of changes in lake and glaciation levels in that region. 'The Pluvial Theory', as it became known, correlated wetter episodes (pluvials) in East Africa with glacials in Europe on the assumption that lower temperatures would result in lower evapotranspiration, and hence higher effective rainfall. A pluvial chronology developed by Leakey and Solomon (1929) was formally adopted by the first Pan-African Congress on Prehistory in 1947 (Deacon and Lancaster, 1988; Table 7.1). Inherent in the Pluvial Theory were two possibilities for the distribution of climatic belts in the tropics; first, that the belts were compressed towards the equator, as envisaged by Penck (1914), the second that pluvials represented wetter conditions throughout the inter-tropical belt (e.g. Wayland, 1934).

In southern Africa, where palaeoenvironmental considerations have long been the preserve of archaeologists and anthropologists, the Pluvial Theory was applied avidly to geomorphological studies of river terraces and the archaeological material contained within them, notably of the Vaal River (Van Riet Lowe, 1930; Sohnge, Visser and Van Riet Lowe, 1937), the Zambezi (Maufe, 1930, 1939; Armstrong and Jones, 1936; Bond, 1957) and other rivers in Rhodesia (Jones, 1944, 1946). It soon became apparent, however, that the fitting of relative dating schemes and stratigraphic correlations into the pluvial framework was something of a circular argument; climatic concepts were dictating the interpretation of the evidence rather than vice versa. Cooke (1958) and Flint (1959) suggested that the hypothesis be abandoned.

Up to this point little information had come from the Kalahari. Although the significance of the Kalahari landforms had been recognised (Livingstone, 1858a; Passarge, 1904), no chronological framework was available for the wetter and drier episodes that they indicated. Schwarz (1920), for example, believed that the Okavango–Makgadikgadi fossil lake system represented desiccation within historical time. Wayland (1954), in his review of the geology and prehistory of the Bechuanaland Protectorate, drew comparisons with the work carried out on the Zambezi, and placed the Kalahari firmly within the pluvial framework. Grove (1969) later proposed a simplified sequence of events following his geomorphological study of Ngamiland, but concluded that a chronology was not possible without absolute dating.

Following the abandonment of the Pluvial Theory, Deacon and Lancaster (1988) note a return to field sites in southern Africa, accompanied by improvements in palaeoenvironmental techniques, particularly the availability of radiocarbon assay as a method of absolute dating. Advances were made in palynological studies, long-term cave sites, and in the development of astronomical models to explain climatic change, in particular the temperature-based Milankovitch model, later confirmed by deep-sea cores. Of more importance to the Kalahari, however, was research carried out in North and East Africa on fluctuations in lake levels (Butzer *et al.*, 1972; Street and Grove, 1976), which indicated arid conditions at the Last Glacial Maximum (c. 18 000 BP), extending through the Late Glacial to about 13 000 BP, and followed by episodes of high lake levels during the early Holocene. The East African sequence has been subsequently refined to a resolution of 1000 year time windows (Street-Perrott and Roberts, 1983), extended to other parts of the tropics where lakes are present (Street-Perrott, Roberts and Metcalfe, 1985), and, in turn, has become the basis for further palaeoclimatic models (Hastenrath and Kutzbach 1983; Kutzbach, 1983). In southern Africa, however, where suitable lakes are not present, the model became difficult to substantiate (e.g. Street-Perrott and Harrison, 1985).

Despite the pioneering work of Butzer *et al.* (1973) on Lake Alexandersfontein in the Kimberley area, and Cooke (1975a) on Drotsky's Cave, the Kalahari became enmeshed in the East African chronology by default. Van Zindren Bakker (1976), in an attempt to reconstruct the late Quaternary palaeoclimate of southern Africa, drew on evidence from adjacent areas to suggest that the Kalahari mirrored conditions in the Sahara, thereby introducing an element of global symmetry to palaeoclimatic models. This opinion was reiterated in subsequent publications (Van Zindren Bakker 1980, 1982a). Heine (1978a, b, 1979, 1982, 1988b) published a series of radiocarbon dates from calcretes found within the Okavango–Makgadikgadi lake system in the Middle Kalahari, which supported Van Zindren Bakker's views. The imprecise context of these dates has led to considerable subsequent criticism (Cooke, 1979b; Butzer, 1984a; Helgren, 1984; Rust *et al.*, 1984), and it would appear, in retrospect, that some researchers are still trying to reconcile facts with the pluvial hypothesis. Subsequent research has attempted to redress the balance.

7.2 Kalahari palaeoenvironments: problems and research methodology

Deacon and Lancaster, in their useful volume on Late Quaternary palaeoenvironments in southern Africa (1988), repeat the dictum of Hecht *et al.* (1979) that palaeoenvironmental reconstruction is only as good as the data on which it is based. Further, such reconstructions form a second order approximation of the original proxy data, while palaeoclimatic modelling forms a third order in the series. Proxy data can yield information on either temperature (e.g. speleotherms) or precipitation (e.g. lake levels, dunes), and can be dated by either relative (e.g. by archaeological association) or absolute methods, of which radiocarbon dating is by far the most common.

The corpus of radiocarbon dates relating to the entire Kalahari numbers approximately 230, of which 50 per cent are from studies of the Middle Kalahari, 10 per cent from the southern and western parts of the Desert, and the remainder from sites on the Kalahari periphery. This, in itself, is both a measure of the low levels of scientific activity in the region and a reflection on some of the difficulties encountered in acquiring data. The primary problem is the Kalahari Group sediments themselves, which, as discussed in Chapter 3, are much altered, are difficult to differentiate, contain few fossils and have low levels of preservation of organic matter. There is very little material suitable for radiometric dating, and the material used most frequently, calcrete, generally yields the 'youngest possible' date as calcrete formation is not necessarily monogenetic (Netterberg,

1980). The environmental significance of the calcrete itself may be open to question, and problems of carbonate contamination also arise. Thus, calcrete dates must be treated with caution, and used as indicators rather than stratigraphic markers.

The Kalahari also has few long-sequence sites, either archaeological or geomorphological, of the type found elsewhere in southern Africa. The paucity of quality archaeological sites (section 8.1) in particular has long delayed the study of Stone Age cultures in the region, and discouraged small-scale detailed studies. Palynology, too, is rendered invalid by the poor pollen preservation and the ubiquity of savanna vegetation types. To this list can be added the effects of poor access, lack of scientific manpower, and the prior claims of other scientific priorities.

The scientific approach in the Kalahari, therefore, has been towards large-scale geomorphological studies with a deductive basis, as opposed to the small-scale, multidisciplinary studies with an inductive basis encountered in the southern periphery of the continent. Thomas (1987b: p. 5) justifies this approach on the grounds that the spatial and temporal unevenness of the data, together with the progress made in contributory disciplines, requires that constant revision and re-evaluation of conclusions must take place.

A result of this approach has been an imbalance in the types of proxy data acquired. Geomorphological data tends to indicate palaeohydrological conditions, whilst biological evidence moves towards palaeotemperature inferences (Thomas, 1987b: p. 15). Within the palaeohydrological realm moister conditions (pans and lakes) are easier to assess and date than the arid conditions indicated by dune systems. These imbalances are ones which remain to be addressed by future, small-scale studies.

7.3 Environmental change from the Cretaceous to the late Quaternary

A general problem of land-based palaeoenvironmental evidence, that the further back in time the more fragmentary, more disturbed and more difficult to date the data becomes, coupled with the problems which the Kalahari itself introduces, means that very little is known about Kalahari environments in the period spanning post-Gondwanaland times to the late Quaternary. Nonetheless, some deductions can be made, using the limited information which the Kalahari Group sediments themselves contain, together with data from a very limited number of sites in the Middle Kalahari and the southern margin of the Kalahari Desert, and some inferred information from lacustrine and aeolian deposits.

The tectonic and sedimentary development of the Kalahari, discussed in detail in Chapters 2 and 3, clearly demonstrates that major environmental changes occurred before the better understood late Quaternary, but the role which climatic changes played is little known. If duricrusts are ignored, given their problematic interpretation in climatic terms, then the presence of basal conglomerates, gravels and marls (sections 3.3.1 and 3.3.2), compared with upper sands (section 3.4) which display significant evidence of aeolian activity, could perhaps be taken as a basic, broad indication of a transformation from wetter to drier conditions in the Kalahari, perhaps during the Tertiary (e.g. Smit, 1977).

Even if such a basic scheme of sedimentation can be accepted, however, its interpretation in climatic terms, whether partially or totally, is somewhat suspect on a number of grounds. First, there is as yet no substantive evidence which allows the presence of basal, probably Cretacaeous to Tertiary, gravels and marls to be interpreted in a climatic context. The post-Gondwanaland evolution of the southern African subcontinent means that these sediments can at the very least be partially explained in terms of drainage development and the presence of an early endoreic drainage system (section 2.3.3) with, for example, the modern course of the Zambezi not being established until the late Pliocene or early Pleistocene (Dixey, 1950; Lister, 1979; Thomas and Shaw, 1988). Second, even if the fluvial sediments do contain a climatic imprint, it is most likely to represent conditions in the source areas, marginal to the Kalahari. Third, the transition to the Kalahari Sand does not automatically indicate a change to aeolian processes operating under more arid conditions, as aeolian imprints in the sand are at least to some degree due to post-depositional reworking rather than the initial deposition (section 3.4.2). The disparate ages assigned to the original deposition of the Kalahari Sand, ranging from Miocene to mid-Pleistocene, further complicate the issue (section 3.4.1).

7.3.1 The Middle Kalahari

Despite these reservations and the general lack of empirical data, studies from the Middle Kalahari do shed some light on pre-late Quaternary conditions, providing some evidence for an end-Tertiary to early Pleistocene transition towards aridity from previously wetter conditions. The basal cave-fill sediments in the Gcwihaba and Nqumtsa Hills, assigned a late Tertiary age, contain windblown sand (Cooke, 1975a) and micromammal remains (Pickford and Mein, 1988) which suggest arid to semi-arid conditions comparable with those of today. Cooke (1980) proposes that the nearby (northern) dunefield was formed at this time, while the caves must have been formed by solution during an unspecified preceding humid period. Cooke (1980) also implies that, during the early and middle Pleistocene, climatic oscillations contributed to fluctuations in the size of Lake Palaeo-Makgadikgadi, while Helgren (1984) used tentative archaeological evidence to suggest that the highest level of the lake occurred before 200 000 BP, in the middle Pleistocene.

7.3.2 The Southern Kalahari margin

Direct evidence of early conditions in the Southern Kalahari is generally lacking but information from the depositional complexes of the Gaap Escarpment (Butzer et al., 1978), on the southeastern margin of the area, and from Kathu Pan, just within the limits of the Southern Kalahari (Figure 6.12), does provide some indication of the conditions which may have prevailed.

The 275 km long dolomite Gaap Escarpment possesses 70 to 120 m high cliffs from which springs have emerged during the Cainozoic. The spring waters have led to the deposition of tufas in the form of cave fills, aprons and carapaces, which not only contain remains of considerable archaeological significance (e.g. Dart, 1926) but provide some indication of late Tertiary to Pleistocene climates over the Gaap Plateau. The tufas are also related to angular breccias which are regarded as a consequence of frost shattering affecting the face of the escarpment.

The Gaap depositional complexes have been shown to be cyclical, with six cycles identified, each containing up to six complex facies changes that are interpreted as representing changing climatic conditions which include shifts in both precipitation and temperature characteristics (Butzer et al., 1978). The evidence indicates periods of high intensity rainfall, cold climates with frost shattering, perennial stream flow from higher rainfall and spring activity from recharged aquifers, alternating with reduced rainfall under semi-arid conditions. The four earliest depositional complexes have been assigned ages ranging from the late Tertiary to the mid-Pleistocene (Butzer et al., 1978; Butzer, 1984a, b).

The earliest depositional cycle contains thick frost-shattered basal breccias along the escarpment foot which may date from the Miocene or early Pliocene (Butzer, 1974) and are indicative of very cold conditions at this time (Butzer et al., 1978), comparable with evidence from elsewhere in the Southern Hemisphere. Times when total rainfall was insufficient to lead to spring activity and tufa deposition, but when high intensity, run-off effective rainfall events occurred, are marked by the reworking of the breccias and the development of scarp-foot debris fans. Such conditions appear to have prevailed during part of the early-Lower Pleistocene. Major tufa deposits indicate long periods of significant spring discharge and humid conditions, and, by inference from areas of South Africa with active tufa formation today, annual rainfall in excess of 600 mm, though this figure could be altered if accompanied by temperature changes. Conditions with tufa formation have been identified for the Late Pliocene, mid- or late-Lower Pleistocene (with the massive 60 m thick tufas containing remains of *Australopithecus africanus*), mid-Middle Pleistocene, following a cold period of breccia formation, and late-Middle Pleistocene (Butzer, 1974). The base of the last of these phases dated to approximately 250 000 BP using U/Th dating (Vogel in Butzer, 1984a). Some periods of enhanced rainfall, estimated as up to 150 per cent of current levels or 600–650 mm, gave rise to corrosion of fissures in the tufas due to the emerging spring waters having lower bicarbonate concentrations. At other times, periods of limited geomorphological activity, which to some extent represent the hiatuses in the tufa–breccia complexes, are seen as comparable with present day conditions (Butzer, 1984a).

The cyclical nature of the deposits and the interpretations of Butzer and his co-workers give an overall picture which suggests that climatic conditions on the Southern Kalahari margin experienced notable fluctuations during the late Tertiary to mid-Pleistocene. The lower units from the wetland sediment sequence at Kathu Pan (or vlei), located some 150 km west of the Gaap Escarpment (Figure 6.12), and described as 'perhaps the best paleoenvironmental sequence from the Kalahari Basin' by Butzer (1984a), provide some supporting evidence for fluctuating climates from within the southern extremity of the Kalahari itself, though given the considerable lack of resolution of the data and the broad time periods in which individual sediment units are bracketed, attempts to correlate events from the two areas should be avoided.

Kathu Pan is an ephemeral wetland area underlain by basal Kalahari Group sediments that are capped by 40–50 m of karstic calcrete. The height of the watertable in the calcrete (currently 1 m below the surface due to modern groundwater pumping), with current mean annual rainfall between 350 and 450 mm) is determined by artesian water from the neighbouring Kuruman Hills and Korannaberg. Fluctuations in pan sedimentation therefore represent long-term groundwater trends rather than short-term geomorphic events (Butzer, 1984a). The sediments from a doline fill in the calcrete have been described by Butzer (1984a), with the lower units spanning the Lower to Middle Pleistocene. Sedimentation on the pan floor has consisted of periods with a high, near-surface watertable leading to the deposition of fine sediments and peaty organic material, disrupted by times with a substantially lower watertable, periods with a fluctuating, intermediate watertable, and episodes of rapid groundwater incursions giving rise to the eruption of springs at suitable points within the karstic system (Butzer, 1984b).

The development of Kathu doline represents a long period of subsurface solution operating under very low watertable and therefore dry conditions, prior to the late-Lower Pleistocene but after the formation of the calcrete, which postdates the basal Kalahari beds and is therefore ascribed to the Tertiary

(Beaumont *et al.*, 1984; Butzer, 1984*a, b*). Basal peaty deposits 4 m thick in the doline are interpreted as the outcome of perennially wet conditions during the late-Lower or early-Middle Pleistocene. This was followed by drier conditions in which the watertable fell, allowing periodic peat fires, indicated by an ash layer, to affect the pan. The ash layer is overlain by sediments containing Middle Pleistocene faunal and archaeological remains, with the sediments themselves indicating an initial pan phase with a fluctuating watertable allowing the accumulation of both organic and aeolian sediments, followed by a period of groundwater recharge and spring-eye eruptions indicative of conditions wetter than today. A subsequent, mid-Upper Pleistocene, period with the watertable at least 4.5 m below the surface, is marked by the incursion of flash-flood deposits, containing Middle Stone Age artefacts, on to the pan surface. This period must have been substantially more arid than today's conditions (Butzer, 1984*a*).

7.4 The Kalahari in the late Quaternary

Sites of palaeoenvironmental significance are shown in Figure 7.1. The distribution and quality of the sites is uneven; large gaps exist north of the Zambezi and in the central Kalahari, while closed sites, with the potential for long-series chronologies, exist only at Drotsky's Cave and Wonderwerk Cave in the

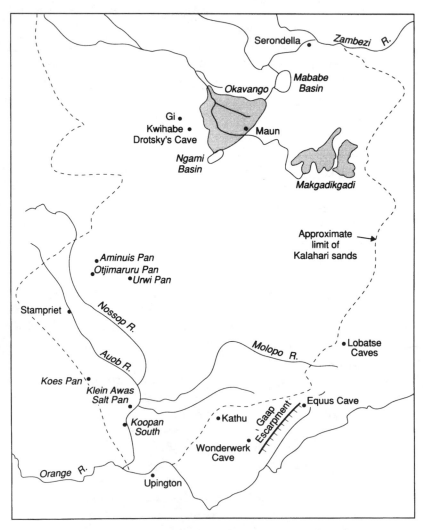

Figure 7.1. The location of sites yielding important palaeoenvironmental data for the Kalahari. The map includes data from Deacon and Lancaster (1988).

west and southeast respectively. The geomorphology of most of these sites has been discussed in Chapters 5 and 6; here the main palaeoenvironmental findings are indicated for each group of sites, before summarising the palaeoenvironmental data in section 7.5.

7.4.1 Drotsky's Cave and adjacent sites

A series of radiocarbon dates for stalagmites in Drotsky's Cave (section 6.5.2; Figure 6.10) was published by Cooke (1975a) and later adjusted (Cooke and Verhagen, 1977; Cooke, 1984). These are shown in Appendix 1A. When viewed in association with dates for calcretes in the adjacent Gcwihaba Valley, Cooke suggested that they represent a cycle of cave development from phreatic through vadose to arid conditions, followed by valley incision and return to the phreatic state within the cave. This cycle appears to have occurred on a number of occasions in the Late Quaternary, accounting for the four generations of sinter so far identified.

Cooke and Verhagen (1977) identify major periods of sinter deposition, representing vadose conditions in the cave, and periods of increased local rainfall, at 45 000–37 000 BP, 34 000–29 000 BP and 16 000–13 000 BP, the latter particularly well represented in three stalagmites. Holocene sinter formation is indicated at 2000 BP and 750 BP.

Dating of a further three stalagmites by ^{14}C and U/Th determination (Shaw and Cooke, 1986) confirmed the humid episode at 16 000–14 000 BP, and suggested further humid intervals in the Holocene at 6000–5000 BP and 4000 BP. Good correlation was achieved between the ^{14}C and U/Th methods.

Cooke interprets the results to indicate increases in local precipitation. Episodes of rapid sinter deposition may represent phases of rainfall up to 300 per cent of the present mean.

Drotsky's Cave is the only closed site known within the Kalahari, and is of major importance in palaeoenvironmental reconstruction. Work in progress (Cooke, pers, comm.) includes palaeotemperature measurements from speleothems, the extention of U/Th dating back to around 300 000 BP, and thermoluminescence (TL) dating of inblown cave sediments.

Archaeological investigations at Gci (Gi) and Quangwa, in the Dobe Valley 70 km northwest of Drotsky's Cave (Helgren, 1978; Helgren and Brooks, 1983; Appendix 1B) have indicated the past existence of a lake, between 4 m and 9 m deep, within the valley on two occasions. This lake is taken to represent humid conditions, degenerating into a series of ephemeral pans during periods of semi-arid climate. The first lake has been equated with Middle Stone Age cultures: the second may have existed at 23 000–22 000 BP, prior to the arid phase of the last glacial maximum at around 19 000 BP, with semi-arid conditions persisting since this time. Helgren and Brooks suggest that the calcrete date of 31 000 BP may postdate the earlier humid period, but note a number of problems arising from interpretation. Pleistocene tectonic activity seems to have occurred at the Dobe Valley, as well as in the Gcwihaba Hills.

7.4.2 The Okavango–Makgadikgadi palaeolakes

The Okavango–Makgadikgadi complex (Figure 5.2) forms one of the most extensive palaeoenvironmental sites in Africa, including the shorelines and terraces of the 945 m Lake Palaeo-Makgadikgadi and 936 m Lake Thamalakane stages, as well as individual levels in the Makgadikgadi, Mababe and Ngami Basins (section 5.3). Palaeoenvironmental reconstruction is hindered, not only by the extent of the landforms, but also by the tectonic history of the region, and the climatic signal from distant as well as local sources; much of the inflow into the system has its origins in the Angolan Highlands (Cooke, 1980).

Apart from a solitary date provided by Street and Grove (1976), the first suite of dates for the Okavango–Makgadikgadi complex was published by Heine (1978b) (Appendix 1C). Fifteen dates were taken from unspecified locations and stratigraphic contexts in the Ngami and Makgadikgadi basins, and interpreted to represent the existence of a large lake between 30 000 and 19 000 BP, followed by a period of aridity until the onset of a minor pluvial at around 12 000 BP. Holocene conditions were initially humid, followed by semi-aridity at around 9000 BP. As already noted, the observations attracted criticism. Subsequent papers (Heine, 1979, 1982) did little to clarify matters; the locations of the dated sites along the main Ngamiland road were not published until 1987 (Heine, 1987), still without altitudinal controls and geomorphological relationships. Meanwhile, the interpretation of the dates became increasingly confused and contradictory, culminating in a paper (Heine, 1988a) attempting to fit Passarge's original stratigraphy into a series of lake stages from 30 000 BP to the present. Such an exercise clearly lies beyond the scope of the available radiocarbon dates. These dates, particularly those derived from shell deposits, are potentially of great value, but require re-evaluation in the field.

Subsequent work (Cooke, 1984; Cooke and Verstappen, 1984; Shaw, 1985a, 1986; Shaw and Cooke, 1986; Shaw and Thomas, 1988; Shaw et al., 1988) has produced a suite of 48 radiocarbon dates (Appendix 1C) related to accurately located landform features such as fossil shorelines, while Shaw (1988a) has provided a recent overview of the palaeohydrology of the complex. Although calcrete has been the primary dating medium, no palaeoclimatic interpretation has been placed upon it. Instead, it has been assumed that calcrete is most likely to form during phases of lowering watertables, and therefore postdates a specific lake level. Where radiocarbon dates indicate contemporaneous calcrete formation on different shoreline levels, obviously the lower level is more significant.

The early part of the record (50 000–20 000 BP) must be treated with caution due to the limitations of the dating technique. It suggests that early arid conditions in the Makgadikgadi Basin at around 46 000 BP were succeeded by the 945 m Lake Palaeo-Makgadikgadi level between 40 000–35 000 BP, although Helgren (1984), on the basis of the distribution of Early Stone Age artefacts, suggests that this level may be mid-Pleistocene. Low levels were attained from 35 000–26 000 BP, with calcrete formation on all three main strandlines. Between 26 000 and 10 000 BP a succession of lakes occurred at the 920 m level, interspersed with lower levels, as at 21 000 and 19 000 BP. The group of dates around 19 000 BP, including one from a peat deposit at Tsoi (GrN 9677), indicate cool, as well as dry conditions.

Dating of the 945 m level in the Mababe and Ngami Basins has proved unsatisfactory so far. The record here begins with low levels at 26 000–24 000 BP, followed by a lacuna until around 17 000–13 000 BP, when a series of dates from shell deposits (section 3.6.2) on the 936 m shoreline and associated terraces throughout the system as far north as Serondela on the Chobe indicate a transgression to the Lake Thamalakane stage. Investigation of the palaeohydrology of the Boteti River (Shaw et al., 1988) suggests that the Lake Thamalakane stage was coincident with the 920 m level in the Makgadikgadi Basin at this time. Lake levels fell from 12 000–9500 BP as evidenced by a series of calcrete dates throughout the lake system.

Initially the Holocene was characterised by low lake levels, though a rise in Ngami to 932 m is noted at around 6000 BP (Shaw, 1985a). Of more widespread importance is return to the Lake Thamalakane Stage at 2500–2000 BP, accompanied by overflow to the 912 m level in the Makgadikgadi and increased levels in the vicinity of the Zambezi–Chobe confluence, suggested by a series of dates from shells at terrace sites, fossilised *Acacia tortillis* at Phatane Gap, Ngami, and calcretes at a variety of localities.

Both of the major palaeolake stages must have derived input from inflow, primarily from the catchment in the Angolan Highlands, as well as local sources, the latter indicated by delta development in the local stream networks at the 945 m and 936 m levels. However, the calculation of palaeohydrological budgets (Grove, 1969; Ebert and Hitchcock, 1978) suggests that the Lake Palaeo-Makgadikgadi stage would require some 50 km^3 yr^{-1} of inflow, a figure that can be met only by a major input from the Zambezi River. Thus, the Lake Palaeo-Makgadikgadi stage may have been caused by a set of unique tectonic and climatic conditions, in which the climatic input cannot be estimated satisfactorily. Helgren (1984) goes so far as suggesting that this stage may have been a short-lived catastrophic event, although this is unlikely given the massive scale of the Gidikwe Ridge.

The 936 m Lake Thamalakane stage, however, has recognised hydrological controls (section 5.3.6), a shoreline sequence undisturbed by tectonic activity, and evidence of multiple transgressions. There is also evidence for increased local precipitation in the terraces and delta of the Ngwezumba and other tributaries on the eastern sides of the Mababe and Ngami Basins. Shaw (1985a) estimated that an increase in precipitation in the Middle Kalahari (including the Okavango) of 160 per cent of the present mean would sustain Lake Thamalakane, while 225 per cent would also allow for the Boteti overflow and the 912 m level in the Makgadikgadi Basin. However, given the 'amplifier effect' (Street, 1980) inherent in the Okavango Delta, and the hydrological relationships with the Zambezi at the 936 m level (Shaw and Thomas, 1988), in which small percentage increases in the Zambezi input would generate large quantities of water, it seems likely that no more than a 100 per cent increases in the mean would be sufficient (Shaw, 1988a).

7.4.3 Etosha Pan

Rust (1984, 1985) has produced a suit of 33 ^{14}C dates, based mostly on calcretes, for Etosha Pan (Appendix 1D). Three of these (Pta 3035, 3036, 3038) are of stromatolites at Insel, representing an 8 m lake level within the pan complex. Unfortunately, the maximum dates (more than 40 000 BP) do not resolve the age of the shoreline, which Wilczewski and Martin (1972) assume to be of Pliocene age.

Rust, wary of palaeoclimatic inference from calcretes (see Rust *et al.*, 1984), interprets his data in morphodynamic terms, with periods of stability, characterised by calcrete development, at 33 000–28 000 BP, 22 000–18 000 BP, and 10 500–9000 BP. Conversely, geomorphic activity, including slope denudation, aeolian activity and basin sedimentation, took place at 18 000–10 500 BP and 9000–3500 BP. Higher groundwater tables are indicated from the lacustrine calcretes and spring tufas at Namutoni at 11 900 ± 120 BP (Pta 3401) and 9310 ± 90 BP (Pta 3043).

Heine (1982) provides four ^{14}C dates (Appendix D) in the range 13 680–10 670 BP which are taken to be representative of a lacustrine phase equivalent to one in the Makgadikgadi. Again, no locational or stratigraphic data is offered, and the fact that date Hv 9493 is taken from the land snail *Xerocerastus* makes the interpretation suspect.

7.4.4 Pans and rivers of the southern and western Kalahari

In the southwest Kalahari most palaeoenvironmental evidence has come from the study of pans, although some data has also been derived from the Molopo River system (Appendix 1E).

Lancaster (1979) provides evidence of a lacustrine phase at 17 000–15 000 BP from stromatolites 1.5 m above the level of Urwi Pan, indicating permanent, wave-agitated waters, preceeding a phase of drier conditions characterised by the deposition of saline clays. He suggested that rainfall 1.5 to 2 times the present mean, coupled with lower temperatures, would be sufficient to maintain the pan by run-off. Subsequently, Lancaster (1986b) has stated that the pans were controlled by higher groundwater tables, following the findings of Heaton, Talma and Vogel (1983) that groundwater recharge occurred in the nearby Stampriet aquifer at 35 000–26 000 BP and 14 000–8000 BP. Lancaster (1986b) also reports increased spring discharge at the Aminuis and Otjimaruru Pans at around 33 000–30 000 BP. Deacon and Lancaster (1988), interpret the body of evidence from the southwestern pans as indicating higher groundwater tables and discharge at 33 000–30 000 BP and 17 000–11 000 BP.

The stratigraphic framework for the Molopo system put forward by Heine (1981, 1982) has been described by Butzer (1984a: p. 48) as 'very cursory . . . do[ing] little to resolve later Pleistocene geochronology', largely because Heine has confused terraces within the sequence, and again used *Xerocerastus* as a humid indicator. However, most of the dates are based on molluscs of the genera *Unio*, *Corbicula* and *Bulinus* which, derived from fluvial sediments in the river bed, would indicate perennial or semi-perennial flow conditions in the Molopo, Auob and Nossop Rivers within the dated range 16 600–12 500 BP. Minor fluvial episodes are indicated from the mid-Holocene (around 4000 BP). On the other hand, the stromatolite-like concretions dated at 23 410 BP (Hv 9885) from Klein Awas Pan, and assumed to have a humid context, are probably calcretes.

7.4.5 Landforms of the Southern Kalahari margin

A group of intensively studied sites occurs at the Kalahari periphery in the northern Cape, and includes Kathu Pan (Beaumont *et al.*, 1984; Butzer 1984a), the Wonderwerk and Equus Caves (Avery, 1981; Beaumont *et al.*, 1984; Butzer, 1984a, 1984b; Thackeray *et al.*, 1981; Van Zindren Bakker, 1982b) and the tufa carapaces of the Gaap Escarpment (Butzer *et al.*, 1978). Descriptions of the morphology and early palaeoenvironmental status of these landforms appear in Chapter 6 and section 7.3.2, and summaries of their palaeoclimatic evidence, drawn from archaeological, geomorphological and palynological studies, have been published by Butzer (1984a, 1984b), and Deacon and Lancaster (1988). The published radiocarbon dates are shown in Appendix 1F, and are particularly valuable for a consideration of Holocene changes.

Research at Kathu Pan (Figure 7.2) has concentrated on infills, partly organic, of dolines in the calcrete base of the pan, and indicates changing groundwater conditions. Low but stable watertables with minimal organic accumulation are indicated from 32 000–11 000 BP, followed by an episode of dry conditions between 11 000–7400 BP, characterised by low organic matter and episodes of sheetwash accumulation. Peat accumulation and the pollen spectra indicate conditions wetter than present from 7400 to 4500 BP, followed by a further arid phase to 2700 BP. Current conditions were established at this latter date. At Equus Cave cooler conditions are indicated prior to 7500 BP, with Kalahari thornveld similar to the present between 7500 and 2400 BP.

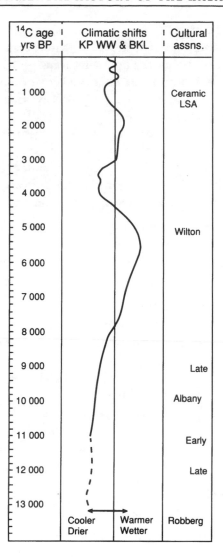

^{14}C age yrs BP	Climatic shifts KP WW & BKL	Cultural assns.

Figure 7.2. Climatic change in the southeastern Kalahari from 13 000 years BP to the present. After Beaumont et al. (1984).

The record at Wonderwerk has been taken from two series of strata, that in the interior of the cave (the WA section) covering the mid-Pleistocene to 12 000 BP, and the WB section at the cave entrance dated from 13 000 BP to the present. Following discrepancies between the interpretation of micromammalian fauna (Avery, 1981) and pollen (Van Zindren Bakker (1982*b*), a revised palaeoenvironmental framework has been put forward by Beaumont *et al.* (1984) and Butzer (1984*a*).

Within this framework a moist and cold climate was experienced at 30 000–26 000 BP, accompanied by frost-spalling within the cave. This pattern was repeated at 13 000–11 800 BP, and wetter episodes were recorded at 10 000–5000 BP and 1900–800 BP, with conditions similar to today at 4000–1900 BP. Vegetation changes included a shift from open scrub savanna to savanna woodland at 6000–5000 BP, with reversion to treeless grassland in the mid-Holocene.

Beaumont *et al.* (1984) have summarised the conditions from all three of these sites as a shift from a cool dry Pleistocene climate to the moist warm Holocene optimum from 8000 to 5000 BP, followed by cooler, drier and more variable conditions from 4500 BP onwards.

The geomorphological studies of the Gaap Escarpment by Butzer *et al.* (1978) have indicated the existence of six major cyclic depositional phases of breccia and tufa accumulation at four sites (see section 7.3.2), of which the two youngest cycles, from the Blue Pool site, cover the late Quaternary. Blue Pool Tufa I, was formed at 94 000–30 000 BP, separated by an erosional hiatus from Blue Pool Tufa II, dated 33 000–14 000 BP. Holocene depositional episodes are indicated at 9800–7600, 3200–2400 and 1300–250 BP (Butzer, 1984b). As tufa deposition currently takes place only during years of above-average rainfall, an annual mean of 600–1000 mm (unadjusted for any accompanying temperature changes) is indicated for these episodes.

Work has recently commenced on the dating of stalagmites and sinters at the Lobatse Caves in southeast Botswana. Initial dates from Cave II, reported by Shaw and Cooke (1986), indicate sinter formation at 26 000–22 000 BP and 18 000–17 000 BP.

7.4.6 Aeolian landforms

The extensive fields of vegetated linear dunes which occur throughout the Kalahari have been described by Deacon and Lancaster (1988: p. 62) as 'one of the most impressive pieces of evidence for the widespread nature of late Quaternary climatic change in the Kalahari region.' Unfortunately, they have also proved to be the most elusive to date with any certainty; at present, episodes of widespread aeolian activity are fixed by the most imprecise of all types of relative dating – the absence or presumed absence of evidence to the contrary.

The distribution, morphology and characteristics of the Kalahari dunes have already been discussed (section 6.1), as has their classification into dunefields or ergs which may be of different ages. As already noted, the Kalahari dunefields have counterparts in other semi-arid parts of the world, including Australia, South America, India and North Africa. Research in some of these regions, notably Australia and North Africa, has indicated that aeolian activity occurred during an arid phase coincident with the Last Glacial at 18 000–12 000 BP, with the core desert areas subject to the longest episodes of activity (Thomas and Goudie, 1984).

Two urgent questions arise in respect to the Kalahari dunes: the first concerns the nature of the palaeoenvironmental conditions which can be deduced from the dunes, either as a whole, or as a set of dunefields; the second involves the dating and correlation of the episodes of aeolian activity.

On the former, Lancaster (1980, 1981) suggested that the limits of dune activity within southern Africa today are coincident with the 150–200 mm isohyet, which must have moved some 1200 km northeast to permit aeolian activity in the northern-most relict dunes which he identified, and 250–300 km for movement in the Southern Kalahari. Differences in dune morphologies were used to infer different ages and, in turn, the concept of multiple phases of aridity of varying intensities. Lancaster also concluded that the three dune systems which he identified had different patterns of sandflow, relating to prevailing anti-cyclonic circulation conditions at the times of dune formation (Figure 7.3), Lancaster's (1981) 'group A northern dunes' (our northern dunefield in Chapter 6) corresponding to a colder climate with greater wind intensities, 'group B northern dunes' (eastern dunefield) corresponding to a shift in the winter circulation 200 km north of its present locus, and the southern dunefield representing a stronger circulation analogous to the present-day October pattern. Greater palaeo-wind intensities, on more than one occasion, are clearly a prerequisite of Lancaster's model.

Thomas (1984a) pointed out that dune mobility in the Northern Kalahari,

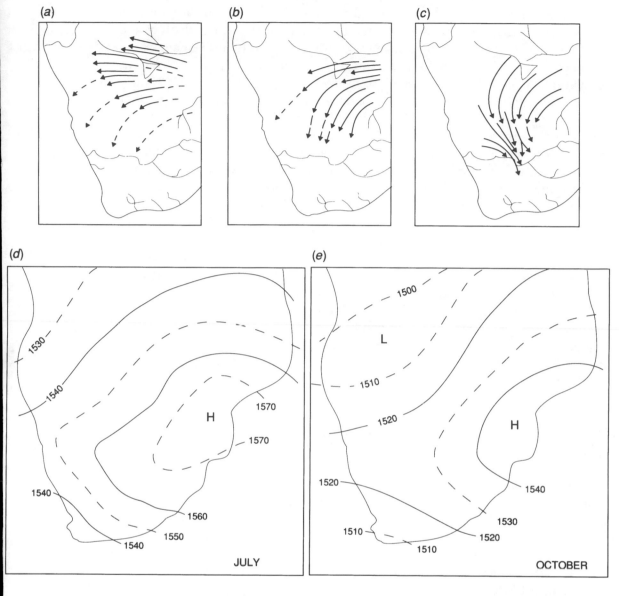

Figure 7.3. Palaeo-sandflow conditions inferred by Lancaster (1981a) from the orientation of Kalahari dunefields. (a) Northern dunefield; (b) eastern dunefield, (c) southern dunefield. A stronger anticyclonic circulation is interpreted from (a), (b) suggests a circulation comparable to that common during July today (d), but located about 200 km further north, and (c) suggests a pattern achieved today during October (e).

particularly as far as northern Angola, would require an isohyet shift of 2300 km under Lancaster's model, far greater than that proposed for analogous dunefields elsewhere, and suggested instead that this field may become active under a higher precipitation regime should greater wind strengths result in higher evaporation rates.

Recently, these extreme scenarios have been modified. Lancaster (1988) notes that dune mobility is related to vegetation cover as well as wind strength and gross precipitation, and has reassessed dune mobility using an index based on precipitation, potential evaporation and wind strength. He suggests that, given a probable 5° C mean temperature decrease during late Glacial conditions, and the wind velocity increase of 117 per cent of the present mean proposed by Newell, Gould-Stewart and Chung (1981), a fall in precipitation of 30–40 per cent would be sufficient to mobilise dunes throughout the Southern Kalahari, with a comparable figure of 30–50 per cent during Holocene arid events, which would have experienced temperatures above the present mean. The movement of the northern dunes, however, would require a greatly enhanced anticyclonic circulation, and were not implicated in this scenario.

Thomas (1988c), applying the recent findings of Ash and Wasson (1983), Wasson and Nanninga (1986) and Tsoar and Møller (1986) which suggest that aeolian activity can be significant on linear dunes with even up to a 35 per cent vegetation cover, concludes that present vegetation densities are sufficiently low on dune crests in the southern dunefield to permit significant aeolian activity. The linear dunes of the Southern Kalahari may not therefore be palaeoforms in the usual sense, though their morphology may have changed (see section 6.3.3) in response to changing environmental and vegetation conditions. Thus, true relict dunes resulting from Quaternary climatic shifts are restricted to the eastern and northern dunefields of the Middle and Northern Kalahari and adjacent areas of western Zimbabwe and Zambia. This also implies that dune mobility in these areas would occur at precipitation levels well above 150 mm yr^{-1}.

The dating of the envisaged arid phases has also been problematic. Mostly it has been based on identification of windows in the humid chronology, and by comparison with arid regions elsewhere (e.g. Wasson, 1984, in Australia). Inevitably it led Lancaster (1981: p. 344) to the conclusion that 'the evidence from southern Africa thus agrees basically with that from Australia and the southern Sahara and provides further support for a worldwide glacial/arid correlation in subtropical latitudes', a cause also espoused by Heine (1982). As a chronology of humid episodes has developed, the arid windows have become narrower and more clearly defined.

An alternative suggestion (Lancaster, 1986b) has been that increased wind strengths and higher watertables are not mutually exclusive, indicating climates that are more changeable and extreme. Following Bowler's (1973) model of environmental conditions for the formation of clay lunette dunes, Lancaster re-interprets his 16 000 BP dates at Urwi Pan, immediately preceding a lunette phase, to represent such a period of low precipitation and high ground water. This tends to ignore the evidence of higher precipitation from a variety of other sources already mentioned.

There have evidently been a number of Quaternary arid episodes, as evidenced, for example, by the complex relations between dunes and the shorelines of the Palaeo-Makgadikgadi system, which have affected the dune subsystems to different degrees. In northern Angola, dune mobility probably took place before 38 000 BP (Van Zindren Bakker and Clark, 1962), while the southern dunes may have been mobile (Deacon and Lancaster, 1988; Lancaster, 1988) prior to 33 000–28 000 BP, between 20 000 and 17 000 BP, coincident with the calcrete formation episode at 12 000–9000 BP, and again in the mid-Holocene. A recent paper by Lancaster (1989) adjusts these arid periods in the southwest Kalahari to prior to 32 000 BP, between 19 000 and 17 000 BP, between 10 000 and 6000 BP, and finally, between 4000 and 3000 BP, thereby underlining the fragility of the evidence on which these assessments are made.

With the exception of the dune-base peat date of 19 680 ± 100 (GrN 9019) at Tsoi, in the Makgadikgadi, the only absolute dates available for dune forms at present are of Holocene age. Lancaster reports a clay lunette at Koopan Suid (Lancaster, 1986b) which had ceased formation prior to 1000 BP (Pta 3865), while Rust (1984) reports a date of 3510 BP (Pta 3044) from calcrete within a small, stabilised barchan in the Andoni Bay dunefield of Etosha.

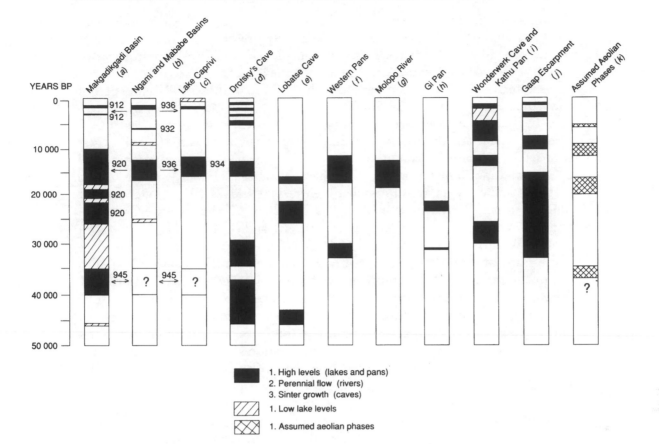

Figure 7.4. Summary of palaeoclimatic data for the Kalahari. Sources: (a) Cooke and Verstappen, 1984; Cooke, 1984; (b) Shaw, 1985a; Shaw and Cooke, 1986; (c) Shaw and Thomas, 1988; (d) Cooke and Verhagen, 1977; Cooke, 1984; Shaw and Cooke, 1986; (e) Shaw and Cooke, 1986; (f) Deacon and Lancaster, 1988; (g) Heine, 1982; (h) Helgren and Brooks, 1983; (i) Beaumont et al., 1984; Butzer, 1984a; (j) Butzer et al., 1978; (k) Deacon and Lancaster, 1988.

7.5 The chronology and climatic scenarios of the late Quaternary

The palaeoenvironmental information for all of the major Kalahari sites is shown in Figure 7.4. A first glance will show that it primarily indicates precipitation changes, with an emphasis on episodes of greater, rather than lesser, precipitation than at present. The evidence tends to be fragmentary, with apparent conflict between sites.

That there should be such conflict is not surprising, for it is inherent in the Kalahari environment itself, which occupies a transitional position between the tropical summer rainfall and temperate winter rainfall belts, both of which play a part in the present-day distribution of rainfall. Thus, there need not be regional correlation in palaeoclimatic events, although Heine's (1982) proposed extreme of aridity in the Middle Kalahari simultaneous with a humid episode in the south is highly unlikely. This caveat will also apply to sites which indicate simultaneously local and regional conditions, such as Lake Palaeo-Makgadikgadi.

Conflict will also arise through the nature and quality of the evidence. The long periods of humidity indicated by cave sinters and the spring tufas at the Gaap Escarpment are due more to lack of discrimination within the deposits than to climatic conditions, while the uncertainty over early lake levels is often the result of the masking or obliteration of the evidence by subsequent lacustrine events.

Again, some sites are more responsive than others. Butzer (1984*a*: p. 51) perceives the problem as '. . . in part one of incompletely preserved records, in part of differential thresholds for response, and, finally, of field and laboratory work insufficiently detailed to discern the true complexity of local records'.

Despite this criticism, a number of conclusions can be drawn. The early part of the radiocarbon record, from 50 000 to 20 000 BP, contains the most discrepancies. In the Middle Kalahari there appears to be conflict from the sites at Gci, Drotsky's Cave and the Makgadikgadi, although Shaw and Cooke (1986) suggest four sets of geomorphic conditions likely to arise from varying inputs from local conditions and input from the Northern Kalahari. Humid conditions are certainly indicated at most sites between 35 000 and 22 000 BP, although at different times, particularly between 35 000 and 28 000 at Drotsky's Cave, Gci Pan, the Western Pans and the Gaap Escarpment, and 24 000 and 22 000 BP in the Makgadikgadi, Lobatse Cave and Gci Pan. Within this broad belt there are suggestions of drier episodes, certainly the Okavango-linked lakes appeared to dry out at around 25 000 BP, representing conditions at least as dry as at present.

From about 20 000 BP through to the Last Glacial Maximum at 19 000–18 000 BP cold and dry conditions prevailed, possibly accompanied by aeolian activity, with dry lakes and some peat formation. The only evidence to the contrary comes from the Gaap Escarpment, where tufa accumulation seems to have continued at the Gorrokop site. This was followed by a phase of greater moisture availability throughout the Kalahari, with the exception of the southeast sites, from 17 000–12 000 BP, with a well established core at 16 000–13 000 BP. This episode was probably of greater extent and humidity than envisaged by Deacon and Lancaster (1988), and extended from the Zambezi as far as the Orange–Vaal drainage, as shown by the remarkable lake level changes at Alexandersfontein (Butzer *et al.*, 1973) and Swartkolkvloer Pan (Kent and Gribnitz, 1985) in the northwest Cape.

At 12 000 BP drier conditions and falling watertables set in. Diminished river flows are represented by the calcretisation of fluvial sediments in the Okavango lakes and the Gcwihaba valley, while cooler, drier conditions are found in the south. Conditions similar to the present were established throughout the region by the early Holocene. Subsequent fluctuations have been of relatively low amplitude and duration, and include more humid episodes at Drotsky's Cave at 6000–5000, 4200–3600 and 2500–2000 BP. The first and last of these is also reflected in lake levels, and the 2000 BP episodes also saw a return to the Lake Thamalakane stage, and is probably of archaeological significance (section 8.1.5). In the Southern Kalahari these fluctuations were not quite synchronous, with humid episodes at 3200–2000 and at around 500 BP. Smaller fluctuations are indicated in the subsequent archaeological record.

7.6 The Kalahari and palaeoclimatic models

On a larger scale Quaternary climatic change can be seen as the effects of variations in the earth's energy budget, expressed through variations in global temperature and atmospheric circulation patterns. The modelling of these changes is based on the generalisation of available sets of regional data, and, inevitably, the extension of conclusions to areas for which no data are available, with its attendant risks. As with the interpretation of proxy data, this third level of approximation is limited by the quality of the data on which it is based.

Palaeoclimatic modelling can be carried out at the regional or global level ('Global Climatic Models' or GCMs), though inevitably the preference for the global tends to prescribe the assumptions of the regional, for comparisons with

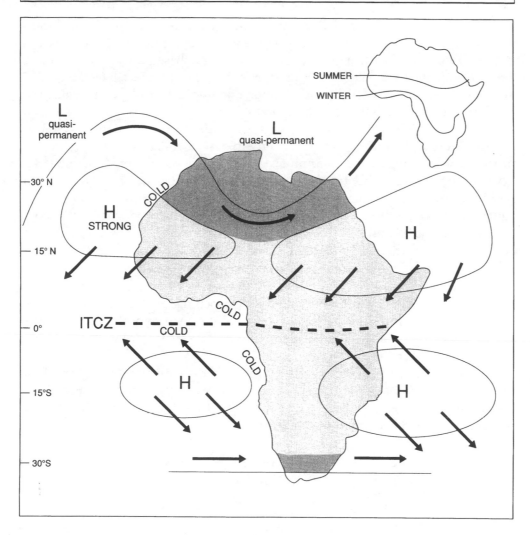

Figure 7.5. Nicholson and Flohn's (1980) model of atmospheric circulation over Africa during the Last Glacial Maximum. Shaded areas are those regarded as wetter than at present.

other parts of the world are inevitable. Most models have been created for the Last Glacial Maximum and the Holocene, the periods for which data is most readily available.

Two approaches require examination; first, the ability of models to explain the palaeoclimatic patterns observed in the Kalahari, and second the contribution of the Kalahari to the refinement of palaeoclimatic models.

A primary question has been the source of moisture in humid periods. For the humid period 16 000–13 000 BP Lancaster (1979) proposes greater summer rainfall caused by a more southerly position of the Inter-tropical Convergence Zone, while Butzer *et al.* (1973) suggest more effective winter cyclonic rainfall. Rust *et al.* (1984), noting the transitional nature of the Kalahari climate, suggest a compromise of variable interaction on a season-to-season basis. As the precipitation source represents the contemporary subcontinental circulation, its understanding is fundamental to the reconstruction of the climate of the subcontinent.

Deacon and Lancaster (1988) provide an excellent review of palaeoclimatic models of the Last Glacial Maximum in the regional context, and suggest they fall into three groups: those that indicate an equatorward movement of circulation belts in both hemispheres (e.g. Van Zindren Bakker, 1967, 1976; Nicholson and

Flohn, 1980; Figures 7.5 and 7.6); those which indicate intensified circulation patterns without latitudinal displacement (e.g. Butzer *et al.*, 1978; Van Zindren Bakker, 1982*a*), and a third group, using an analogue based on the atmospheric circulation patterns during present drought cycles (e.g. Tyson, 1986; Cockcroft, Wilkinson and Tyson, 1987), which involve latitudinal displacement. For the Kalahari region these approaches involved a range of scenarios. Van Zindren Bakker (1980, 1982*a*), for example, by strengthening the circulation systems, brought increased winter rain to the southern desert whilst extending the ITCZ further south during the summer months. Nicholson and Flohn (1980) inferred dry conditions throughout the Kalahari, while Cockcroft *et al.* (1987) and Tyson (1986) believe that an enhancement of the Southern Oscillation and an assumed equatorwards migration of the subtropical convergence brought wetter conditions in the winter rainfall belt, and aridity further north. However, they also conceded that, as there is antipathy between summer and winter rainfall in their model, there would also be episodes when the pattern is reversed.

Deacon and Lancaster (1988), in summary, suggest that despite the major improvements in models in recent years, palaeoclimatic evidence in southern Africa is still insufficient to test the validity of the rival hypotheses. The only available method of testing rainfall seasonality, xylem analysis of grass stems, has not yet been utilised in the region.

However, the data presented here suggest that conditions during the Last Glacial were neither as arid nor as uniform as suggested by Nicholson and Flohn (1980), and the great latitudinal extent of the 16 000–13 000 humid episode requires input from both winter and summer rainfall sources. Whether these

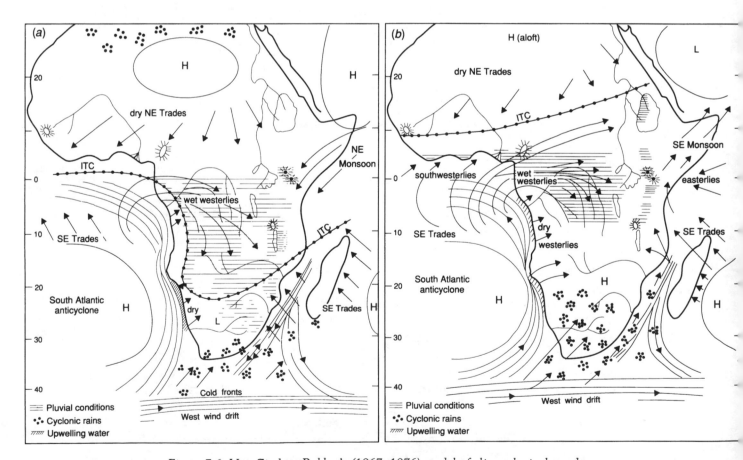

Figure 7.6. Van Zindren Bakker's (1967, 1976) model of climatological trends over Africa during episodes of glaciation. (a) Southern summer and (b) northern summer. Latitudinal contraction of climatic belts is assumed.

inputs were alternating, as suggested by Tyson (1986), or contemporaneous, as implied by Van Zindren Bakker (1976), is beyond the resolution of the data at present.

The Kalahari data also have implications for global climatic models in that it is not coincident with evidence from other tropical arid areas. Although it conforms with the concept of tropical aridity at the Last Glacial Maximum (Williams, 1985), subsequent late glacial humidity, and events during the Holocene, do not conform with the pattern established for North Africa from the study of lake levels (Street-Perrott and Roberts, 1983). Thus, models which imply universal aridity, or 'global symmetry' between hemispheres (e.g. Nicholson and Flohn, 1980) are suspect.

Modelling of the lakes of North Africa and Asia Minor (Hastenrath and Kutzbach, 1983; Kutzbach, 1983) suggests that the lacustrine episodes of the early Holocene in these areas are due to an enhancement of the monsoonal circulation in the northern hemisphere, itself due to an orbitally induced increase in solar radiation and enhanced thermal contrast between land and sea at this time (COHMAP, 1988). This, in turn, may have led to a weakening of the ITCZ in Africa south of the equator, and subsequent drier conditions.

An alternative explanation is suggested from the analysis of palaeo-temperature data (Harrison et al., 1984), which indicates sufficient variability in the meridional temperature gradient to permit the southwards displacement of climatic belts in both hemispheres during the last glacial, with subsequent displacement of the ITCZ into the southern hemisphere. This, in turn, would account for the observed disparity between the hemispheres. The present generation of models, of increasing complexity and resolution, explore further the hemisphere gradients which would explain these observed differences.

Part Three

THE HUMAN FACTOR

8 The Kalahari in the archaeological record

DESPITE THE LIMITATIONS of the Kalahari environment, there is a long, but not necessarily continuous, history of human occupation, extending from the Early Stone Age to the present day. Within this span societies have evolved from a hunter–gatherer existence to pastoralism and agriculture and on to multiple land use, exhibiting increasing degrees of technological sophistication within each stage.

These societies have adapted to the Kalahari environment, and have, in turn, produced changes within the environment itself. In general terms, the evolution of the social groups, accompanied by growth in population numbers, has led to increasing independence from environmental factors, accompanied recently by marked, frequently detrimental, impacts upon the environmental base.

Recent years have seen much research activity in the fields of archaeology, ethnology and history, which have permitted a much clearer view of human activity in the distant past, and have challenged many of the popular notions of 'primitive' societies in the region. This chapter examines the early history of human settlement and assesses its impact upon the Kalahari Desert in the light of this research. In Chapter 9 we pursue this theme into the nineteenth and twentieth centuries, during which impacts gain in momentum and variety.

8.1 The scope and limitations of the prehistorical record

The term 'prehistory' covers the time before the existence of continuous written or oral records; in the Kalahari this essentially means before the sixteenth to nineteenth centuries, depending on locality. It covers the Stone Age, in which small groups pursued an environmentally opportunistic life-style of foraging, or hunting and gathering, using stone implements. The Stone Age is conventionally divided into Early, Middle and Later periods (ESA, MSA and LSA; Goodwin, 1926). These subdivisions are based on the associations and typologies of stone artefacts and may not have chronological significance; indeed, the use of the terms is based more on convenience than scientific classification (Volman, 1984).

In the Kalahari the Late Stone Age ended as late as 400 years ago (Robbins, 1984). The foraging culture has been perpetuated, in part, by the Khoisan, who can thus be considered as direct descendants of LSA peoples. Although this life-style is rapidly disappearing, it offers ethnologists a valuable analogue for the behaviour of Stone Age cultures which cannot be gained from archaeological evidence alone.

Prehistory also incorporates the Iron Age, which saw the development of ceramics, metallurgy, pastoralism, agriculture and intra-regional trade, accompa-

191

nied by the growth of large and socially complex settlements at the Kalahari periphery, exhibiting a high degree of political organisation on a regional scale. The migration of further Bantu-speaking agro-pastoralists from surrounding areas of southern Africa from about AD 1500 onwards crosses the ill-defined boundary into the historical period.

The presence of long-term human occupation, particularly in the earlier stages of prehistory, has not always been obvious. Although southern Africa has a rich archaeological heritage extending back to the early hominids and the evolution of 'anatomically modern man' (Deacon, 1988), the Kalahari has not received much consideration. From a quantitative point of view the relative scarcity of sites has been frequently noted (e.g. Ebert et al., 1976; Helgren and Brooks, 1983). This has tended to perpetuate the long-held view (Van Riet Lowe, 1935) that the Kalahari was avoided by early peoples as being too inhospitable, or at least occupied only in selected localities when climatic conditions were favourable. This notion has been countered by Ellenberger (1972), who points out that stone artefacts have a wide distribution in the region, and are particularly common in the vicinity of water courses and pans, while Hitchcock (1982) reports that the New Mexico Kalahari Project of 1975 identified over 100 archaeological sites in the Makgadikgadi area alone.

Qualitatively, the Kalahari has also been found wanting; recent reviews of the Stone Age by Volman (1984) and Deacon (1984) consider only three Kalahari and peri-Kalahari sites (Gci, Kathu Pan, Wonderwerk) to be of archaeological importance on the subcontinental scale.

A parallel debate as to the timing and extent of the penetration of Iron Age cultures into the Kalahari has followed a similar path. The theory that such penetration was delayed until the last two centuries by environmental factors has also recently been dispelled (see Denbow and Wilmsen, 1986).

A more important cause of low archaeological visibility has been lack of research. Although archaeological studies started with the Vernay-Lang Expedition of 1930, progress has been slow. Campbell (1982) notes that serious archaeological research in Botswana did not commence until 1968; by 1982 only 20 sites had been properly excavated, of which 5 had been dated.

There is no doubt, however, that the sites themselves are not ideal. As noted in Chapters 6 and 7, the Kalahari offers few closed sites, such as caves, that would preserve archaeological materials in stratigraphic sequence, while the high rate of decay of organic matter through biological activity reduces its availability as both archaeological material and a dating medium. Most archaeological material occurs as metachronous surface accumulations of resistant artefacts, principally stone tools and pottery fragments, exposed by deflation and surface wash. Localities where artefacts appear to be in situ may also be suspect; research on the prehistoric succession in the Kalahari Sand at Gombe, Zaïre (Cahen and Moeyersons, 1977), backed by laboratory simulations (Moeyersons, 1978), indicates that lithic artefacts may become redistributed through the profile by creep and bioturbation, thus challenging the concept of stratigraphy in Kalahari sediments.

Despite these problems, however, much has been achieved in archaeological and anthropological studies during the past two decades. In particular the timespan of human occupation has been greatly expanded, and fixed by a series of radiocarbon dates (Appendices 1B, 1C and 2).

8.1.1 The Early Stone Age (ESA)

The ESA is characterised by the presence of large bifacial cutting or digging tools, termed handaxes, and flaked pebbles, thought to be cleavers (Figure 8.1). These

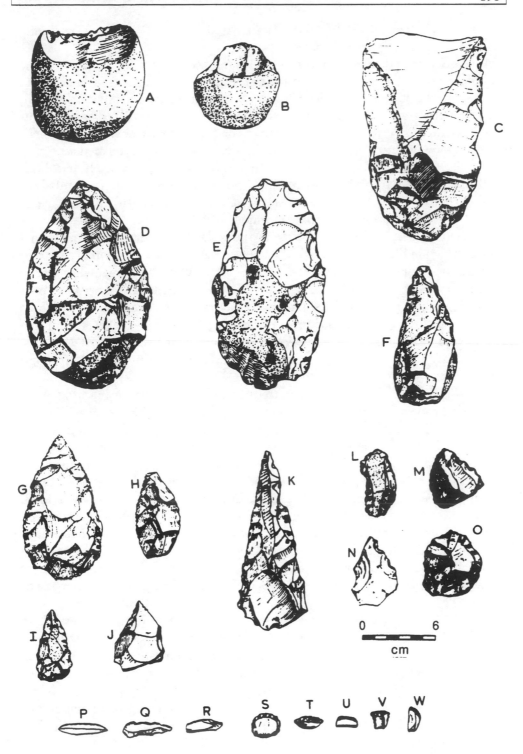

Figure 8.1. Some Stone Age tools from the Kalahari. After Cooke (1979).
ESA (Acheulian) A–B: pebble tools, C: cleaver, D: handaxe. (Charama) E:
handaxe, F: pygmy handaxe.
MSA G–I: bifaced points, J–K: unifaced points, L–M: side scrapers, N: flake,
O: core.
LSA (Microliths) P–R: backed blades, S–W: scrapers.

implements are found throughout southern Africa, usually within river gravels,
where they have frequently been reworked and redistributed. Their nature and
context makes them difficult to correlate and almost impossible to date in
absolute terms. Early attempts to create chronological sequences based on the
refinement of implements have long since been abandoned (see Mason, 1967;

Sampson, 1974), and cultural subdivisions based on variations between assemblages are very much open to debate.

Most ESA artefacts recovered in sub-Saharan Africa are now placed in the Acheulean. With the possible exception of the Zambezi Valley sites (Clark, 1950, 1975), only the Upper Acheulean is represented in southern Africa (Deacon, 1975), covering a timespan from approximately 1 million years BP to about 200 000 years ago (Volman, 1984). North of the Limpopo a transitional phase to the MSA may be represented by assemblages of pick-like tools termed core axes (Clark, 1970) classified as Sangoan, while the late Sangoan itself has been subdivided into Bambata and Charama industries after sites in Zimbabwe (Cooke, 1966). The Charama industry is thought to be not less than 115 000 years old on the basis of present evidence (Volman, 1984).

The distribution of Stone Age sites in the Kalahari Desert and surrounding areas is shown in Figure 8.2 and Table 8.1. The greatest concentration of reported ESA sites lies along the hardveld rivers in the vicinity of Gaborone, where stone implements are common and can easily be spotted by the casual observer. Cooke, in his review of Stone Age localities in Botswana (1979), examined Wayland's collection of 'zone fossils' accumulated during his time as Director of the Bechuanaland Geological Survey Department, and identified a total of 159 sites, of which 36 were ESA sites in the vicinity of the Taung, Ngotwane and Limpopo Rivers.

Sites within the Kalahari occur less frequently and are invariably found along valleys or in the vicinity of pans. Yellen (1971) has noted ESA artefacts in the Quangwa Valley, part of the long but discontinuous record in this area that runs through to the LSA (Helgren and Brooks, 1983). Aldiss (1987b) reports 25 axes of Acheulean type from the Okwa Valley near Tswaane which appear to be unrolled (i.e. *in situ*), and are made from local siltstones, sandstones and silcrete. A factory site has also been reported from Samedupe Drift on the Boteti River (Wayland, 1950a, b; Cooke, 1979) of transitional Charama (Late ESA to MSA) age, where tools were manufactured from the silcrete which outcrops at that point. Recently, Campbell (1988a) has reported the widespread presence of handaxes along the Nchabe and Boteti river beds between Kumana and Samedupe, at Chanoja on the Boteti, and at several sites in the lower Thamalakane, all made of silcrete.

The relative scarcity of ESA materials and lack of chronological controls in both the Kalahari and other parts of southern Africa underlines how little is known of these early cultures, particularly of their subsistence patterns. Although faunal remains at ESA sites are few on the subcontinental scale (see Klein, 1979 and 1980 for reviews), it would appear that, in comparison with MSA sites, they contain a high proportion of remains from large and dangerous animals, such as rhinoceros, indicating a propensity for scavenging rather than organised hunting.

The ESA represents a long time span of technological conservatism, in which the techniques of tool production changed little, and experiments with other materials, such as bone or wood, have not been preserved. Ebert *et al.* (1976), in reporting the find of 447 Acheulean artefacts at Serowe, of an estimated age of 500 000–100 000 years, question the interpretation of such artefacts from a purely functional view. They suggest either that the quantities of tools suggest a prolonged period of sedentary occupancy, which seems unlikely in view of subsequent developments in the Stone Age, or that the tools were not used in a functional manner, i.e. only replaced once they had been worn out. In this case it is possible that tool manufacture was repetitive behaviour, very similar to that found in higher apes, and that Acheulean men had not reached the stage of carrying tools with them.

8.1.2 The Middle Stone Age

The succeeding Middle Stone Age lasted from around 130 000 to around 20 000 BP (Volman, 1984). Artefact assemblages from the MSA contain a variety of flake, core and blade tools, smaller and more finely worked than the ESA implements. Many of the tools have been retouched, and may, towards the end of the MSA, have been hafted on to handles.

MSA sites are more common in the Kalahari than those of the ESA (Hitchcock, 1982), but are still found mainly in the vicinity of pans and valleys. Most of these sites have been interpreted as locations of tool manufacture or hunting/trapping camps, some covering extensive areas. Again the open nature of these sites has precluded observations beyond comment on the lithic assemblages. Cooke (1979) has noted a tendency towards finer-grained rocks, particularly chalcedony, although silcrete is still the dominant source material, as at Orapa (Cohen, 1974) and the Nata River (Cooke and Patterson, 1960a). The tools have been attributed to various industrial complexes, such as Stillbay (Cooke and Patterson, 1960a), Bambata (Cooke, 1979) and Pietersburg (Tobias, 1967), but the same problems of correlation remain.

The most intensively studied MSA site occurs at Gci Pan in the Quangwa Valley of western Ngamiland (Yellen, 1971; Brooks and Yellen, 1977; Helgren and Brooks, 1983), which has also provided palaeoenvironmental information (section 7.4.1). In the floor of Gci Pan an archaeologically sterile calcrete layer separates and closes a lower MSA layer from the LSA materials in the surface deposits. The MSA layer is composed of interstratified alluvia and colluvia containing up to 1500 pieces of stone, tooth and bone per square metre (Brooks and Yellen, 1977). The lithic artefacts are mostly retouched points and scrapers, which have their nearest typological equivalent in the Bambata complex of Zimbabwe. Faunal remains recovered include over 1000 teeth and fragments, bone and ostrich shell, and include three extinct species: giant zebra (*Equus capensis*) giant buffalo (*Pelorovis antiquus*) and a giant hartebeest (*Megalotragus priscus*).

The dominant faunal species in the Gci MSA layer are zebra and warthog. This is at variance with the general pattern of MSA sites in southern Africa, which yield small to medium ungulates, particularly the eland (Volman, 1984). Klein (1979) notes that warthog are relatively rare in MSA sites, presumably because they are difficult to hunt without elaborate strategies. The dominance of these two animals at the site suggests that ambush strategies had already been developed at Gci.

^{14}C dates from ostrich shell in the MSA layer give infinite dates, while the overlying calcrete has yielded a date of 31 470 BP (see Appendix 1B). Helgren and Brooks (1983) suggest that MSA occupation occurred in brief but intense spells around the ^{18}O stage boundary 4–5, circa 75 000 BP.

8.1.3 The Later Stone Age

The Later Stone Age in southern Africa appeared between 35 000 and 20 000 years ago, and persisted until historic times. It is characterised by assemblages of small bladelets designed for hafting on to shafts, together with a range of convex, circular and elongate scrapers. These are usually termed 'microliths' on account of their size, although, as Deacon (1984) points out, microliths were not produced throughout the LSA; larger tools were revived around the terminal Pleistocene/early Holocene (c.12 000–8000 BP).

Alongside the evidence of these more sophisticated tool-making practices, other archaeological remains indicate human development during the LSA,

Table 8.1. *Stone Age sites in the Kalahari and environs*

	Location	Reference	Site characteristics
1	Victoria Falls	Clark, 1950, 1975	Terrace deposits. Early ESA to LSA
2	Mababe Depression	Lamont, reported by Hitchcock, 1982. Campbell, 1970	ESA on raised beaches, LSA site and rock paintings at Savuti
3	Tsodilo Hills	Junod, 1963; Denbow & Campbell, 1980; Campbell et al., 1980.	LSA sites, rock paintings
4	Gci/Quangwa Valley	Yellen, 1971; Brooks & Yellen, 1977; Helgren & Brooks, 1983	ESA to LSA. Dated MSA/LSA site at pan edge at Gci
5	Caecae (XaiXai)	Yellen, 1971; Wilmsen, 1978	LSA campsite at pan edge
6	Drotsky's Cave	Yellen et al., 1987	LSA (terminal Pleistocene) dated site
7	Gemsbok Pan	Van Riet Lowe, 1935	ESA on pan periphery
8	S.E. Lake Ngami	Cooke & Patterson, 1960a	MSA factory site on raised shoreline
9	Toteng	Cooke, 1979; Robbins, 1984	MSA and LSA (dated) sites
10	Nchabe River	Campbell, 1988c	28 ESA to LSA sites along river
11	Samedupe Drift	Wayland, 1950b; Cooke, 1979	ESA (Charama) quarry and manufacture site on banks of the Boteti River
12	Makalamabedi Drift	Van Waarden, 1988	MSA (Tshangula) in bed of Boteti River
13	Nata	Bond & Summers, 1954; Cooke, 1967	A series of MSA/LSA hunting camps on Nata River and lake shorelines
14	Kedia Hill	Cooke & Patterson, 1960b	MSA hunting camp
15	Toromoja/Gwi Pan	Denbow & Campbell, 1980; Helgren, 1984	LSA settlement and burial. Dated sites
16	Orapa Mine	Cohen, 1974	MSA factory site on beach ridge
17	Letlhakane Well	Cooke & Patterson, 1960b	MSA/LSA site on pan edge
18	Serowe	Ebert et al., 1976	ESA site on terrace of Metsemasweu River
19	Okwa Valley	Aldiss, 1987b	ESA sites on river terraces
20	Nossop Valley	Tobias, 1967	MSA sites on river terraces
21	Limpopo, Taung and Ngotwane Rivers	Wayland, 1950a; Ebert et al., 1976; Cooke, 1979; Hitchcock, 1982	Abundant ESA sites in river beds and terraces
22	Kathu Pan and Wonderwerk Cave	Beaumont et al., 1984	ESA to LSA sequence in pan peats and cave sediments. MSA/LSA dated

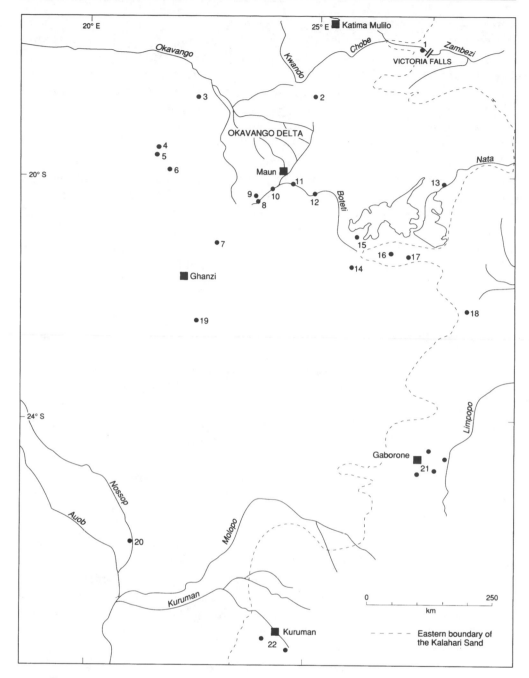

Figure 8.2. Distribution of Stone Age sites in the Kalahari and environs. Numbered sites equivalent to Table 8.1. Minor sites from Cooke (1979).

including bone and wooden tools, personal ornaments (particularly ostrich shell beads), rock paintings and engravings, specialised fishing and hunting implements, including the bow and arrow, and evidence for formal burial of the dead. As many of these implements and practices have been encountered among the Khoi ('Hottentot') and San ('Bushmen') peoples in the Kalahari in historic times, these groups have been seen as extensions of LSA culture, although the manufacture of stone tools is no longer practised. These analogues are discussed in section 8.1.4.

As with earlier assemblages, attempts have been made to subdivide the LSA into industries or industrial complexes, which can be defined as tool-making cultures producing artefact assemblages with particular characteristics. The initial division into Smithfield and Wilton industries (Goodwin and Van Riet

Lowe, 1929) has been overtaken by the increasing availability of sites and data; subsequent attempts to revise the classification (Sampson, 1974; Phillipson, 1977) have become equally outmoded. Deacon (1984) suggests that sufficient dated sites are now available in southern Africa to identify regional trends in LSA development, especially in the Holocene. She proposes a chronological sequence of late Pleistocene, terminal Pleistocene and Holocene cultural/stratigraphic units with distinct industrial complexes. A final category of the Late Holocene covers the period in which pottery and metal artefacts also occur.

Most of the Kalahari LSA sites fall into this latter category. However, there are two confirmed late Pleistocene sites. The first is in the Depression Rock Shelter (previously termed 'Depression Cave') in the Tsodilo Hills, where excavations to a depth exceeding 5 m have yielded evidence of sporadic occupation of the site from recent times back to 19 000 BP, and, on the basis of sedimentation rates, probably to 30 000 BP (Robbins and Campbell, 1988). LSA microliths occur at the level of around 19 000 BP, in association with a small piece of chromite that appears to have been used as a natural colouring pigment. This episode of relatively sophisticated microlith technology was followed by an occupational hiatus to 13 000 BP, at which point the microlith tradition continued.

The second site is at Drotsky's Cave, where Yellen et al (1987) have excavated part of the floor of the northeast chamber (Figure 8.2) close to the entrance. The excavation yielded 848 faunal fragments, 193 fragments of ostrich egg shell, and a few lithic items, most of which have been classified as debitage. Dating of charcoal fragments within the excavation has given a date of 12 200 BP, suggesting that the site forms part of the Terminal Pleistocene/Early Holocene non-microlith industrial complex, one of the few that occurs in the interior of southern Africa. Yellen and Brooks (1987) also suggest that some of the artefacts found above the calcrete layer at Gci may be contemporaneous, but have not been accurately dated. They lie between the calcrete (22 500 BP minimum age) and materials dated to within the last 2000 years close to the surface.

Helgren (1984) describes an LSA settlement close to the Boteti Delta at Toromoja. Here an assemblage of microliths of Wilton type, fashioned from silcrete, are found in association with faunal remains of reedbuck and lechwe, wetland ungulates no longer found in the area, together with zebra and wildebeest remains. Fish remains have also been recorded. The site has suffered greatly from bioturbation and has a questionable stratigraphy. A date of 2960 ± 50 BP on a calcareous soil horizon suggests that settlement may be mid-Holocene.

Most other sites that have been investigated and dated postdate contact between LSA and Iron Age peoples. At Gwi Pan, close to Toromoja, a burial site has been excavated of probable late LSA age, the skeleton yielding a date of 235 ± 370 BP (Helgren, 1984). Further to the west, the surface layers at Gci are of similar age (see Appendex 1B) and contain scattered accumulations of artefacts and waste material associated with two hearths and a pit, which has been interpreted as a game trap (Brooks and Yellen, 1977). At Toteng, on the Nchabe River, large quantities of Wilton-type microliths fashioned from silcrete and chert have been located in association with Iron Age pottery above a layer dated at 400 ± 100 BP, suggesting that stone tools were manufactured in quantity until fairly recently (Robbins, 1984). Sites such as this tell us more about the evolution of societies during this period than about the behaviour of the LSA peoples per se, and are discussed more fully in section 8.2.

8.1.4 Ethnological evidence for Stone Age adaptations

There is a wealth of ethnological data now available on the Khoisan peoples, covering every aspect of their distribution, movement, social relationships, language, beliefs, diet and occupation (e.g. Schapera, 1930; Story, 1964; Silberbauer, 1965, 1972; Lee, 1972, 1979; Tanaka, 1976; Yellen and Lee, 1976, Vossen, 1984). On the assumption that the Khoi and San are direct descendants of LSA peoples, still practising a foraging life-style, but without the use of stone implements, it has been possible to use these ethnological observations as analogues for LSA behaviour. Conversely, the comparison of archaeological and historical findings can, at least, provide temporal depth to the ethnological studies.

Such comparisons are not without difficulties. Deacon (1984) notes that ethnological/archaeological relationships are likely to be closer in the Kalahari than in other parts of southern Africa since the same environment is common to both. However, the ethnological data, both current and historical, relate to a time at which the Khoisan have already had some 2000 years of contact with outside influences, during which their behaviour is likely to have undergone changes which may not be easily detectable in the archaeological record.

Further, conclusions drawn from the archaeological evidence, such as group size and occupational strategies, may not take into account the complexities of the relationship between the San and their environment. Deacon (1984), for example, notes that the seasonal nucleation of social groups is closely dependent upon environment and strategy; the G/wi group of the San tend to congregate in the summer months to utilise standing water, while the !Kung disperse at this time to harvest the Mongongo nut (*Ricinodendron rautanenii*).

The recent past has also encompassed episodes of non-foraging occupation, such as herding (Hitchcock, 1982), and even absorbtion into the 'Bushman' battalions of the South African Defence Force. Ebert (1978) has described these shifts as a continuum of adaptations available to hunter–gatherers, with groups occupying options on the continuum in response to prevailing conditions. The wide range of practices of different San groups is thus a further caveat in the search for analogues.

The general pattern, however, must have similarities. Yellen and Harpending (1972) have noted that the !Kung dry season camps tend to be found at the same locations as LSA sites, suggesting similar aims, although contemporary sites are smaller in area (Hitchcock, 1982) and usually more distant from water sources such as pans (Wilmsen, 1978). It is probable that LSA groups, like their modern counterparts, were small and highly mobile, with group size and range dependent on the availability of food and water resources on both a seasonal and a long-term basis.

Of prime importance was the availability of water, either as surface or near-surface resources. Water-bearing plants, such as the Tsama melon (*Citrullus lanatus*), were also likely to have been used in areas and times of water shortage, but have not been preserved in the archaeological record.

There was probably a high dependence on vegetable foods, although these leave few remains at archaeological sites, and there is some debate as to the degree of change in the relationship between hunting and gathering over the last 2000 years (e.g. Schapera, 1930; Story, 1964). Lee (1979) found that a !Kung group in winter obtained some 70 per cent of their calories from vegetable products, with a strong dependence on a limited number of species, usually those with non-perishable roots, nuts or seeds, as staple foods. Perishable items such as fruits and leaves are consumed in season. Certainly, the remains of Mongongo nuts have been found at a number of LSA sites, for example, in the Tsodilo Hills (Campbell, Hitchcock and Bryan, 1980), and have recently been dated to a number of levels in the Depression Rock Shelter site from the first century AD back to around 7000 BP (Robbins and Campbell, 1988). Other staples have undoubtedly been utilised over time.

Table 8.2. *Kill rates and meat yields in one year among a group of ≠ Xade San*

Species*	Meat yield per animal (kg)	Total yield			Kill rate	
		Per annum	Summer (Dec.–May)	Winter (June–Nov.)	Summer (Dec.–May)	Winter (June–Nov.)
Giraffe	136	136	136	—	1	—
Eland	86	774	688	86	6	1
Kudu	45	135	135	—	3	—
Gemsbok	68	1360	816	544	12	8
Hartebeest	36	396	216	180	6	5
Wildebeest	36	432	252	180	7	5
Springbok	11	352	253	99	23	9
Duiker	4.5	293	112	181	25	40
Steenbok	3.6	245	98	147	27	41
Springhare	0.7	155	119	36	170	52
Fox	1.8	33	8	25	4	14
Jackal	2.3	32	6	26	3	11
Rodents	0.06	18	5	13	180	115
Birds	0.23	37	18	19	81	84
Tortoises	0.12	53	47	6	390	50
Snakes	0.23	11	8	3	34	16

After Silberbauer (1972).

* Other animals: porcupine (2), warthog (1), ants (2.5 litres yr^{-1}), termites (8 litres yr^{-1}), caterpillars.

Faunal remains at LSA sites indicate a high proportion of small to medium ungulates, all of which occur in the Kalahari at present. Smaller mammals and some reptiles, such as tortoises and snakes, are also present. The larger, migratory mammals, such as the zebra and buffalo, are less well represented, while more dangerous creatures tended to have been avoided, although some examples, such as bones of the white rhinoceros (*Ceratotherium simum*) found in the hunting pit at Gci, have been recorded (Brooks, 1978). These observations accord well with the results of a study of kill rates and meat yields amongst a ≠ Xade group (Silberbauer, 1972), summarised in Table 8.2. The study shows that the bulk of meat comes from a few larger mammals during summer months, while in winter the San rely on smaller animals, particularly small buck, rodents and hares. It is in these winter months, when game is scarce, that greater reliance is placed on vegetable products. Silberbauer (1972) also notes that hunting with bow and arrow is prevalent in summer, possibly because the arrow-poisons derived from the chrsysomelid beetle (*Diamphidia nigro-ornata* and related species: Woolard, 1986) can be obtained only at this time. In winter the less effective procedures of snaring are adopted.

As the San now pursue foraging over a more limited range than previously, some of the LSA strategies have fallen into disuse. In riverine areas hunting has now become impractical due to pressure on the land, and fishing or fish trapping, usually involving *Clarias* and *Barbus* spp. (Robbins, 1984) has been discontinued for similar reasons. The San also use three types of structures which have LSA analogues, namely campsites, hunting blinds and pitfalls (Brooks, 1978; Crowell and Hitchcock, 1978). The shelters built at campsites from poles and grass are usually destroyed by termites within the space of a few years, but the associated hearth sites, where most of the communal activity takes place, become the foci of tool and food waste deposits. Hunting blinds, on the other hand, are more substantial structures utilised sporadically over long periods of time. Most of these are of the pit type, such as the one recorded at Gci (Brooks and Yellen, 1977), and accumulate tools, food waste and charcoal. The third type, the pitfall for large game, is no longer excavated by the Khoi, but was observed on a number of occasions by early European travellers (e.g. Burchell, 1822). The decline in the use of pitfalls may be associated with the decrease in large mammals and their subsequent protection.

Present San practices are certainly no guide to the size of LSA communities, nor to the ranges which they covered. Hitchcock (1982) notes that a major change brought about by Iron Age contacts, and accentuated towards the present day, has been a decrease in mobility. Initially, sedentariness leads to a reduction in group and site size, as comparative studies of the Nata River and central Kalahari populations show (Yellen, 1977; Hitchcock, 1978). However, prolonged settlement may lead to overall increases. Wilmsen (1978) suggests that the San population at Caecae is now 50 per cent higher than a century ago, a response to distant political events and subsequent migrations of other groups. These pressures would be unlikely to have affected LSA peoples until the last few hundred years. Ebert (1978) differentiates between hunter–gatherers in high-population, sendentary areas, and those in remoter, less favourable terrain, who have fluid social organisation and low recognition of social and territorial boundaries. It is probable that LSA populations were modelled on these latter lines.

8.1.5 Stone Age peoples and the environment

It is not possible, from the evidence available, to estimate the size of populations during the Stone Age. The discontinuous nature of the sites and high degree of environmental adaptation suggested by ethnological studies indicate a population of small groups constantly moving towards fresh resources, particularly water.

As the water resources themselves are controlled by fluctuations in climate, occupation was discontinuous and population sizes were variable. Hitchcock (1982) notes that San groups responded to the droughts of 1933, 1947 and the 1960s by out-migration, an option that is becoming less available at present. It is reasonable to suppose that Stone Age populations responded in a similar way to the long-term fluctuations described in Chapter 7.

Out-migration can be inferred from gaps in the archaeological record, while the prevailing conditions can be tentatively ascertained from archaeological and associated geomorphological data. The hypothesis that migration follows rainfall pattern shifts has been found applicable in the archaeological records of Bushmanland and Griqualand West, to the south of the Kalahari, for the Last Glacial (Beaumont, 1986), and may apply to earlier periods. A similar pattern of occupation during wetter episodes in the Kalahari should be expected.

In this context, Wayland (1954) suggests that ESA populations were present during periods of higher rainfall and permanent water, and Ebert (1977) proposes cooler temperatures and reduced seasonality of rainfall to generate the savanna grasslands that would have supported the types of game favoured by MSA hunters. Certainly, MSA peoples occupied the Dobe Valley during a lacustrine phase (Helgren and Brooks, 1983; section 7.4.1).

In the Late Glacial this pattern is not evident. The lack of early LSA sites may be due to deficient research, but Campbell (1982) suggests that the Kalahari became depopulated between 16 000 and 8000 BP despite the evidence for more favourable climatic conditions in the first 4000 years of this span. Only at Gcwihaba is there good evidence for occupation at a time when the climatic amelioration was ending. A clearer correlation emerges in the late Holocene, where an episode of higher lake levels and established river flows between 2500 and 1500 BP coincides with the introduction of pastoralism in the region (Denbow & Wilmsen, 1986). This is discussed in section 8.2.

It is difficult to estimate the impact of Stone Age foraging on the environment, although, given low population numbers and resource requirements, it is unlikely to have been great. Klein (1980) suggests that the extinction of some larger mammals at the beginning of the Upper Pleistocene was largely due to environmental factors, although MSA and early LSA hunting could have contributed to their demise. However, Klein's studies (e.g. 1981) of mortality profiles among ungulates, especially the eland, at Stone Age sites suggests that MSA peoples took animals which, on account of age or health, were not reproductively relevant to the herds. Certainly, LSA populations, with improved hunting techniques, exploited the same populations that exist today.

8.2 Later prehistory: the advent of pastoralism

The development of pastoralism among LSA peoples in the Kalahari about 2000 years ago heralds a period of technological change and increasing social organisation. It was followed by the introduction of metal working by the fourth century AD and the development of social hierarchies, based on pastoralism at the Kalahari periphery, towards the end of the first millenium. These 'Cattle Chiefdoms' had considerable influence in the Kalahari, and extensive trade links to coastal areas.

Recent research has covered some 34 archaeological sites of this period in Botswana, together with others in neighbouring countries, and is backed by a suite of approximately 80 radiocarbon dates (Denbow and Wilmsen, 1986). These sites and their chronology are shown in Figure 8.3 and Table 8.3; the dates are given in Appendix 2. To this may be added the 320 Iron Age peri-Kalahari sites identified by the presence of *Cenchrus ciliaris*, a distinctive grass which colonises abandoned cattle kraals (Denbow 1979, 1984). These have added considerable depth to our understanding of an area long considered to be a cultural vacuum. As Denbow (1986: p. 3) notes '. . . the long and complex past of the peoples of this region has often been condensed into an ahistorical and timeless caricature when compared with events in neighbouring countries'. The picture now emerging is not of a frontier of human settlement, but of a region fully integrated into the southern African social milieu.

8.2.1 Pastoralism among LSA communities

Current evidence suggests that domestic animals and ceramics appeared in southern Africa just over 2000 years ago. At Bambata Cave in the Matopos Hills of Zimbabwe sheep remains and a ceramic type known as Bambata ware, representative of a western stream of Early Iron Age (EIA) cultures, have been dated to 200 BC (Walker, 1983), while in Namibia and the Cape Province sheep and/or goats occur with LSA artefacts at around AD 100 (Klein, 1984). The wide dispersal of animal domestication, preceding the introduction of metallurgy by several centuries, suggests its spread from the north among the Khoisan peoples, in particular the Khoi, who were practising animal husbandry at the time of European arrival in southern Africa. These husbandry practices have been well documented (e.g. Elphick, 1977).

Studies of Central Khoisan (Khoe) languages show that cattle-related words have their roots in the Khoe dialects on northern Botswana, and that the division into Khoekhoe ('Hottentot') and non-Khoekhoe (essentially San) languages may

Table 8.3. *Radiocarbon sequence of archaeological sites in the Kalahari and Eastern Hardveld of Botswana for the last two millenia*

Date	Middle Kalahari	Eastern Hardveld	Southeastern Hardveld
	Gci		
1800			
	Kgwebe	Shashe	Magagarape
1700			
	Depression Cave		
1600	Depression Cave		
	Toteng		
1500			
		Domboshaba	
	Depression Cave	Toutswe	
		KM 30	
1400			
		Toutswe	Broadhurst
		KM 30	
1300			
		Toutswe	
1200			
	Serondela	Toutswe	
		Thatswane	
1100			
	Nqoma	Toutswe	Moritsane
		Kgaswe	Magagarape
1000			
	Outpost 1	Toutswe	
	Matlapaneng	Kgaswe B55	
	Nqoma	Maiphetwane	
		Taukome	
		Thatswane	
900			
	Nqoma	Letsibogo	Magagarape
	Society		
	Matlapaneng		
	Hippo Tooth		
	Caecae		
800			
	Nqoma	Taukome	Thamaga
	Divuyu	Bisoli	Magagarape
	Serondela		
	Caecae		
	Qogana		
	Chobe		
700			
	Nqoma	Taukome	Magagarape
	Divuyu	Bisoli	
	Matlapaneng		
600			
	Divuyu	Letsibogo	Baratani
			Magagarape
500			
			Magagarape
400			
		Maunatlala	
300			
200			
100			
	Depression Cave		
0AD			

After Denbow (1986), Campbell (1988*b*, 1989), Robbins and Campbell (1988).

have taken place about 2500 to 2000 years ago (Ehret, 1967; Vossen, 1984). Ehret further suggests that the pastoralist vocabulary of the Bantu of the Angola–Namibia region is derived from the Khoisan, inferring that their animal husbandry practices were acquired in the same manner.

Despite these indications of a central role for the Kalahari region in the dispersal of domesticated animals, no early sites have been directly dated. The earliest site is at Lotshitshi on the edge of the Okavango, dated to around AD 290, where cattle bones are found in association with a variety of game remains and a limited quantity of pottery, possibly of Bambata type. However, undated, but possibly earlier, sites of LSA artefacts and Bambata ware occur along the Boteti River, in the southwest Makgadikgadi Basin, and to the east of Sua Pan (Denbow and Campbell, 1980). These sites suggest a relatively dense pattern of settlement, partially reliant on fishing and riverine hunting. Denbow (1986) concludes that the Okavango–Makgadikgadi axis played an important part in the diffusion of

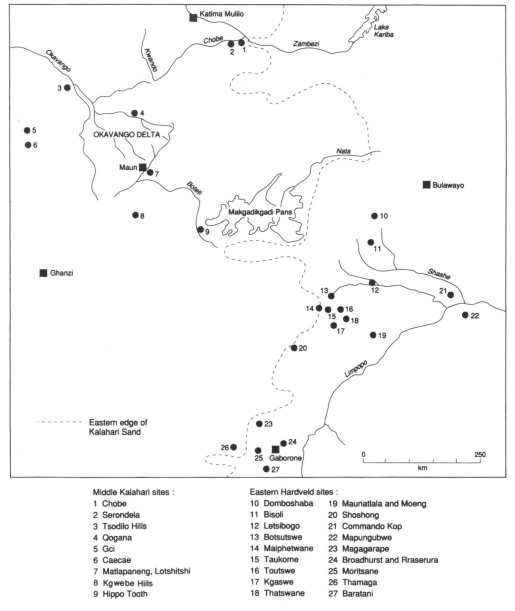

Middle Kalahari sites :	Eastern Hardveld sites :	
1 Chobe	10 Domboshaba	19 Maunatlala and Moeng
2 Serondela	11 Bisoli	20 Shoshong
3 Tsodilo Hills	12 Letsibogo	21 Commando Kop
4 Qogana	13 Botsutswe	22 Mapungubwe
5 Gci	14 Maiphetwane	23 Magagarape
6 Caecae	15 Taukome	24 Broadhurst and Rraserura
7 Matlapaneng, Lotshitshi	16 Toutswe	25 Moritsane
8 Kgwebe Hills	17 Kgaswe	26 Thamaga
9 Hippo Tooth	18 Thatswane	27 Baratani

Figure 8.3. Distribution of EIA archaeological sites in the Kalahari and Eastern Hardveld of Botswana. After Denbow (1986) and Denbow and Wilmsen (1986).

Figure 8.4. An example of the abundant San rock art in the Tsodilo Hills, northwest Botswana. The animals in the upper frieze have been interpreted as domestic cattle. Photograph by Alec Campbell.

domestic animals around 2000 years ago, particularly as climatic conditions appeared more favourable at this time.

Later LSA sites occur in conjunction with EIA artefacts, and it is not always possible to draw firm conclusions on their relationships. At pre-Iron Age sites it would appear that, although livestock were kept, this was subsidiary to the hunting and foraging economy. This distinction becomes less apparent at the Tsodilo Hills sites of Divuyu and Nqoma (Denbow, 1980; Denbow and Wilmsen, 1983) and at Matlapaneng at the base of the Okavango. Both span the period between the sixth to the eleventh centuries AD and contain EIA and LSA artefacts, as well as evidence of cattle husbandry as a primary occupation. Tsodilo contains one of the finest concentrations of rock art in southern Africa, some 2000 paintings at 200 sites, many depicting domestic cattle (Cambell *et al.*, 1980; Figure 8.4). The variety of styles and lack of direct dating make it difficult to assign the paintings to either Khoisan or later haMbukushku artists, but suggest a high degree of involvement by the Khoisan in cattle management and allied developments by the end of the first millenium.

8.2.2 The Early Iron Age

Early Iron Age communities arrived in the vicinity of the Zambezi River about AD 200 (Cambell, 1982) and spread rapidly southwards. They can be identified archaeologically by semi-permanent villages, and the use of domestic stock, ceramics, crops and metals (Maggs, 1984). It is not known whether all these items were introduced in a single economy as, of course, ceramics and domestic stock were already part of the Khoi tradition. However, the EIA tradition was firmly established in southern Africa by AD 500.

The interpretation of this tradition has depended much on the identification of ceramic typologies and the assumption that the EIA correlates with the spread of Bantu-speaking peoples, both assumptions being open to question (Maggs, 1984).

The diffusion and chronology of different 'streams' of EIA industrial complexes in the region (e.g. Phillipson, 1977) are likely to be revised with future research.

In the Kalahari region two centres of EIA activity are apparent (Table 8.3, Appendix 2), one within the Middle Kalahari, centered on lakes, rivers and other water resources; the other in the hardveld between the eastern boundary of the Kalahari Sand and the Limpopo River, extending east and northeast into the Transvaal and Zimbabwe. Strictly speaking, the latter is not part of the Kalahari . (see discussion in Chapter 1), but it has been considered so by many archaeologists (e.g. Kiyaga-Mulindwa, 1983; Denbow, 1986), particularly as the influence of these communities extended far into the sandveld.

Sites in the first group include settlements in the Tsodilo Hills, at the periphery of the Okavango Delta at Matlapaneng, and on the southern bank of the Chobe River at Serondela, falling mostly into the period AD 600–1100. The Tsodilo sites include Depression Rock Shelter, where the earliest ceramics of the region have been dated to the first century AD (Robbins and Campbell, 1988), together with later Iron Age artefacts. The adjacent Divuyu and Nqoma village sites (Denbow and Wilmsen, 1983, 1986) preserve evidence of pole and wattle house structures, iron and copper implements which appeared to have been worked on site, and carbonised seeds of Sorghum (*Sorghum bicolor*), Pearl millet (*Pennisetum americanum*) and melons, which suggest some agriculture. The ceramics are of two chronologically separated types and appear to have their nearest affinities in Angola and southwest Zambia. Cattle appear to have become increasingly important in the economy at Tsodilo, the earlier Divuyu site preserving a higher proportion of sheep and goat remains. Initial local trade links to the Okavango are indicated by the presence of mollusc shells (*Unio* spp.), while the presence of marine shells and glass beads suggests that Nqoma had strong links to the Indian Ocean coast by the ninth century AD.

The contemporary riverine sites (Denbow and Wilmsen, 1986) indicate stronger associations with the Gokomere tradition of Zimbabwe, particularly the Kumadzulo–Dambwa complex found along the Zambezi River. Ceramics of this type are found along the Chobe, Okavango, Thamalakane and Boteti Rivers, all at sites predating the tenth century. Other finds include clay figurines of cattle at the eighth-century Serondela site and hut remains at Qogana and Matlapaneng. The latter is a major seventh- to tenth-century site also preserving evidence of metal working in the form of bloomery waste, grains and cow peas (*Vigna unguiculata*), cattle, as well as cowrie shells and glass beads from the coast. Both the Tsodilo and riverine sites suggest a mixed economy, still partially dependent on wild animals, fish and gathered vegetable products (Turner, 1987a, b).

A different pattern emerges at the eastern hardveld sites. Although early sites have been dated to AD 380 at Maunatlala and AD 440 and 520 in southeast Botswana (Magagarape and Baratani), most indicate settlement between AD 600 and 1300 as a series of villages in a hierarchy of sizes, all closely linked to cattle herding, indicated by the 200 or more kraals containing vitrified cattle dung. These deposits range from 30 to 100 m in diameter and 25 to 150 cm in depth (Denbow, 1986), and have been much reduced by burning. The ceramics of these settlements have affinities with the Gokomere and Zhizo types of Zimbabwe and the northern Transvaal, with variations between the northeastern (Bisoli), eastern (Toutswe, Taukome, Kgaswe) and southeastern (Moritsane) settlements, suggesting migrations at different times.

These EIA communities reached a high level of social organisation. Toutswe (Figure 8.5), like Mapungubwe and K2 in the vicinity of the Limpopo River, was a major town, probably the largest EIA settlement in southern Africa. It has an area of approximately 100 000 m², was connected into the network of trade links

Figure 8.5. Oblique aerial view of Toutswe, eastern Botswana, the site of a large Early Iron Age settlement, which occupied most of the hill summit. The open ground at the foot of the hill towards the right is the remains of an extensive cattle kraal, distinctively colonised by the grass Cenchus cillaris. Photograph by Alec Campbell.

which reached the Indian Ocean, and was probably the focus of considerable political power, with satellites perhaps as far distant as Tswagong in Sua Pan, dated at around AD 1100 (Campbell, 1982). However, the Limpopo settlements were partially based on trade in gold and ivory. The emphasis on cattle at Toutswe represents a continuity of economic conservatism.

Denbow (1986) argues that cattle husbandry increased rapidly at the expense of agriculture between AD 500 and 700, to the point where some 80 per cent of meat consumed in primary centres was of domestic stock (Welbourne, 1975). This may have been a response to low and erratic rainfall, which would have made agriculture unreliable but ensured disease-free conditions for stock. In turn, this would concentrate wealth into the hands of a few individuals, so creating a centralised power base. Certainly, age distribution data of cattle remains from major and secondary centres (Denbow, 1983; Voigt, 1983; Turner, 1987a) indicate that prime animals were slaughtered at the former, while less valuable animals were utilised at smaller centres, indicating a different social stratification.

Denbow (1982, 1986) has further examined the pattern of settlements in the Toutswe region, and identified a three-tier hierarchy of settlements, based on settlement and kraal size. He identified 159 class 1 sites, with a midden size of 1000–5000 m^2. Of these, some 76 per cent were located on the plains, usually at the boundary between sand and clay soils. This corresponds with the present pattern of rural settlements in eastern Botswana, and is a response to availability of water and the choice of soil types to minimise drought risk. The larger class 2 and 3 settlements, with middens up to 100 000 m^2, are found exclusively on hilltop sites, suggesting an emphasis on defence. The spatial relationships of the class 3 sites (Toutswe, Bosutswe and Shoshong) point towards a high degree of autonomy or even competition between centres. The overall image of the Toutswe culture is of a series of centralised 'cattle chiefdoms', reliant on agriculture and pastoralism in satellite settlements, probably with an element of transhumance into more distant regions. It is possible that the thirteenth-century Toutswe complex supported a regional population as high as the present population (Campbell, 1982), with the same degree of environmental pressure.

After AD 1200 the Toutswe area went into rapid decline, followed in about 1250 by the Limpopo settlements. As this decline was not uniform throughout the hardveld, the northeastern and southeastern sites persisting until the fourteenth century, Denbow suggests that overgrazing and drought are the most likely causes of the collapse of the Toutswe culture. Palaeoclimatic evidence is lacking for this period, but the absence of settlements dated to 1200–1500 at the Tsodilo Hills and riverine sites in the sandveld can also be construed as a deterioration in the environment, followed by the abandonment of settlements.

8.2.3 The Late Iron Age

The Late Iron Age (LIA) is characterised by a change in ceramic typology, the introduction of nucleated hilltop settlements, stone walling and the creation of political structures based on cattle-wealth. Throughout much of southern Africa the transition from Early to Late Iron Age took place between 1000 and 1200. In the Kalahari region the changes became apparent after 1500, although Maggs (1984) argues that the Toutswe culture was a transitional form, an EIA culture containing many of the elements that distinguish the later phase.

The full development of the LIA took place in western Zimbabwe (Leopards Kopje) and in the northern Transvaal (Mapungubwe), culminating in the Great Zimbabwe Tradition, comprising a Zimbabwe Phase (1250–1450) and a Khami (Kame) Phase (1450–1686) (Maggs, 1984). The Great Zimbabwe Tradition represented the zenith of the cattle- and gold-based semi-centralised political system, controlling a vast area between the Zambezi and Limpopo Rivers.

The Kalahari appears to have been peripheral to this empire. Following the centuries of decline in the Toutswe culture, settlements of the Zimbabwe and Khami Phases appeared in northeastern Botswana, and into the eastern sandveld. Domboshaba dates to c. 1460, and became associated with gold mining (Molyneaux and Reinecke, 1983). Other minerals exploited include copper at Matsitama, iron in the Tswapong Hills and specularite at Pilikwe. Walled structures on the eastern and western peripheries of Sua Pan were possibly related to cattle, or to trade in salt (Figure 8.6).

Elsewhere, fresh migrations were leading to the populating or repopulating of the Kalahari from the periphery (Figure 8.7). This expansion was, in part, a centrifugal effect of the growth of large states elsewhere, particularly the Sotho-Tswana kingdoms of the South African plateau, and the growth of the Lozi Empire on the middle Zambezi (Campbell, 1982; Ngcongco, 1982). This pattern of migration, fission and fusion of groups, and assimilation of existing communities was accentuated and accelerated by the *Difaquane*, the 'crushing' which affected southern Africa in the early nineteenth century. By 1800 Bantu-speaking peoples had settled throughout the eastern hardveld, around the Okavango, Boteti and northern rivers, and along corridors provided by the Molopo network and the line of pans centred on Kang and Tshane. The Khoisan and early migrants, such as the baKgalagadi, were pushed back towards the Kalahari Desert, assimilated into or dominated by the Tswana communities.

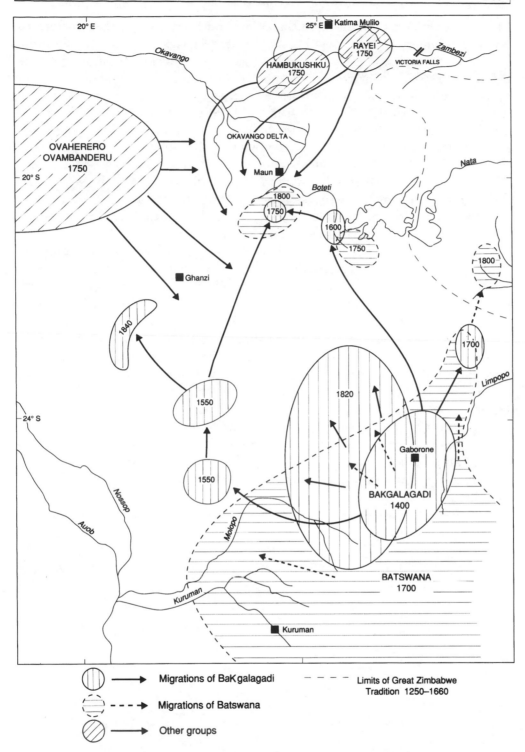

Figure 8.6. Settlement in the Kalahari in the Late Iron Age. After Tlou and Campbell (1984), Tlou (1985), Botswana Society (1988).

The history of these migrations is extremely complex, and, on the basis of oral record, crosses the boundary from prehistory to the historical period. Detailed histories of these migrations, and their consolidation into Tswana states, have been published elsewhere (Campbell, 1980, 1982; Ngcongco, 1982; Tlou and Campbell, 1984; Tlou, 1985).

Some of the earliest arrivals were Kgalagadi groups who settled along the

Molopo River and southeastern Kalahari around 1500, with further migrations into the Kalahari via Mabuasehube, reaching the Kgwebe Hills at c. 1700. In the southeast the baKgwatheng group of the Kgalagadi were closely followed by the baKwena. Despite conflict between the two groups, both expanded rapidly between 1650 and 1750; by this time the baKwena settlement near Molepolole may have contained 10 000 people (Campbell, 1982). Further fission sent off new migrations: the baNgwato branch of the baKwena moved north to Shoshong, then divided again, with the baTawana moving via the Boteti River to Kgwebe and Ngamiland (Tlou, 1985). Meanwhile pressures in the southeast led to fresh relocation, the eventual submission of the Kgalagadi groups to the Batswana, and the migration of some of these lesser Tswana groups, notably the baKhurutshe, into the Boteti region. By 1800 Tswana groups had settled the entire eastern hardveld, the Boteti and Molopo corridors and southern Ngamiland.

Similar events were occuring on the middle Zambezi and around the Zambezi–Chobe confluence in the early eighteenth century, where expansion of the Lozi Empire pushed first the baYei, then the haMbukushku groups along the network of waterways to the Okavango, which was entirely settled by 1800. These groups came into contact with the baKgalagadi and baTawana from the south and east, resulting in the formation of a Tawana state in the nineteenth century. Smaller groups came sporadically eastwards from Namibia, ranging from the ovaMbanderu pastoralists who settled around 1500 at the Kunene River, moving on to Ngamiland between 1600 and 1800, to the ovaHerero refugees from the 1904–5 German–Herero war.

The unrest of the 1820s mostly affected the southeastern region, where civil war and the immigration of waves of dispossessed people fragmented the existing communities. It was followed by the re-establishment of Tswana states, and the on-migration, assimilation or subjugation of minority groups, notably the baKgalagadi.

The result of these processes in many ways parallels the events of the first

Figure 8.7. Walled site on Kubu Island, western Sua Pan. Although the provenance of this site is not clear, it may represent the furthest extension of the Great Zimbabwe Tradition into the Kalahari.

millenium; the establishment of cattle-based political units based at the periphery of the Kalahari sandveld, the baTawana state close to its core. However, the populations involved were far larger and their influence was greater, in both penetration and effect, in the desert regions. There was also greater diversity in occupation. Some groups, such as the baYei, introduced new fishing methods to the Okavango, while the haMbukushku peoples relied primarily on fishing and dryland agriculture. In the western Kalahari some Khoisan and ovaMbanderu peoples practised nomadic pastoralism, and the 'bushman' life-style observed among the San by early European travellers may have been a response to pressure from the new migrant groups. The human population had greatly increased; its impact on the environment was about to become substantial.

9 The Kalahari environment in the nineteenth and twentieth centuries

THE PRESENCE of large settled populations in centralised Late Iron Age states provided the impetus for the intensive exploitation of the Kalahari environment, a process which has accelerated to the present day. Exploitation was also fuelled by a demand for Kalahari products outside the region, and the introduction of suitable technologies for their exploitation, both of which came with the arrival of Europeans in southern Africa. External influences have led inevitably to closer links with the world beyond southern Africa, and diversification of the range of resources and services demanded from the Kalahari.

9.1 The nineteenth century: trade, technology and environmental change

Initial resource exploitation during the nineteenth century involved hunting and trading those natural products for which a demand had arisen in the industrialising countries of the northern hemisphere, notably ivory, furs, ostrich feathers and hippo teeth, as well as introducing livestock to an increasing proportion of the land. Both were to benefit from the availability of new technologies. The trade goods received in return included tobacco, sugar, tea, cloth, clothing and a range of manufactured European goods. Above all it involved the trading of firearms and ammunition, which were viewed not only as a means of defence and increasing production but as a source of political power. So great was the apprehension over the trade in firearms in the African states that the Boers in South Africa sought to limit this trade by the Sand River Convention of 1852, to no avail.

Trade expanded rapidly in the middle years of the nineteenth century. The mechanisms of trade and exploitation, however, were not entirely in the hands of European travellers, as is generally believed. Trade routes still followed the networks developed during the Iron Age, with the production and intermediate transfer of game products and cattle still in the hands of separate ethnic groups. As Denbow and Wilmsen (1986: p. 1513) note: 'Relations of production and exchange were thus not strictly bounded by ethnic and linguistic divisions but cut across them.'

9.1.1 The impact of hunting

The reduction in wildlife numbers which accompanied the arrival of Europeans was not unique to the Kalahari but a feature of European expansion through the subcontinent as a whole; as early as 1657 the Dutch in the Cape had sought restrictions in the hunting of elephants, rhino and hippo in view of their declining

212

numbers (Tree, 1989). The difference lies in the rapidity with which the slaughter proceeded. Larger mammals such as the elephant were eliminated over most of the Kalahari in the space of about three decades. Their survival in areas of the Middle Kalahari was greatly dependent on the protection afforded by the tsetse fly. This is because effective hunting relied on the introduction of not only guns but also the horse and the ox-wagon. The former allowed greater mobility and range, the latter permitted the transport of heavy loads. Both horse and ox were liable to sleeping sickness (*nagana* in Swahili or *kotsela* in Setswana) spread by the insect.

The horse was introduced into the region by Europeans and their Khoi assistants. This introduction appears to have been a gradual process: the first horses recorded in Ngamiland were those which accompanied Livingstone in 1849, while Letsholathebe of the baTawana purchased his first animal from the trader Joseph McCabe in 1852 (Tlou, 1985). Today, their use for cattle herding is widespread in Ngamiland. The ox-wagon, too, became the principal means of transport in the nineteenth century, first reaching the Kalahari from the Orange River in about 1806. An estimate of its value can be gained from the exploits of the hunter Gordon Cumming (1850), who passed through the southeastern Kalahari en route to the baNgwaketse kingdom in 1843. On one hunting trip he took 50 000 percussion caps, 16 000 bullets, 400 lb (180 kg) of lead and 500 lb (225 kg) of gunpowder, in addition to other supplies (Tree, 1989), a load clearly not practicable by human porterage. The wagon traffic also tends to be underestimated; A.W. Eriksson, a Swedish compatriot of C.W. Andersson, built up a large hunting and trading empire on the Walvis Bay–Ngamiland route in the 1870s, employing up to 40 Europeans, and with an estimated capital of £200 000 (Tabler, 1973). In 1877 Richard Frewen (Tabler, 1960) noted that the concentration of travellers and their retinues at pans on the Zambezi Road towards the end of the dry season was having a noticeable impact on water, grazing and game availability.

The total number of animals slaughtered in the pursuit of sport and profit is difficult to estimate. Denbow and Wilmsen (1986) suggest that 100 000 lb (45 400 kg) of ivory left the Kalahari during the mid-nineteenth century. It is probable that the figure was greater than this: George Westbeech, trading from Mpandamatenga, is thought to have exported between 20 000 and 30 000 pounds (10 000–14 500 kg) of ivory a year between 1871 and 1875 (Tabler, 1963; Sampson, 1972), representing the slaughter of 500 to 700 elephants annually. The majority of these animals were killed by hunters, mostly 'Cape Coloured', in Westbeech's employ, or by hunting parties sent out by Sepopa of the Barotse. Even though elephant numbers probably declined westwards and southwards into the desert, the scale of slaughter was still awesome in these areas. Hendrik Van Zyl, a Boer settler in the western Kalahari, accounted for 400 elephant in the Ghanzi area in 1877, and the following year, in a party of six guns, killed 103 elephant in a single afternoon at Olifants Pan (Main, 1987). The decline in smaller animals followed a similar pattern. On the basis of the £15 000 worth of wildlife products sold annually by Francis and Clark, traders of Shoshong, Campbell (1980) suggests that the total trade by 1865 amounted to 5000 elephant, 3000 leopard, 3000 lion, 3000 ostriches, 250 000 small fur-bearing animals and about 100 000 meat animals a year. Restriction of European hunting became an inevitable, if belated consequence; for example Khama of the baNgwato had banned the killing of ostrich by Europeans by 1875 (Tabler, 1960).

Campbell and Child (1971) have charted the decline in game numbers as Europeans moved north. The Molopo was traversed at various points between its headwaters and Werda by Bain in 1826 (Lister, 1949), Cornwallis Harris in 1836

(Cornwallis Harris, 1852), Gordon Cumming in 1843 (Cumming, 1850) and Cotton Oswell in 1845 (Oswell, 1900). Their reports suggest that the Molopo, at least in its upper reaches, was a permanent stream of modest dimensions, surrounded by thickets of reeds and long grass. The Nossop, too, contained permanent springs and seasonal flow in its upper reaches (Andersson, 1857). On the Molopo elephant, buffalo, hippo, rhino, waterbuck and reedbuck were present in the 1850s, but are now extinct, while elephant, buffalo and giraffe were common on the Nossop as far south as Twee Rivieren. The giraffe seem to have died out in the area around the turn of the century. Further north, the Letlhakeng mekgacha are described as well-watered, and large animals are known to have penetrated along the Okwa–Mmone drainage system into what is now the Central Kalahari Game Reserve. The decline in game seems to have been accompanied throughout by a diminution of surface water, particularly wetland habitats, and a change in rangeland type (section 9.1.3).

9.1.2 The spread of pastoralism

Increased human populations inevitably led to permanence of settlement, and to the spread of domestic animals. The value of cattle, and, to a lesser extent, of sheep and goats, was enhanced by their transition from a source of domestic wealth to an object of trade. Between 1860 and 1890 some 12 000 head of cattle were exported annually from the northwestern Kalahari alone to the expanding markets of the Cape (Denbow and Wilmsen, 1986).

The increase in herd size led first to a form of transhumance, where cattle were trekked from the home village to areas where grazing or water was plentiful (Campbell, 1981). This, in turn, was replaced by attempts to ensure permanent water supplies by digging wells, usually on pans or in the beds of mekgacha. These had the adverse effects of lowering the local watertable, leading to a diminution of surface springs and pools, and creating concentric rings of range degradation around water points by encouraging permanent cattle populations. Deep, hand-dug wells became common; Farini (1886) recorded a well 100 feet deep at the Nossop–Auob confluence in 1885, and other wells of antiquity have been recorded in the area south of the Makgadikgadi Basin (Hitchcock, 1985).

Control of water points thus became incorporated into the Batswana system of land tenure, with the chiefs (dikgosi) responsible for the allocation of land, including pasture, to the tribal authorities, who then subdivided on the basis of wards. Although grazing lands were subject to the scrutiny of overseers, whose role encompassed the location of cattle posts to prevent overgrazing, the growth in herd size rendered their role ineffective.

Demand for grazing land and provision of watering points led eventually to changes in land use. Hitchcock (1985) notes that provision of wells in the 'Western Sandveld' part of the baNgwato territory led to its designation as grazing area rather than royal hunting area, the usual role of peripheral land. Conflict was also common: serious disagreement broke out in 1886 at Lephepe between the baKwena and baNgwato over water rights, and Frederick Barber lamented the tribulations encountered in negotiating the use of a waterhole near Mmashoro in 1875, an area which was then used as an ostrich hunting ground by baNgwato royalty (Tabler, 1960).

By the end of the nineteenth century cattle were grazed throughout the Kalahari sandveld, save those areas infested by tsetse fly. Even this control was not reliable; the rinderpest epizootic which swept through the region in the 1890s decimated game and livestock, thereby depriving the fly of its diet. By 1900 the tsetse fly had been reduced to small pockets in the riverine areas of the Chobe and Okavango, although its subsequent recovery was swift (Potten, 1976).

9.1.3 Changes in the environment

Major changes occurred in the Kalahari environment in the nineteenth century, not all of which can be directly related to changes in the balance between domestic and wild herbivore populations. The change most frequently noted from travellers' records is the diminution of surface water supplies and the reduction of associated wetland habitats. This became associated in time with the idea of the progressive desiccation of the interior of southern Africa, an idea which persisted well into the twentieth century, and formed the basis for the ambitious schemes of Professor Schwarz (section 1.3.3) in the 1920s. Despite several analyses of travellers' records (e.g. Kokot, 1948) and available meteorological data, the evidence for this desiccation remains ambiguous.

In part this results from great emphasis on the history of Lake Ngami, a goal of European exploratory ambitions in the mid-nineteenth century, which shrank from a perennial lake of some 800 km^2 to an ephemeral body, at most of 250 km^2, between 1850 and 1880, as a result of the drying up of the Thaoge inflow from the Okavango (Shaw, 1985b). Although this has been interpreted in climatic terms, the hydrological change is more likely to be the result of human interference (Shaw, 1985b), or of channel shifts in the upper Okavango (Snowy Mountains, 1987: vol. 2). Nevertheless, the Okavango as a whole underwent a series of hydrological shifts, with considerable reduction in outflow by 1920 (Shaw, 1986), and the shrinking of the wetland habitats of the lower Okavango, Boteti and Savuti Rivers.

Springs feeding pans also dried up. Livingstone (1858a: p. 78) noted on his travels that 'In every salt-pan in the country there is a spring of water to one side.' Large numbers of these ceased to function, such as those at Khakhea, Tshabong and probably at Kang, which had a large population around 1820 (Campbell and Child, 1971). Similar changes occurred in the Molopo and Auob–Nossop valleys, and the central Kalahari mekgacha. In part this may have been caused by well digging at these localities. Equally likely is the destruction of the reedbeds themselves, which were trampled by stock and burned by local inhabitants, both to flush out game and to provide nutrients for the growth of pasturage. The burning of *Phragmites* was practised extensively by the baTawana around Lake Ngami, and was mentioned by a number of European travellers (e.g. Baines, 1864).

A further change coincident at this time was bush encroachment; the spread of unpalatable shrubs and bushes into the grassland (see also Figure 4.10a). Bain noted open grassland stretching for several hours' travel on either side of the Molopo in 1826, and, after passing through a belt of thorn forest, travelled eastwards through similar open country in the headwaters of the Moselebe mokgacha (Campbell and Child, 1971). This area now supports mostly *Acacia/Terminalia* scrub savanna. In the Middle Kalahari a large belt running westwards from the Nata River through the Mababe Depression is now covered in dense Mopane scrub, but was described by Selous (1893) as grassland in the 1870s. Similar processes have been operative in the vicinity of the Boteti River, where dense stands of Mosu bush (*Acacia tortillis*) have replaced riverine forest. Observations based on more recent cattle introductions suggest that the replacement of grassland by scrub and woodland may be completed in as little as 20 years (Child, 1968).

Associated with vegetation changes were shifts in the balance between animal species (Campbell, 1981). Large, economically valuable species with a high dependency on water, such as elephant, rhino and hippo, were the first to disappear. Animals sensitive to grassland quality, such as sable, roan, tsessebe and zebra, once common throughout the southern and eastern Kalahari, became restricted to grasslands north of the Makgadikgadi Basin. In turn, less specialised herbivores, such as buffalo, kudu, impala, springbok and possibly gemsbok, extended their ranges. The balance between springbok and impala is of particular interest in that both these species occupy very similar ecological niches, with the former tending to occupy the drier areas. The springbok has increased in numbers and distribution over the past century, and in particular since the 1920s, when large migrations of this species have taken place. A similar pattern has emerged in the central Kalahari where wildebeest have increased as zebra have disappeared from the area.

Campbell (1981) further notes that, although game were common throughout the Kalahari Desert in the nineteenth century, herd size was restricted to a few tens or hundreds of animals, frequently of mixed populations. There were no reports, even as late as Lieutenant Hodson's (1912) travels, of the massive populations of springbok, wildebeest and hartebeest which occurred between the 1920s and 1970s, sometimes numbering tens of thousands. These herds, migrating in search of food and water, were to die in large numbers once fences were erected (section 9.4.2), and were themselves a symptom of the ecological imbalance that had arisen in the Kalahari. This is in contrast to the region south of the Kalahari, where large migrations of *Trekbokken*, mostly springbok and wildebeest, were reported in the mid-nineteenth century (Cumming, 1850; Tabler, 1960) and were easy targets for hunters.

9.2 The growth of the cattle industry in the colonial period

The end of the nineteenth century saw a low point in the fortunes of the Kalahari. Both game and cattle stocks had been greatly reduced by hunting or disease, the markets for game products had collapsed, efforts to find economic minerals had failed, and the region itself had passed from an active frontier to an impoverished backwater of the British Empire (section 1.3.3). The inhabitants of the region, with the exception of the white settlers in southern and western parts, were forced back to subsistence activities, or became migrant labourers to the mines of South Africa. The cattle industry itself experienced mixed fortunes as a political issue between the Union of South Africa, Southern Rhodesia and the Bechuanaland Protectorate. Nevertheless, cattle became the economic mainstay of the region, and the changes brought about in the Kalahari in the twentieth century are closely linked to the development of the cattle industry.

Following the rinderpest epizootic, which devastated cattle populations throughout most of southern Africa, there was an initial recovery in the beef market associated with the Anglo-Boer War, with Bechuanaland cattle stocks being replenished from breeding herds obtained from the Luzi people of Barotseland, now the Western Province of Zambia, which had been unaffected by the rinderpest outbreak (Prins, 1980; Wood, 1988). Thereafter, however, the Kalahari cattle industry was hindered by many problems. These included the economic recession of the interwar years, the re-expansion of the tsetse fly in the Middle Kalahari, and a series of severe droughts, in particular that of the 1930s. Foot-and-mouth disease also became a serious problem, with outbreaks in 1933–4, 1937, 1944, 1947–9, 1957–8, the 1960s and 1977 (Hitchcock, 1985).

Even the cattle herders of the Upper Zambezi flood plain in the Northern

Kalahari did not go unscathed. The export of cattle from this area had grown considerably after 1896 and by 1910 was averaging nearly 8000 per year (Wood, 1988). In 1916, however, it was halted by a ban imposed by the colonial administration following an epizootic of contagious bovine pleuropneumonia, which reduced the region's cattle numbers from 350 000 in 1915 to 72 000 in 1926 (Wood, 1983). The ban lasted until 1947 when it was lifted following a livestock vaccination programme and the extension of a veterinary cordon fence along the border with Angola. Exports resumed but fell far short of pre-1916 levels, even into the early 1950s (Hellen, 1968). Several factors probably accounted for this (van Horn, 1977; Wood, 1988): for instance, the desire to increase herd sizes before recommencing large-scale exports, a shortage of grazing lands and high wet season cattle mortality due to the demise of drainage systems on the Zambezi flood plain, caused, in turn, by a labour shortage due to out-migration to South African mines. The colonial authorities also showed little interest in the Barotseland cattle industry, preferring instead to promote attempts to raise crop production. By way of contrast, further to the south in the Kalahari core, the administration of the Bechuanaland Protectorate, despite their own financial and administrative shortcomings, still believed that cattle ranching was, scientifically and financially, the only viable land use. Such resources as were available went this way.

The first veterinary officer in the Protectorate was appointed in 1907 (Falconer, 1971), and attempts were made both to improve the quality of the cattle and to extend the available rangeland. Attempts were also made to establish markets for the beef in the face of competition from the Imperial Cold Storage Company, which held the monopoly in beef marketing in southern Africa, and a generally hostile attitude from the South African and Southern Rhodesian governments, who sought to protect the interests of their own, white stock ranchers. The attempts by Colonel Rey, Resident Commissioner of the Protectorate in the 1930s, to establish a market for the Middle Kalahari cattle in the Congo, make fascinating study (Hubbard, 1981; Rey, 1988).

Quality control inevitably involved the prevention and control of foot-and-mouth disease, as cattle could not be exported from infected areas. Outbreaks of the disease were convenient reasons to block the trade; a total embargo was placed on Bechuanaland cattle from 1933 to 1935, and led to frantic efforts to inoculate cattle and to restrict movement of animals. The logical extension of this was the construction of cordon fences. The first of these was erected in 1954 from Maitengwe on the Zimbabwe border, through Dukwe and Tlalamabele to Makoba, and has been followed in the 1960s and 1970s by a series of fences (Figure 9.1) which effectively divide northern and eastern Botswana into a series of large paddocks. These fences have been a source of much environmental concern on the grounds that they disrupt the migration of wild animals, an issue brought to the fore by the massive die-off of wildebeest in the vicinity of the Khukhe Fence in 1962–3 (section 9.4.2).

The extension of rangeland also called for the elimination of tsetse fly in northern Botswana. From the small pockets of infestation present after the rinderpest epizootic, the fly re-established itself rapidly in the Ngamiland and Chobe Districts, increasing the area of infestation in the former by 25 000 km^2 between 1902 and 1962 (Potten, 1976). So acute was the situation by 1942 that cattle were evacuated from the western fringe of the Okavango Delta to Lake Ngami, and plans made to evacuate the Maun–Ngami corridor. A Tsetse Fly Control (TFC) Department was founded in the Bechuanaland Protectorate in 1942, which concentrated initially on the eradication of the game reservoir around Maun, leading to the slaughter of over 60 000 animals, mostly kudu and buffalo, in the following 20 years. Spraying with Dieldrine, and later, Endosulphan, proved more effective, and has been the policy of TFC from 1966 to the present, though experiments are now being conducted with odour-baited traps (Bowles, 1989). As a consequence, tsetse fly is now restricted to small areas within the Okavango Delta.

This policy, while laudible in itself, has had environmental consequnces. The first, and as yet unquantified, is the effect of insecticides upon the Okavango

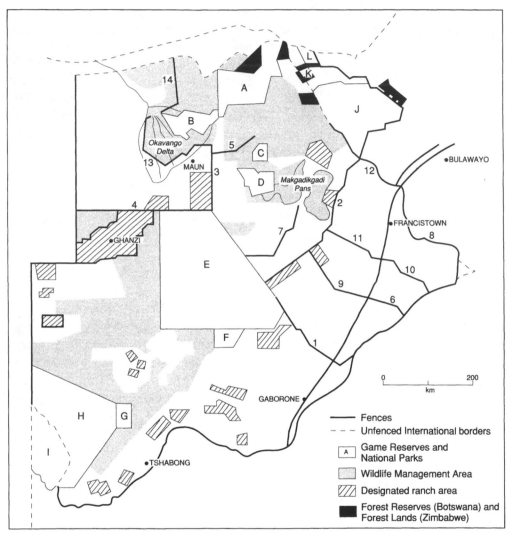

National Parks and Game Reserves	Veterinary Cordon Fences
A Chobe National Park	1 (1954) Dibete Cordon Fence
B Moremi Wildlife Reserve	2 (1955) Central Ngwato
C Nxai Pan National Park	3 (1955) Ngamiland
D Makgadikgadi Pans Game Reserve	4 (1958) Kuke
E Central Kalahari Game Reserve	5 (1968) Shorobe–Chobe
F Kutse Game Reserve	6 (1977) Palapye–Sherwood
G Mabuasehube Game Reserve	7 (1979) Orapa
H Gemsbok National Park	8 (1979) Vakaranga–Tuli
South Africa	9 (1981) Makoro–Makoba
I Kalahari Gemsbok National Park	10 (1981) Serule–Limpopo
Zimbabwe	11 (1982) Serule–Thalamabele
J Hwange National Park	12 (1982) Vakaranga–Tomasarka
K Kazuma Pan National Park	13 (1982) Gomare–Shorobe ('Buffalo Fence')
L Zambezi National Park	14 (1989) Northern Fence

Figure 9.1. Veterinary fences, designated land use areas and proposed Wildlife Management Areas in the Kalahari.

ecosystems. The second is the eradication of the fly in the Okavango to the extent that an invasion of that fragile ecosystem by cattle has become possible, and has been prevented only indirectly by the construction of yet another cordon fence, the so-called 'Buffalo Fence', around the southern and western perimeters of the Delta in the 1980s. This was, in fact, erected to protect cattle, by stopping them coming into contact with buffalo which might carry foot-and-mouth disease, but it has possibly been of even greater benefit to game by effectively halting the advance of large numbers of cattle into the Delta. The policy and cost of replacing natural controls on cattle movement with artificial limitations is questionable, and the environmental consequences are hotly debated.

A more important policy, as far as the sandveld regions are concerned, was the implementation of water point schemes. Even though movement of cattle into the Kalahari had been underway for decades, as late as 1943 it was observed that the majority of Bechuanaland's 1.5–2 million cattle were to be found in hardveld areas (Schapera, 1943). As a response to the drought of the 1930s, a borehole programme was started under government auspices, with an intensification of that programme after 1956. The diesel pump (Figure 9.2) proved to be even more effective than manual labour in the provision of water, and boreholes sprang up throughout the sandveld, each controlling a zone of grazing of 10 km radius or more. The costs of drilling are high, the more so given a high proportion of blank wells, so inevitably the control of these water points became invested in fewer, and richer, individuals. The total number of boreholes in the Kalahari is unknown, but Campbell (1981: p. 114) notes that it is 'very considerable', and that they are ubiquitous outside game reserves.

The enthusiasm of the colonial administration for the cattle industry remained undiminished, despite climatic and economic setbacks. The failure of the large ranches established in 1949 by the Colonial Development Corporation at Nata, and Bushman Pits near the Boteti River, provided lessons concerning environmental deterioration that went unheeded. Hitchcock (1985) notes that, at independence in 1966, Botswana inherited a cattle industry beset by land shortage, ecological deterioration and economic inequality, trends which have continued, through two major droughts, to the present day.

9.3 The winds of change: political independence and economic growth

The 1960s saw the start of a new political era in southern Africa, with the beginnings of the process of territorial independence from colonial powers. This process has inevitably had a significant impact on the Kalahari environment as economic bases have developed and changed in the post-independence period, patterns of land use and ownership have changed, and natural resources have been sought and exploited as part of the quest for economic, as well as political, self-sufficiency.

The impact of these political and economic changes has inevitably been uneven in both time and space. Whereas Zambia and Botswana achieved independence in 1964 and 1966 respectively, it came later in Zimbabwe (1981) and Namibia (1990). This has inevitably resulted in a change of focus. In the colonial era the Bechuanaland Protectorate and adjacent territories were physically and economically peripheral to South Africa (section 1.3.3), and the Kalahari sectors of South West Africa and the Rhodesias peripheral within these territories themselves, as illustrated by the case of Barotseland (Western Province, Zambia) which had demonstrated its agricultural potential in the late nineteenth century but experienced little infrastructural development until after independence (e.g. Wood, 1988). The latter aspect still persists, and is reflected in the high proportion of land given over to freehold farms, forest reserves and game parks in states in which the Kalahari impinges on peripheral areas, types of land use not liable to sudden change. It includes game reserves which have subsequently become important National Parks: Etosha in northern Namibia (established in 1907) and Hwange in western Zimbabwe (1927).

The location of the economic core, however, has shifted. South Africa has

Figure 9.2. A vintage diesel pump, of the type introduced in the early twentieth century, which increased yields from many Kalahari boreholes.

increasingly come to view her northern border, including the Kalahari of Gordonia, as a *cordon sanitaire* against the rest of Africa, exemplified by the fence that runs along the thalweg of the Molopo Valley, and extends along the Namibia–Botswana border on the twentieth and twenty-first lines of longitude. In turn the scale of post-independence change has undoubtedly been greatest in Botswana. This partly reflects the underdevelopment inherited from its protectorate status, but also the generation of wealth from a variety of sources over the past two decades.

The following sections focus on some of these changes, assessing their impacts on the peoples, wildlife and environment of the Botswana Kalahari, and summarising the environmental issues that arise therefrom.

9.3.1 Administrative change

It has been noted by Yeager (1989) that two activities carried out by the colonial administration of Bechuanaland were to have a profound influence on the political economy of Botswana and, by extension, on the Kalahari environment. First, the traditional Batswana cattle culture was subjected to a degree of commercialisation, and second, large areas of land were set aside for the protection of wildlife before and after independence. To these a third can be added, for, as already noted, the last years of colonial control saw the commencement of borehole drilling to supply water in the Kalahari sandveld. Though initially a drought-relief measure, this was also carried out to promote cattle production (Mazonde, 1988) in what Debenham (1952) had envisioned as vast areas of untapped grazing in the central and western areas of the country. These activities indicate that it is not possible to examine the changes which have been effected upon the Kalahari environment since the 1960s without some consideration of certain aspects of the political evolution of post-independence Botswana, since the two have been inexorably linked (Picard, 1980; Parson, 1981; Mazonde, 1988; Yeager, 1989).

Without detailing the intricacies of the administrative and political changes that took place (see Picard, 1980; Hitchcock, 1985, and Yeager, 1989 for summaries), independence saw a restructuring of the role of local decision-makers, chiefs and village headmen with their advisory councils (*dikgotla*), who fulfilled the role of determining land and water allocation and stocking levels. As Yeager (1989) has noted:

> Their purpose was to guarantee that the potentially dangerous mixture of communal land tenure and individual cattle ownership did not create a pastoral 'tragedy of the commons' invoking overgrazing and soil loss.

After independence the role of dikgotla and chiefs was reduced. District Land Boards were established to determine land allocation and, although chiefs sat on these, all decisions which they made now had to be ratified by central government. In environmental terms, this has meant that in areas where cattle keeping has been a traditional activity, local groups with a long-term interest in maintaining the environment have lost much of their control to centralised administration. Inevitably, the latter have major interests in shorter-term national and financial goals, particularly as independence generated the need to participate in a global economic system.

These changes have had their greatest direct effect in the populated areas of eastern Botswana, marginal to the Kalahari. Whereas the Kalahari core is not an area where cattle-keeping has been a traditional activity, it has been increasingly influenced by the effects of both the changes in land allocation procedures and the post-independence goal of boosting cattle productivity, which have resulted in the westward expansion of the cattle industry into the sandveld on an unprecendented scale (Cooke, 1985). The growth has been made possible only by an accelerated programme of borehole drilling, and as such has not, in the main, involved traditional small- and medium-scale cattle herders. Benefit has accrued to an entrepreneurial group which has evolved since independence, largely within the political elite that has grown up simultaneously (Mazonde, 1988; Yeager, 1989).

9.3.2 Economic development

The Kalahari environment has contributed significantly to the economic growth which Botswana has experienced since independence. In 1966 Botswana was one of the poorest nations on earth, partly because of lack of investment during the colonial period and partly because the environment appeared to offer limited agricultural potential and virtually nothing in the way of non-renewable mineral resources. Since independence the transformation has been extraordinary in both its speed and its scale, and has been based on both cattle and minerals, together with infrastructural developments which have expediated the opening up of the Kalahari. The consequential annual per capita income growth of between 8 and 11 per cent in Botswana has consistently been among the world's highest (Yeager, 1989).

Cattle in Botswana numbered about 1.1 million in 1966, after dropping below a million in the early 1960s due to drought (Cooke, 1985). By 1984 this total had grown to over 3 million, and though official agricultural statistics are too aggregated to give precise details (Arntzen and Veenendaal, 1986), it seems than an increasing proportion of them is found in the Kalahari. Despite having a culture in which pastoralism is a traditional pursuit, an increasing proportion of the national herd is owned by a decreasing proportion of the population. It is currently estimated that 45 per cent of rural households in Botswana own no cattle (Arntzen and Veenendaal, 1986) and that 60 per cent of all cattle are owned by 5 per cent of the population (Yeager, 1989). Allowing for population growth, however, it is estimated that the actual number of individuals with livestock (predominantly cattle) using grazing resources rose by 28 per cent in the 11 years prior to 1984.

Rising cattle numbers alone do not contribute to economic growth. The major market for Botswana cattle is within the European Economic Community (EEC), where meat is purchased at heavily subsidised prices for both political and aid purposes. Beef sales to the EEC rose from 6821 tonnes in 1966 to 16 285 tonnes in 1984, but this, in fact, represented a significant drop in the percentage of total beef exports – from 97 per cent to 55 per cent – as markets have been diversified. In environmental terms, the huge growth in the national herd over the same period has not been matched by a rise in the annual percentage off-take, which despite fluctuating around 10 per cent a year over this period (Arntzen and Veenendaal, 1986), fell from 13.5 per cent in 1966 to 10.7 per cent in 1984.

Though cattle, increasingly derived from within the Kalahari, have made an important contribution to economic growth in Botswana, livestock and other agricultural products have formed a decreasing proportion of the Gross Domestic Product since independence, falling from 39 per cent in 1966 to 7 per cent in 1983. Over the same period the contribution of mining activities grew dramatically, from virtually nothing in 1968 to nearly 50 per cent in 1984 (Government of Botswana, 1985a).

Farming activity is almost negligible in the Kalahari areas, with the exception of the Okavango, Boteti and Chobe flood plains. The government has promoted a search for suitable irrigable and dryland farming sites with the aim of achieving self-sufficiency in agricultural production. Investigation of possible sites has indicated little potential for irrigation, a result of erratic water supplies, poor soils and distance from markets. In fact, the arable sector in Botswana as a whole, despite the involvement of two-thirds of the populace, has declined from 0.6 ha per person in 1940 to 0.3 ha per person in 1981 (Arntzen and Veedendaal, 1986), producing as little as 10 000 tonnes of cereals annually during the 1980s drought.

The mining sector is dominated by diamonds, copper/nickel and coal, the first amounting to some 80 per cent of the total mineral value in 1988. In the same year Botswana produced some 15.2 million carats, 21 per cent of the world's diamond output, from three open cast mines located within the Kalahari sandveld, at Orapa, Letlhakane and Jwaneng (Figure 9.3). The diamonds are found in kimberlite pipes located 50 to 100 m beneath the Kalahari Sand surface, with that at Orapa, where mining started in 1971, being one of the world's largest. Further prospecting for diamonds has been in progress in the Central Kalahari since the early 1980s (section 9.3.4).

Reserves of bituminous coal, estimated at being the eighth largest in the world, are found in Karoo strata less than 100 m below a sand cover at Morupule, near Palapye. At present, coal is mined only for the domestic power industry. Construction started in 1988 on a soda ash plant at Sua Pan, to be completed in 1991 at an estimated cost of US$ 370 million. The plant aims to produce 300 000 tonnes of soda ash (sodium carbonate) and 650 000 tonnes of table salt (sodium chloride) per year, sufficient to meet the southern African demand, from a resource which is estimated to be unlimited (Figure 9.4).

The deciduous woodlands of the Kalahari in northeastern Botswana and northwestern Zimbabwe support a small but significant forestry industry, which was somewhat curtailed during the 1920s by the establishment of Hwange National Park. In contrast, a major growth industry is tourism. This is an important earner of foreign currency but its expansion has to be reconciled with the detrimental environmental impacts which it can create if expansion is uncontrolled or unmonitored, especially within fragile and unique environments such as the Okavango Delta. The tourist industry is largely, though by no means exclusively, centred upon a corridor in the Middle Kalahari extending from Victoria Falls and Hwange National Park in Zimbabwe, through Chobe and Okavango to Etosha in Namibia. As much of the activity occurs within National Parks, the infrastructures necessary to control and monitor the industry exist, at least in theory (but see section 9.4.2).

9.3.3 Demographic patterns

In 1904 Hodson (1912: p. 21) estimated the population of the Protectorate, including the Kalahari, to be '120 000 natives and 1000 Europeans'. This had increased to 941 027 by the 1981 census, and an estimated 1.2 million in 1988. Projections for the year 2011 based on present trends suggest a population approaching 3.1 million (Central Statistics Office, 1987; Table 9.1).

Botswana has a relatively small, though unevenly distributed (see Figure 1.4) population, but one that is growing rapidly: with a net increase of 3.6 per cent per annum, it is second only to Kenya on the African continent. Tumkaya (1987) describes the main characteristics of this population: a high proportion (48.3 per cent) of children below the age of 15, a phase of decreasing mortality coupled with increasing fertility, and a disproportionate number of females (sex ratio = 100:89), all factors which contribute to rapid population growth.

Population densities in rural Botswana, which is primarily composed of areas within the Kalahari, have also been increasing despite an overall trend for rural-

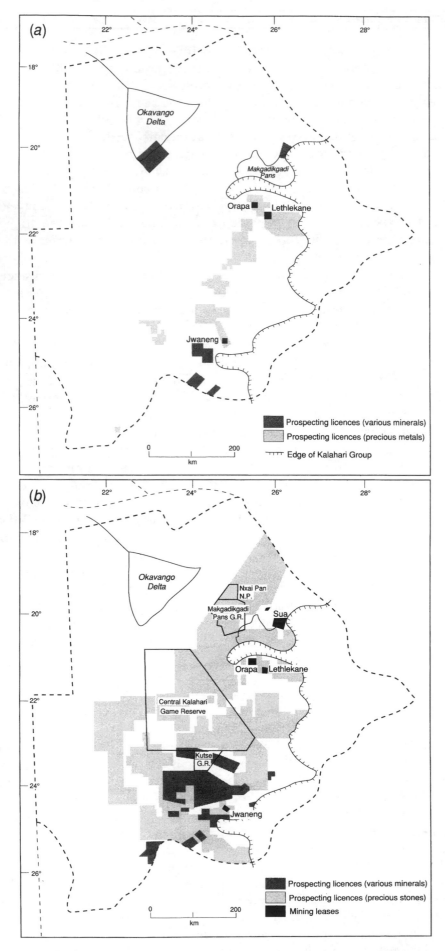

Figure 9.3. The growth of mining leases and prospecting licences in the Botswana Kalahari between (a) 1 January 1985 and (b) 1 July 1989. After Government of Botswana (1985b, 1989).

Figure 9.4. The pilot soda ash plant operated by BP Ltd on Sua Pan in the early 1980s. The new plant, due for completion in 1991 at a cost of US $370 million, will be able to satisfy the regional demand for soda ash and common salt.

urban migration during the last two decades. The latter is a typical developing world phenomenon, with economic advantages perceived to be greater in towns, though improvements in infrastructure, such as roads, clinics, schools and reticulated water supplies, have to some extent encouraged the growth of villages. The overall rural density rose from 0.85 to 1.33 persons per km² between 1964 and 1981, with some marked increases in Administrative Districts covering the Kalahari core (Table 9.2). Tumkaya (1987) notes that unchecked population growth will inevitably bring about deteriorating standards of living and have an adverse impact on the physical environment as the carrying capacity of the land is reached. The necessary educational and health programmes aimed at reducing fertility have been proposed but not pursued with any conviction. Population

Table 9.1. *Returns of population censuses and future population projections for Botswana*

Census year	Enumerated population
1904	120 776
1911	125 350
1921	152 983
1936	265 756
1946	296 310
1956	309 175
1964	514 876
1971	596 944
1981	941 027
1988 (est.)	1 200 000
1991 (est. – high variant)	1 373 000
2001 (est. – high variant)	2 026 000
2011 (est. – high variant)	3 029 000

After Central Statistics Office (1987) and Tumkaya (1987).

Table 9.2. *Population increases in Kalahari-centered administrative districts in Botswana in the 1964, 1971 and 1981 censuses and 2011 projections*

District	1964 Pop'n/Density*	1971 Pop'n/Density*	1981 Pop'n/Density*	2011 Pop'n/Density*
Kweneng	68 106 (1.90)	65 951 (1.84)	117 127 (3.26)	330 538 (9.21)
Ngamiland	41 855 (0.38)	51 323 (0.47)	68 063 (0.62)	124 564 (1.14)
Chobe	4 982 (0.24)	5 097 (0.25)	7 934 (0.38)	25 377 (1.22)
Kgalagadi	15 351 (0.14)	16 337 (0.15)	24 059 (0.22)	69 668 (0.65)
Ghanzi	16 137 (0.14)	16 658 (0.14)	19 096 (0.16)	51 885 (0.44)

Data from Tumkaya (1987).
* Density expressed as persons per square kilometre.

estimates for peri-Kalahari states suggest that the same trends of high population growth rates, urban migration and slower rural increases are present, leading to similar population pressures.

9.3.4 Opening up the Kalahari

An inevitable consequence of economic and demographic growth has been an improvement in the transport and communications infrastructure. At independence, Botswana inherited a single-track railway line, completed in 1897, in the extreme east of the country and one of the world's poorest road systems, with less than 25 km of tarmac surface (Campbell, 1980). The Kalahari was especially poorly served. Since then there has been considerable investment in road improvement programmes. The major villages of the eastern hardveld are now connected by tar roads, with extensions into the eastern Kalahari at Jwaneng, Letlhakeng and Orapa. The main road from Francistown to the Zambezi at Kazangula was completed in 1984, with upgrading of the Tshabong–Bokspits and Nata–Maun–Nokaneng roads and construction of a new road from Jwaneng to Namibia to be completed in the early 1990s.

The provision and upgrading of airstrips has followed at a similar pace. Maun, for example, has one of the busiest airports in southern Africa in terms of take-offs and landings, albeit of light aircraft. A new international airport is under construction at Kasane to serve the growing Victoria Falls–Chobe tourism industry. Telecommunications have also been improved greatly by the introduction of a microwave network.

The improvement and growth of primary links and centres has been obvious; less so has been the opening up of the sandveld by a myriad of tracks and 'cutlines' used for forestry, mineral prospecting and pastoral extension. The provision of veterinary fences, for example, which require constant supervision, has led to a primary network of well-maintained tracks throughout the Middle Kalahari. There are now few areas in the Kalahari which cannot be reached within a few hours given a four-wheel-drive vehicle and the will to travel.

This process has less desirable aspects: in the Central Kalahari Game Reserve diamond exploration has left in its wake 2500 km of temporary cutlines and tracks (Murray, 1988), opening up previously inaccessible areas to depredations by poachers. Recommendations to ensure reclamation and restoration by mining companies, easily enforceable through the granting of prospecting leases, have not yet been implemented.

9.4 Major environmental issues

The expansion of economic activity in the Kalahari has led, eventually, to a realisation that it forms a finite resource which can be managed only by the principles of sustainable development. This is increasingly emphasised in both public and government forums (Kalahari Conservation Society, 1983; Botswana Society, 1984, 1986), and will be enshrined in the proposed National Conservation Strategy for Botswana. The acceptance of these principles was summarised in the 1989 Budget Speech by the Vice-President of Botswana, P.S. Mmusi (1989, p. 2):

> Concerning the environment, it is vital that our development be sustainable. Botswana has only a finite quantity of groundwater, grazing land and forests. If we use up these resources faster than nature can replenish them, production that relies on them will be lower in future years, lowering the standard of living of our sons and daughters.

The acceptance of the principle of sustainable development is a major step forward. Whether the political will exists to carry it through, in the face of growing economic demands on the environment, is another matter. In theory, cattle ranching is an activity based on a renewable resource, which, if carefully managed, should allow the environment to be economically productive. In practice, the management and conduct of the cattle industry suggests that it is primarily exploitative (section 9.4.1).

Alongside this exploitation, efforts have been made to conserve the environment and wildlife through the creation of National Parks and Wildlife Reserves which cover, in terms of percentage land area, far larger areas than in most 'developed' nations. In Botswana virtually all of these lie within the Kalahari (Figure 9.1), while others of great importance are found in neighbouring peri-Kalahari territories including Hwange and Kazuma in Zimbabwe, Etosha in Namibia and the Kalahari Gemsbok Park in the Northern Cape. Although these reserves should justify their existence in economic as well as conservational terms, through income from tourism and wildlife utilisation, none of them is a complete ecological unit, as noted in section 1.3.4. Problems therefore arise in their conservation and management, and in the utilisation of the wildlife within and without their borders (sections 9.4.2 and 9.4.3).

The environmental picture is further complicated by the activities of the mining sector. Commercial mining has only recently entered the Kalahari, and its environmental impact to date has been localised (Campbell and Cooke, 1984, p. 19; Arntzen and Veenendaal, 1986). Given the mineral potential of the area, further expansion and environmental impacts are inevitable, quite simply because developing countries tend to give a high priority to mining as a form of economic land use in order to generate foreign currency, creating conflict with other possible types of utilisation. For example, the frenzy of diamond prospecting activity in the Kalahari led in 1984 to the Botswana government questioning the status of the Central Kalahari Game Reserve, at 52 000 km² the largest area of reserved land in the Kalahari region and second largest game reserve in the world. This, in turn, led to the rapid instigation of a fact-finding mission to investigate the function of the Reserve (Kalahari Conservation Society, 1984, 1985), and the establishment of a management plan pending a decision on its future (Kalahari Conservation Society, 1988a).

The establishment of mines creates a range of adverse environmental impacts, including pollution, disposal of waste, urbanisation and land use conflict. Whereas these impacts can be minimised within the area of a mining lease, as, for example, in the proposed Sua Pan Soda Ash Project, frequently there is no consideration of impacts on surrounding areas, for which no land use may be gazetted (Shaw, 1989b).

9.4.1 The impact of cattle ranching in the Kalahari

Several large-scale projects have been implemented since 1970 in an attempt to expand the utilisation and exploitation of the Kalahari Desert as a major grazing resource, not only in order to facilitate the growth of the Botswana cattle industry, but also to relieve environmental pressures in the eastern hardveld. The Livestock Development Project (LDP 1), which commenced in 1972, was the first attempt to formalise the expansion of the cattle industry into the Kalahari. It aimed to relieve pressure on communal lands in the east by encouraging the owners of large herds to relocate their activities to state-owned lands in the western sandveld (Odell, 1980; Cooke, 1985), where 30 ranches demarcated for breeding and 10 for 'stock growing' were to be established with new boreholes. Access to markets and the abattoir at Lobatse was facilitated by a series of finishing ranches and improvements to a network of cattle trek routes. The project also aimed to expand sheep farming in the Kalahari, but this, by comparison, has only been carried out to a limited extent in the Kgalagadi District.

LDP 1 was beset by considerable logistical, economic and technical problems (Odell, 1980) and in 1975 was subsumed into the even more ambitious and wide-ranging Tribal Grazing Lands Policy (TGLP). This project was multi-faceted in its aims, setting out to overcome social inequalities by extending the benefits of the cattle industry to a wider sector of the population while at the same time tackling the environmental problems associated with ever-growing cattle numbers and densities in the east of the country. It was also the aim of the project to establish better management practices. Overall, therefore, TGLP aimed to allow effective exploitation of the environment at the same time as implementing a degree of conservation by reducing environmental pressures in the most utilised areas.

The mainstay of TGLP was a major reform of land tenure, with all tribal land, i.e. that which was neither state land nor privately owned, totalling about 75 per cent of the country, being placed in one of three categories (Cooke, 1985). These were commercial, containing leasehold ranches to be rented by major cattle owners (Figure 9.1), communal, to be utilised in the traditional manner, and finally reserved land, to be held over for future use. The commercial lands were primarily established in the Kalahari Desert where, under LDP 2 which was set up in 1977, 100 fenced cattle ranches were to be created by 1981 at a cost of US$13 million. By 1984, 218 borehole-dependent ranches had, in fact, been established, most with an area of 6400 hectares but some larger (McLeod, 1986).

TGLP has been widely criticised because of its detrimental social (Hitchcock, 1985; Mazonde, 1988; Yeager, 1989) and environmental (Cooke, 1985; McLeod, 1986) consequences, which have occurred because the project was based on a number of erroneous assumptions and miscalculations. As Cooke (1985: p. 81) has noted:

> If one single misconception can be singled out as critical it was the belief that there were empty blocks of grazing in the Kalahari sandveld, where commercial ranches could be established.

Such blocks were absent from a social viewpoint because the area was already occupied by Khoisan hunter–gatherer groups, at the low densities necessary for their life-style, and in some cases by small groups of cattle herders dependent on borehole-supplied cattle posts (Hitchcock, 1977, 1978; Cooke, 1985). These groups are known collectively by the non-ethnocentric acronym RADs, or Remote Area Dwellers. The designation and installation of ranches has therefore to some degree exacerbated social inequalities in the Kalahari and, with applications for ranch leaseholds restricted to owners with 400 or more head of cattle, the commercial benefits of the scheme were available only to those who were already wealthy (Mazonde, 1988).

From an environmental perspective, a major issue is just how sustainable cattle ranching is in the Kalahari. Verbeck (1968) noted that livestock production in semi-arid areas is beset by a number of problems and limitations including water shortages and the poor quality of grazing. The quality of Kalahari grazing is no exception in this respect, with Skarpe and Bergstrøm (1986) reporting that the grasses are deficient in all the essential nutrients required by livestock with the exception of calcium. Once reliance is placed upon grazing in one particular area on a long-term basis, an inherent characteristic of pastoralism in the Kalahari which relies all year round on water points (boreholes), rangeland degradation not only becomes a possibility but a reality (Figure 9.5). As previously noted (section 4.5.2), this leads to the eradication of the most palatable grass species and ultimately to bush encroachment (e.g. Tolsma et al., 1986; Thomas, 1988e).

In many instances this has been exacerbated by the absence of modern management practices so that in some areas, for example the Ncojane ranches established under LDP 1, the fenced farms were almost completely overgrazed in less than 10 years (Cooke, 1985). To some extent, this has occurred because official stocking rates were either not established, not implemented or not monitored. In other cases, where official stocking levels have been reached, ranchers have grazed their excess cattle on neighbouring communal lands, creating further and more widespread pressures on grazing resources (Yeager, 1989).

One important issue to consider is the potential carrying capacity (PCC) of the Kalahari environment, which represents the volume of livestock which can be supported on a sustainable basis without a reduction in range quality. The PCC of the Kalahari has been estimated using two approaches, mean annual rainfall (Field, 1977) and a basket of environmental parameters including rainfall, evaporation, vegetation type and soil characteristics (Field, 1978; Figure 9.6, Table 9.3). The concept of carrying capacity and its application to arid and semi-arid environments has received criticism, but largely in the context of its

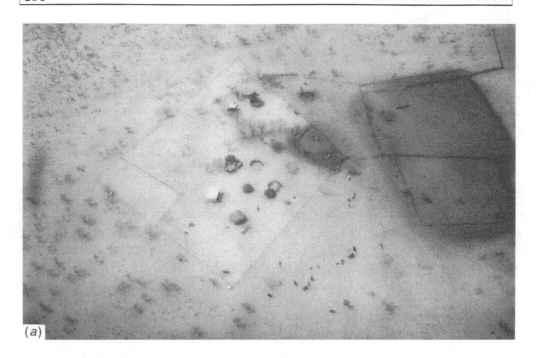

(a)

(b)

(c)

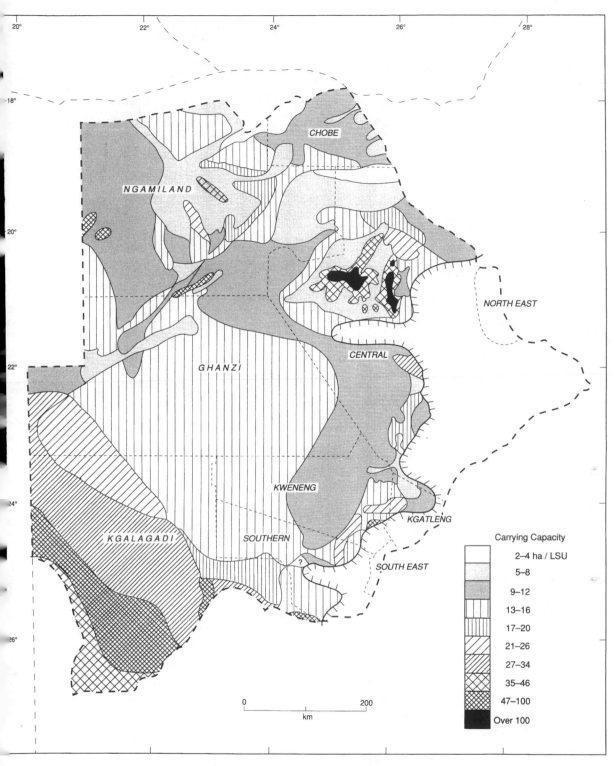

Figure 9.6. *Potential carrying capacities in the Kalahari, expressed as hectares per livestock unit. After Field (1977).*

Figure 9.5. *Environmental degradation occurs increasingly around the point water sources upon which cattle ranching in the Kalahari is based. The aerial photograph (a), taken in September 1988, highlights the degradation and the area of bare ground around a typical borehole development in the central Kalahari. The pump house is the small white-roofed hut towards the top of the photograph, to the right of which is the circular tank into which the water is pumped. The cattle are retained in kraals (fenced enclosures) at night. These show as areas darkened by high concentrations of dung. (b) A similar degraded area around another water point, but in this case the pump (centre) has broken down and the cattle are without water. (c) An older water point in the same area, where bush encroachment has followed the removal of grass species. Photographs by Jeremy Perkins.*

Table 9.3. *Livestock stocking rates and potential carrying capacities (PCC) in parts of the Kalahari*

District	PCC[1,3]	Stocking rates[1,2,4]	
		1980	1984
Southern			
Baralong	12	4.2	6.6
Ngwaketse S	16–21	8.9	15.8
Ngwaketse N	16–21	12.9	10.0
Molopo freehold ranches	12–21	4.7	
Kgalakgadi	21–27	7.0	13.5
Ghanzi Freehold ranches	16–21	9.6	

In all cases stocking rates exceed PCC, i.e. there are insufficient hectares per LSU to allow prolonged use of the environment without considerable degradation.

[1] All expressed as ha/LSU where
     LSU = Livestock unit (450 kg live weight).
[2] Includes all livestock:
     1 cow = 0.7 LSU
     1 goat/sheep = 0.1 LSU
     1 donkey/mule = 0.4 LSU
     1 horse = 0.6 LSU.
[3] From Field (1978).
[4] From Carl Brothers (1982).

application to traditional nomadic pastoralism (Homewood and Rodgers, 1987) rather than the type of borehole-dependent ranching which has expanded in the Kalahari. Though theoretical, the PCC figures may give some indication of the low densities of domestic livestock which the Kalahari is likely to be able to support at sustainable levels. Even though data are highly aggregated, Table 9.3 indicates the considerable disparity between actual stocking levels and PCC values, and how widespread overstocking, in environmental terms, appears to be. Although Skarpe (1981) has shown that cattle are able to browse as well as graze, thereby overcoming grazing shortages due to over-use of the range or to drought, these figures suggest that at current stocking levels the Kalahari is being used as a non-renewable grazing resource, limiting its future utility and, even without the additional impact of drought, causing desertification and long-term environmental degradation.

There is clearly a need for a reduction in cattle numbers, an increased rate of annual stock off-take, which even in times of drought is often less than the cattle mortality rate, and the introduction of better range management practices. The need for such moves has been recognised as the impact of desertification processes has spread, but with considerable short-term gains to be made from the cattle industry, especially with the subsidies which Botswana beef receives on the European market, the incentives to implement reforms are not always apparent, the more so given current patterns of cattle ownership. In this respect, it is interesting to note that in 1983, at the request of the Botswana Government, UNEP (United Nations Environment Programme) produced a report detailing actions needed to improve the condition of the environment through a series of short- and long-term projects (UNEP, 1984). The 15 major proposals contained in the report included four which dealt directly with improving the relationship between livestock and the environment. These called for the establishment of an environmental authority within central government; research into the precise impact of livestock and agriculture on the Kalahari environment; an inventory of the environment and its wildlife resources with proposals on how to promote conservation and the sustainable use of wildlife resources; and research into the restoration of degraded environments (UNEP, 1984). While the majority of the proposals in the report were accepted and have been implemented with foreign aid, these four were rejected (Yeager, 1989).

9.4.2 Wildlife in the Kalahari

The expansion of cattle production into the Kalahari has inevitably affected wildlife populations and generated a number of environmental issues. These include the potential of wildlife utilisation as an alternative to cattle ranching and the direct and indirect impacts of cattle and veterinary fences on wildlife numbers.

On the first point, Murray (1978) has estimated that hunting for the provision of food is a major activity for up to 39 per cent of the population of Botswana, and much of this occurs in the Kalahari, either through licensed tribal hunting or through unlicensed traditional subsistence activities. This probably accounted for about 40,000 animals in 1976, with illegal poaching adding an estimated further 13 per cent to this total (Murray, 1978); these figures are, however, considered to be underestimates by Arntzen and Veenendaal (1986).

Given that indigenous game animals are well adapted to the Kalahari environment, with some species able to survive for long periods without the need for surface water and most species showing grazing and browsing behaviour which is not detrimental to plant communities (Cumming, 1982; Williamson and Williamson, 1983), it has been argued that an alternative strategy to cattle ranching in the Kalahari is to utilise wildlife more widely and effectively (DHV Consulting Engineers, 1980; Cooke, 1985). Additionally, from the point of view of optimising meat production, many wildlife species gain weight faster than cattle and reach maturity more quickly (Johnstone, 1973).

Despite the ravages of nineteenth-century European hunting, the Kalahari still supports a large wildlife population (Cooke, 1985), though it is uneven in distribution and difficult to enumerate. A pre-drought survey by DHV Consulting Engineers (1980) provided estimates of game populations in the Kalahari Desert, together with the numbers which could be harvested annually at sustainable levels, equivalent to 10 per cent or less of the total population for most species (Table 9.4). Again, game ranching may prove an alternative, having proved successful in other parts of Africa, such as Zimbabwe (IUCN, 1988). A recent consultancy (Lawson, 1989) has proposed increased wildlife utilisation in the forms of improved subsistence hunting, game cropping, game ranching and community partnerships with commercial safari and tourist companies, following the Zimbabwe model.

Wildlife utilisation beyond game ranching requires a measure of wildlife protection. Some official recognition has been given to the economic and subsistence potential offered by wildlife in public forums (Kalahari Conservation Society, 1988b) and through the Fauna Conservation Act of 1986 which specifies the establishment of Wildlife Management Areas (WMAs), covering some 20 per cent of Botswana, mostly in the Kalahari (Figure 9.1). Cattle would not be permitted in these areas, where the aim would be to establish rural economies based on the sustainable management of wildlife products and tourism (Carter,

Table 9.4. *Estimated wildlife populations in the Kalahari Desert and possible annual off-take at sustainable levels*
Numbers are based on aerial surveys, which can be very inaccurate.

Species	Population	Possible off-take
Wildebeest	315 058	25 718
Hartebeest	293 462	26 918
Springbok	101 408	9 058
Zebra	100 295	10 029
Ostrich	92 286	5 693
Gemsbok	71 423	5 412
Eland	18 832	1 250
Duiker	6 594	
Kudu	6 429	484
Giraffe	4 406	170
Steenbok	2 122	
Warthog	878	

Data from DHV Consulting Engineers (1980).

1983). In turn, the WMAs would act as corridors for wildlife migration between parks and reserves (Hannah, Wetterberg and Duvall, 1988). At the time of writing, these WMAs have yet to be gazetted.

Opposition to their institution comes from the cattle lobby. As cattle, both ranched and based on traditional cattle posts, are already present in many of these areas, the concept of WMAs has aroused considerable opposition from cattle-owning groups (Yeager, 1989). At the village community level this arises from a long culture of cattle ownership, with hunting as a subsidiary activity, which questions the necessity of WMAs. On the commercial level the profitability of the beef industry is an effective deterrent to change. Thus, both the WMAs and occupations which could benefit from their existence await implementation.

Meanwhile, many of the attributes that make wildlife 'environmentally friendly' have been reduced in their effectiveness as the Kalahari has been opened up and the infrastructure associated with the cattle industry has grown. For example, dispersal and migration are seen as important strategies utilised by many species to reduce competition for grazing and water resources. The construction of fenced ranches and veterinary cordons has obstructed many migration routes (Figure 9.7), preventing animal dispersal during the annual dry season and during droughts, contributing both to the death of individual animals (Thomas, 1986c) and large population groups (Child, 1972; Owens and Owens, 1980; Williamson and Williamson, 1981, 1985) and generating fierce and often emotive debate in international forums.

A recent review of available data (Patterson, 1987) concludes that the construction of artificial barriers, in combination with drought and the expansion of human activity, has had dire consequences for migratory herds. The numbers of animals involved reveal the extent of these catastrophes; Campbell (1981) suggests that up to a quarter of a million animals died in the 1962-3 migrations, while Yeager (1989) reports a further 50 000 wildebeest dying on a single fence in 1983. Campbell (1981) believes the die-offs were not necessarily related to the fences, as the cause of death frequently appears to have been starvation resulting from exhaustion of the grazing resource, in which case the fences merely acted as barriers against which carcasses could accumulate and be easily seen by observers. In support of this contention he cites a similar wildebeest disaster in the 1930s preceding fence construction.

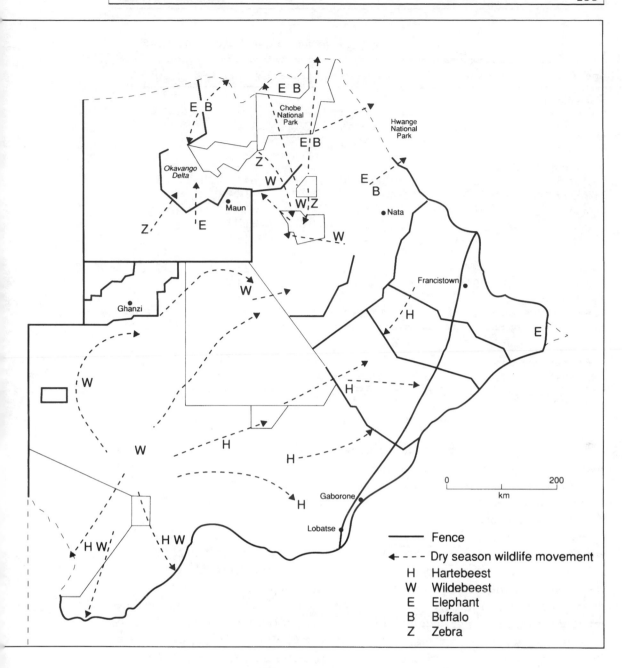

Figure 9.7. Major wildlife migration routes, indicating their disruption by veterinary fences. After Carter (1983).

9.4.3 Wildlife in National Parks

Whereas wildlife populations have been diminishing in many areas of the Kalahari, in others they have been expanding during the twentieth century (Cumming, 1981; Thomas, 1988e), due at least in part to the establishment of wildlife reserves and National Parks and the protected environments which such areas provide, where competing demands for land are excluded. Even the construction of veterinary fences can have some benefits to game (Taylor and Martin, 1987), for example, in preventing a major intrusion of cattle into the Okavango Delta, though it is very doubtful whether these outweigh the substantial disadvantages for game which fences in general create.

Some National Parks have seen staggering increases in the populations of certain animals, of which the elephant is probably the best-known example. This

Table 9.5. *Population estimates for some large mammals in Northern Botswana*

Species	Wet season 1987	Dry season 1987	%change wet–dry	April 1984*
Domestic animals				
Cattle	139 970	140 880	+0.7	
Sheep/goats	27 920	42 250	+51.3	
Donkeys	9 620	7 600	−5.9	
Wild animals				
Buffalo	20 810	133 270	+540.0	35 190
Elephant	50 000	40 530	−18.9	26 121†
Gemsbok	7 120	8 160	+14.6	
Giraffe	6 800	7 570	+11.3	5 201
Hippo	1 780	1 000	−44.0	1 749
Impala	34 050	32 550	−4.4	17 453
Lechwe	28 670	36 320	+26.2	17 703
Tsessebe	10 070	15 580	+54.7	5 501
Wildebeest	11 830	18 300	+51.0	7 218
Zebra	17 760	44 890	+154.0	63 446

After Calef (1988*b*).

* Animal census from aerial survey. The 1984 survey covered 80 000 km, the 1987 census 120 000 km².

† The 1984 elephant count did not include concentration of animals along the Chobe River bank.

is due not only to natural population growth but increasingly to the removal of access to surrounding land, causing a concentration of existing populations in smaller areas (Cumming, 1981). Furthermore, considerable migrations take place within and across park borders in response to grazing and water availability (Calef, 1988*a*; Table 9.5).

Although large indigenous mammals have always interacted with their environment, with some species causing dynamic changes in plant community structures, it is unlikely that large-scale environmental degradation occurred under natural conditions as many plants have efficient defence systems to prevent over-utilisation (Cumming, 1982). The increasing concentration and confinement of the range of large mammals has created situations where natural balances have probably been upset (Figure 9.8). Taylor and Walker (1978) note that in these circumstances overgrazing by wild ungulates can be greater than by domestic livestock because the former are able to feed from a wider range of plants. The problem is exacerbated by the need to provide permanent pumped water sources in many of the Kalahari's National Parks due to the absence of natural all-year-round supplies and the deprivation of access to those outside the parks (Knight, Knight-Eloff and Bornman, 1988; Thomas, 1988*e*). Mitchell (1961) noted that excessive pressure around some boreholes in Hwange National Park had caused the elimination of perennial grasses and created areas of compacted bare ground.

The greatest and most publicised impacts are probably those caused by elephants on tree populations (Anderson and Walker, 1974; Guy, 1976; Cumming, 1982; Lewin, 1986; see section 4.5.2 and Figure 4. 10*b*) and are most noticeable where elephant populations have grown rapidly. In the Hwange National Park elephant increased in numbers from 1000 in 1930 to around 20 000 in 1981 (Cumming, 1981), while the population in Chobe National Park has increased from a few hundred in 1950 to 15 000–20 000 in 1983 (Campbell, 1983) and 55 000 in 1989 (Department of Wildlife and National Parks, quoted in the

Figure 9.8. The increasing confinement of large wildlife species in National Parks and Game Reserves is attractive to tourists, as in this example from Chobe National Park (a), but it can result, (b), in pressure on the environment and severe vegetation disruption, especially around artificial water sources, in this case around a pumped pan in Hwange National Park.

Botswana Guardian, 13/10/89), a current population increase of more than 4000 elephants a year. The two parks together contain more than 10 per cent of the entire remaining African elephant population, as estimated by the Convention on International Trade in Endangered Species (CITES).

Whether the resulting environmental changes are cyclical, and therefore self-correcting (Lewin, 1986; Moss, 1989), or require careful management to prevent permanent environmental damage, has not yet been resolved. In Hwange National Park a policy to regulate elephant numbers by culling has been implemented since 1981 in an attempt to preserve existing vegetation communities and maintain the elephant population at between 12 000 and 14 000 (Cumming, 1983). Culling of 4000 elephant a year in Chobe could also yield US$5 million per annum in income for the park. With recent attempts to place the elephant on Appendix 1 of CITES, thereby banning international trade in elephant products, the question of elephant control has shifted to the arena of international politics.

The increasing confinement of the Kalahari's large wild animals in parks and reserves and the use of such areas for tourism has therefore generated the need for careful management within as well as beyond such areas. The Department of Wildlife and National Parks (DWNP) in Botswana, in contrast with neighbouring countries, is, however, hopelessly undermanned and underresourced, with a ratio of one staff member per 2095 km² of reserve land (Hannah et al., 1988; p. 15), the worst such ratio in Africa. As such, the Department is unable to cope with the basic duties of game and tourist management, and has been identified as a major obstacle to the implementation of satisfactory environmental management strategies (Cooke, 1985; Kalahari Conservation Society, 1988b).

9.4.4 Water resources

Economic growth has inevitably increased the demands for water. In Botswana the primary consumers are livestock (36 per cent), irrigation (35 per cent), mining (12 per cent) urban (12 per cent) and rural (5 per cent) supply (Government of Botswana, 1985a). In the Kalahari areas pastoralism and mining dominate.

With the exception of the Okavango and Chobe networks, nearly all water in the Kalahari is derived from groundwater sources. The historical expansion and increased utilisation of the borehole network has been discussed in section 9.2. This almost uncontrolled sinking of boreholes has been widely criticised as a cause of range degradation for, as noted by Yeager (1989), even a low-output pumping station can usually water more cattle than the surrounding land can support. To avoid the potential environmental effects of providing too much water at one point, boreholes should be a minimum of 8 km apart and support no more than 400 head of cattle (Jarman and Butler, 1971); some boreholes in the past actually supported over 2000 head (Martens, 1971).

To meet the growing demands from livestock in the Kalahari some 2200 boreholes were initially provided by the Government under LDP 1, with a similar number funded privately (Fosbrooke, 1971). In 1975 the Government of Botswana applied a moratorium on the sinking of new boreholes in an attempt to apply greater control on the use of water resources and the environment, but the TGLP provided a loophole to the scheme, making the sinking of boreholes permissible on new ranches.

More recent experience of locating water points for mining, wildlife or construction use has indicated that they are soon unofficially appropriated for livestock, suggesting that tight controls are necessary. Nowhere is this better illustrated than in the Central Kalahari Game Reserve, set up in 1961 to preserve both wildlife and the life-style of hunter-gatherers. Here, the institution of 19 water points for game became the foci of large settlements of Remote Area Dwellers, engaged mostly in pastoralism and illegal hunting. The settlement at Cade, on the western boundary of the reserve, attracted over 1000 inhabitants between 1983 and 1986 (Sweet, 1986). To restore ecological balance the resettlement of the RAD populations outside the reserve was recommended, with the closure of water points as a controversial option to discourage sedentariness (Hannah et al., 1988).

The large number of borehole water sources within the Kalahari inevitably raises questions about the ability of supplies to meet future demands, especially as recharge rates are poorly understood (section 3.7). No data have been collected on rates of borehole depletion, but cases of borehole exhaustion were noted in the 1980s drought, notably in the mekgacha of the Kweneng District, and in the Bodibeng area to the west of Lake Ngami, where increasing salinity in wells led to a migration of the ovaHerero and their livestock to the Western fringes of the Okavango Delta.

In recent years attention has again been focused on the perennial rivers of the Middle Kalahari, notably the Okavango, as a water source. Although massive water transfer schemes (Section 1.3.3) are far in the future, the Orapa diamond mine has been taking water from the Okavango via the Boteti River since 1968, and a series of dams are presently being constructed on the Boteti, Nchabe and Thamalakane Rivers to create 'Lake Maun' as the first step in establishing a controlled water supply for the Maun-Orapa corridor. From an environmental perspective, the extraction of water from the base of the delta is less important than other issues affecting the integrity of the Okavango, notably the incursion of cattle, encouraged by the elimination of tsetse fly (section 9.2), the pressure of uncontrolled tourism, and the recent colonisation of parts of the eastern delta by Kariba Weed, Salvinia molesta (Procter, 1983; Smith, 1985; Kalahari Conservation Society, 1986a). However, it does focus attention on the issue of international water rights. Although Botswana has sole rights to the delta below Mohembo, the Okavango River also passes through Angola and Namibia, and the latter already has the capacity to extract water for the Eastern National Water Carrier, a conduit running from the Okavango to the environs of Windhoek via the Grootfontein wellfield (Kalahari Conservation Society, 1986b). This indicates a need for an Okavango Basin Commission to negotiate areas of potential conflict, in much the same way as ZACPLAN, a consultative body set up in 1987 by the Zambezi states and UNEP, oversees the management of the Zambezi catchment.

9.4.5 Climatic variability and economic development

All human activity in the Kalahari is conducted in a region in which the climate varies around a semi-arid mean. Botswana has experienced two major droughts since independence, in the 1960s, and from 1982 to 1987. These droughts appear to form part of a quasi-18-year cycle apparent in southern Africa's rainfall records (section 4.2.4), with prediction of the 1980s drought made many years earlier (Tyson, 1979).

Some of the effects of the last drought have been noted earlier, many accentuated by other linkages in the man–environment chain. The national herd was reduced by one-third from 3 million to 2 million, though at the same time the carrying capacity probably fell even faster (Yeager, 1989). Wildlife was wiped out en masse, with an estimated loss of half a million herbivores overall, and a reduction in the wildebeest population of 90 per cent (Hannah et al., 1988). In agriculture Botswana's output fell to less than 10 per cent of projected yields, requiring the implementation of drought relief schemes to ensure the survival of the rural populace. Wells dried out or became saline, leading to migrations of people and livestock into areas already under stress (Figure 9.9).

Less known, and notoriously difficult to assess, have been the changes to the soil and vegetation resource during these years. An attempt to estimate land degradation (including reduction of grass cover, bush encroachment and initial soil erosion) from the 1984 LANDSAT imagery was made by Ringrose and Mathieson (1987a,b; Figure 9.10), which showed that range degradation was universal throughout the country, verging on desertification, defined as the large scale disappearance of vegetation, prevalent around major villages. Rates of soil erosion have also been difficult to estimate. Although low topography and high soil permeability limit sheet erosion in the Kalahari, bare soils, particularly the fine-textured soils of pans and lake beds, are prone to wind erosion. The loss of 7 cm of topsoil in the space of a few months has been recorded in the western Okavango floodplain (Snowy Mountains Engineering Corporation, 1989; Vol.1).

Given the scale of the drought tragedy, and the emphasis placed on sustainable development, particularly in the livestock industry (e.g.Arntzen and Veenendaal, 1986), the assumption that drought conditions should form the lowest common denominator for land management should be adopted. However, memories are short. By 1989 the national cattle herd had reached pre-drought levels, and the numbers of smallstock (sheep and goats), which normally fluctuate in inverse proportion to cattle during climatic fluctuations, had reached a record level (J. Arntzen, pers. comm).

9.5 The Kalahari in the future

It would be presumptuous in a few lines to assess the future development of an area of the earth's surface as large as the Mega Kalahari, or indeed the Kalahari Desert, itself far larger than the British Isles. Nevertheless, this survey of the environment through science and human interaction warrants some conclusions.

Figure 9.9. Lake Ngami during the drought, August 1982. The cattle, weakened by lack of grazing, drink water to excess, and are unable to extract themselves from the shallow mud. Ngami dried out some months later.

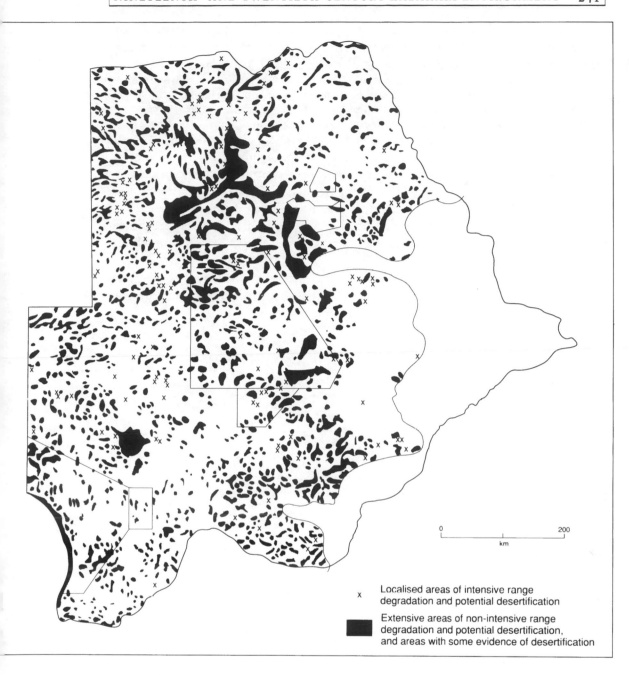

x Localised areas of intensive range
degradation and potential desertification

Extensive areas of non-intensive range
degradation and potential desertification,
and areas with some evidence of desertification

Figure 9.10. Provisional map of overgrazing and desertification in the Botswana
Kalahari in 1984, derived from Landsat 4 data. After Ringrose and Mathieson
(1987a).

The first is that scientific knowledge of the Kalahari has expanded substantially
over the past few decades, yet even as growth occurs, enormous gaps in the
scientific fabric are apparent. In common with other 'wilderness' areas, we are
uncertain what zoological and botanical resources are present in the Kalahari, and
we possess only the most rudimentary ideas as to the basic ecological linkages
which sustain it. This scientific uncertainty extends upwards to the atmospheric
processes, particularly in terms of long- and short-term climatic change, and
downwards into the Kalahari Group sediments upon which the ecosystem is
based.

In turn, the scientific *corpora* support a substantial body of consultative and
advisory reports, from a wide range of organisations, both national and

international. Without exception, these indicate that the present patterns of economic development and activity are not sustainable, that the environment cannot absorb these increasing impacts. Patterson (1987; p.11) summarises:

> It is obvious that the Kalahari is not what it was . . . the days of unlimited free range are gone forever. But with careful planning, regular monitoring and professional standards of management a semi-artificial but viable and extremely valuable Kalahari ecosystem can be maintained through the nineties.

Some factors militate in favour of this. Improvements in the political climate encourage the prospect of co-operation between Kalahari states, while in Botswana economic prosperity, together with the commitment to democratic government and open debate, are optimistic signs. On the other hand, the prevailing traditional attitudes to land, cattle and wildlife, framed in an earlier period of territorial expansion and authoritarian rule, are no longer appropriate to current ecological realities. It is these attitudes among the electorate which generate ambivalence in a democratically elected government. Thus, it will require an even stronger commitment to policies of environmental education, population control, and wildlife and cattle management to reverse the current trends. As Cooke (1985) notes, there is no time to delay.

APPENDICES

Appendix 1 Radiometric dates from palaeoclimatic sites in the Kalahari, including those with geomorphic contexts

Appendix 1A: Radiometric dates from Drotsky's Cave and environs

Cave sinters (from Cooke and Verhagen, 1977; Cooke, 1984)

Lab. no.	^{14}C date	Sample
Wit 309	> 45 000	SII sinter
Wit 310	> 45 000	SII sinter
Wit 308	41 900 ± 5500	SII sinter flowstone
Wit 386	37 000 ± 1700	Repeat
Wit 306	34 000 ± 1700	SII stalagmite
Wit 377	29 300 ± 1000	Repeat
Wit 166	16 000 ± 200	SIII/IV stalagmite
Wit 164	15 200 ± 300	Upper slice SIII/IV stalagmite
Wit 165	15 100 ± 200	Repeat
Wit 168	14 000 ± 200	SIII/IV stalagmite
Wit 170	13 100 ± 200	Upper slice SIII/IV stalagmite
Wit 169	13 000 ± 200	Repeat
Wit 123	2 550 ± 100	SIV stalagmite
Wit 122	2 200 ± 100	Upper slice SIV stalagmite
Wit 195	750 ± 100	SIV stalagmite
Wit 163	Modern	Nodule on stalagmite

Cave sinters (from Shaw and Cooke, 1986)

Lab. no.	^{14}C date	U/TH date	Sample
Hv 13288	16 190 ± 200	19 400 ± 900	Stalagmite 1: sample *a*
Hv 13289	15 600 ± 190	18 000 ± 800	: sample *b*
Hv 13290	14 125 ± 175	15 600 ± 1000	: sample *c*
Hv 13291	5 860 ± 105	6 900 ± 400	Stalagmite 2: sample *a*
Hv 13292	5 060 ± 100	5 800 ± 400	: sample *b*
Hv 13293	3 665 ± 125	4 500 ± 400	: sample *c*
Hv 13294	5 375 ± 100	6 900 ± 400	Stalagmite 3: sample *a*
Pta 4015	5 500 ± 70		: sample *d*
Pta 4021	4 260 ± 60		: sample *c*
Hv 13295	1 185 ± 105	1 800 ± 400	: sample *b*

Appendix 1A: (*cont.*)
Calcretes: Gcwihabe Valley (from Cooke and Verhagen, 1977; Cooke, 1984)

Lab. no.	¹⁴C date	Sample
Wit 313	>45 000	Hard calcrete CII
Wit 385	34 700 ± 2000	Hard calcrete CI
Wit 315	22 700 ± 500	Repeat
Wit 314	11 000 ± 100	Soft calcrete CIV
Wit 376	10 800 ± 100	Repeat
Wit 307	10 000 ± 200	Soft calcrete CIII/CIV
Wit 375	9 800 ± 200	Repeat

Appendix 1B: Radiocarbon dates from the Dobe Valley and Gci Pan, Western Kalahari (from Helgren and Brooks, 1983)

Lab. no.	¹⁴C date	Material	Location
Si 4095	>43 000	Ostrich egg shell	Unit 4
Si 4097	>42 000	Ostrich egg shell	Unit 4
Si 4094	>40 000	Ostrich egg shell	Unit 4
Si 4648	31 470 ± 1010	Calcrete	Unit 3
Si 4090	23 980 ± 590	Calcrete	Unit 2B
Si 4647	22 250 ± 290	Calcrete	Unit 2B
Si 4689	10 255 ± 80	Calcrete	Unit 3
Si 4091c	810 ± 60	Carbonaceous soil	Unit 1A
Si 4091a	495 ± 45	Carbonaceous soil	Unit 1A
Si 4098	110 ± 50	Charcoal	Unit 1A

Appendix 1C: Radiocarbon dates from the Lake Palaeo–Makgadikgadi system

Uncontrolled dates (from Street and Grove, 1976; Heine, 1978b, 1982, 1987)

Lab. no.	¹⁴C date	Material	Location
Hv 8387	31 750 ± 500	Calcrete	?
Hv 8382	30 250 ± 520	Calcrete	Boteti River
Hv 8370	27 350 ± 550	Mollusca	Nata
Hv 8379	27 050 ± 450	Calcrete	?
Hv 8371	25 910 ± 1210	Mollusca	Nata
Hv 8364	24 330 ± 270	Calcrete	Boteti River
Hv 8688	23 750 ± 250	Calcrete	Nata
Gak 4310	20 990 ± 1100	Calcrete	Makgadikgadi
Hv 8365	20 835 ± 100	Calcrete	Boteti River
Hv 8366	19 170 ± 660	Mollusca	?
Hv 8381	14 620 ± 90	Calcrete	Ngami
Hv 8386	14 300 ± 190	Calcrete	Nchabe River
Hv 8380	13 275 ± 110	Calcrete	Boteti River
Hv 8367	11 920 ± 1630	Mollusca	Western Makgadikgadi
Hv 8378	9 390 ± 80	Calcrete	?
Hv 8383	8 720 ± 95	Calcrete	Thamalakane River
Hv ?	4 445 ± 60	Calcrete	?
Hv 8689	4 025 ± 110	Calcrete	Western Makgadikgadi
Hv 8373	2 220 ± 220	Diatom gravel	Nata

Appendix 1C: (cont.)

Surveyed sites (from Cooke, 1984; Cooke and Verstappen, 1984; Helgren, 1984; Shaw, 1985a; Shaw and Cooke, 1986; Shaw et al., 1988; Shaw and Thomas, 1988; Shaw (unpublished))

Lab. no.	^{14}C date	Material	Location	Association
Grn 11156	>52 000	Calcrete	Ngami	945 m shoreline
Grn 11155	46 200 ± 1700	Calcrete	Rysana (Mak)	911 m
Grn 11159	40 200 ± 900	Calcrete	Moremaoto (Boteti)	920 m terrace
Grn 9016	38 200 ± 650	Calcrete	Gidikwe Ridge	945 m shoreline
Grn 9018	35 250 ± 550	Calcrete	Toromoja (Mak)	912 m shoreline
Grn 10369	34 600 ± 1300	Calcrete	Mmatsumo (Mak)	920 m shoreline
Grn 10371	31 800 ± 850	Calcrete	Sua Pan	920 m shoreline
Grn 10372	29 350 ± 650	Calcrete	Sua Spit	912 m shoreline
Grn 10370	29 300 ± 700	Calcrete	Sua Pan	945 m shoreline
Grn 11160	26 590 ± 200	Calcrete	Moremaoto (Boteti)	934 m terrace
Grn 10374	26 050 ± 550	Calcrete	Sua Spit	920 m shoreline
Grn 13712	25 860 ± 200	Calcrete	Tatamoge (Boteti)	932 m terrace
Grn 12627	25 850 ± 500	Calcrete	Goha Pan (Mababe)	932 m shoreline
Grn 9015	25 630 ± 180	Calcrete	Toromoja (Mak)	912 m shoreline
Grn 12294	23 900 ± 550	Calcrete	Setsau (Ngami)	928 m lake floor
Grn 13710	22 540 ± 140	Calc/mollusc	Moremaoto (Mak)	925 m terrace
Grn 10375	21 930 ± 310	Calcrete	Sua Spit	945 m shoreline
Grn 10373	21 920 ± 260	Calcrete	Sua Spit	912 m shoreline
Grn 9019	19 680 ± 100	Calcrete	Tsoi (Mak)	910 m dune base
Grn 9719	19 420 ± 160	Calcrete	Tsoi (Mak)	910 m pan bed
Grn 9677	19 160 ± 250	Peat	Tsoi (Mak)	910 m pan bed
Grn 12627	17 190 ± 210	Calcrete	Ngwezumba (Mababe)	936 m terrace
Grn 14788	15 570 ± 220	Mollusca	Ngwezumba (Mababe)	936 m terrace
Grn 13192	15 380 ± 140	Mollusca	Serondela (Chobe)	933 m terrace
Grn 14787	14 490 ± 150	Mollusca	Okwa Gorge (Mak)	approx. 920 m valley bed
Grn 14786	14 070 ± 150	Mollusca	Gidikwe Ridge	approx. 920 m on ridge
Grn 12623	13 070 ± 140	Calcrete	Ngwezumba (Mababe)	936 m terrace
Grn 15536	11 980 ± 130	Stromatolite	Gidikwe Ridge	approx. 920 m on ridge
Grn 12625	11 950 ± 110	Calcrete	Savuti (Mababe)	934.5 m terrace

Appendix 1C: (cont.)

Surveyed sites (from Cooke, 1984; Cooke and Verstappen, 1984; Helgren, 1984; Shaw, 1985a; Shaw and Cooke, 1986; Shaw et al., 1988; Shaw and Thomas, 1988; Shaw (unpublished)) (cont.)

Lab. no.	^{14}C date	Material	Location	Association
Grn 13191	11 550 ± 110	Calcrete	Serondela (Chobe)	933 m terrace
Grn 13711	10 970 ± 50	Calcrete	Tsogobe (Boteti)	930 m terrace
Grn 12258	10 230 ± 150	Calcrete	Lentswane (Ngami)	943 m
Grn 9017	10 070 ± 60	Calcrete	Rakops (Mak)	920 m shoreline
Grn 13194	9 580 ± 90	Calcrete	Savuti (Mababe)	930 m terrace
Grn 12285	8 920 ± 140	Calcrete	Dautsa Flats (Nga)	928 m lake floor
Grn 11157	6 445 ± 35	Calcrete	Dautsa Ridge (Nga)	approx. 930 m on ridge
Grn 12295	6 440 ± 110	Calcrete	Dautsa Flats (Nga)	931 m lake floor
Si 4646	4 445 ± 60	Calcrete	Gwi Pan (Mak)	912 m terrace
Grn 11161	3 140 ± 50	Calcrete	Xhumo (Mak)	912 m channel
Si 4645	2 960 ± 50	Calcrete	Toromoja (Mak)	912 m terrace
Har 8200	2 290 ± 140	Mollusca	Chobe River	931 m terrace
Grn 12626	2 020 ± 60	Calcrete	Savuti (Mababe)	928 m terrace
Grn 12259	1 970 ± 70	Calcrete	Phatane Gap (Nga)	934 m channel
Grn 13193	1 900 ± 40	Calc. wood	Phatane Gap (Nga)	934 m channel
Grn 9676	1 710 ± 35	Mollusca	Gwi Pan (Mak)	912 m on beach
Grn 9038	1 590 ± 70	Bone	Toromoja (Mak)	912 m on terrace
Grn 12296	1 460 ± 80	Calcrete	Phatane Gap (Nga)	934 m channel
Si 4650	235 ± 370	Bone	Gwi Pan (Mak)	912 m terrace

**Appendix 1D: Radiocarbon dates from Etosha Pan
(from Heine 1979, 1982; Rust, 1984, 1985; Rust *et al.*, 1984)**

Lab. no.	^{14}C date	Material	Location
Pta 3047	46 300 ± 4060	Calcrete	Oshingambo
?	> 44 700	Calcrete	Poacher's Point
Pta 3060	> 43 500	Calcrete	Okatjangee
Pta 3104	> 43 300	Calcrete	Okatjangee
Pta 3038	42 400 ± 1950	Stromatolite	Insel
?	> 42 000	Stromatolite	Poacher's Point
?	> 41 700	Stromatolite	Poacher's Point
Pta 3036	> 40 000	Stromatolite	Insel
Pta 3035	39 300 ± 1470	Stromatolite	Insel
Pta 3066	37 900 ± 1550	Calcrete	Logan's Island
?	37 600 ± 940	Calcrete	Poacher's Point
Pta 3052	33 900 ± 730	Calcrete	Ondongab
Pta 3074	32 500 ± 800	Calcrete	Enguruvau
Pta 3133	32 300 ± 820	Calcrete	Okatjangee
Pta 3100	32 200 ± 770	Calcrete	Insel
Pta 3134	32 000 ± 780	Calcrete	Oshingambo
Pta 3065	31 000 ± 710	Calcrete	Ondongab
Pta 3048	30 300 ± 720	Calcrete	Insel
Pta 3140	28 700 ± 550	Calcrete	Oshingambo
Pta 3058	28 600 ± 480	Calcrete	Enguruvau
Pta 3141	28 200 ± 890	Calcrete	Oshingambo
Pta 3131	26 900	Calcrete	Insel
Pta 3039	22 700 ± 240	Calcrete	Okatjangee
Pta 3059	22 400 ± 240	Calcrete	Insel
Hv 9883	22 250 ± 330	Tufa/mollusc	NW of Homob
Pta 3042	21 400 ± 230	Calcrete	Ondongab
Pta 3053	19 800 ± 180	Calcrete	Insel
Pta 3046	18 100 ± 190	Calcrete	Kapupuhedi Well
Hv 9494	13 680 ± 175	Calcrete	
Hv 9492	12 720 ± 165	Calcrete	
Pta 3041	11 900 ± 120	Calcrete	Kapupuhedi Well
Hv 9493	10 670 ± 465	Mollusca	West Etosha
Pta 3050	10 400 ± 90	Calcrete	Chudop Well
Pta 3040	9 540 ± 100	Calcrete	Kaross
Pta 3043	9 310 ± 90	Sinter	Namutoni Well
Pta 3130	9 210 ± 90	Calcrete	Enguruvau
Pta 3044	3 510 ± 120	Calcrete	Andoni Bay

Appendix 1E: Radiocarbon dates from the pans and rivers of the Southern and Western Kalahari (from Heine, 1978b, 1982; Lancaster, 1979, 1986b, 1989)

Lab. no.	^{14}C date	Material	Location
Pta 4157	33 500 ± 880	Spring tufa	Aminuis Pan
Pta 4153	30 400 ± 600	Spring tufa	Otjimaruru Pan
Pta 4221	23 700 ± 320	Spring tufa	Otjimaruru Pan
Hv 9502	28 000 ± 4900	Organic soil	Auob River
Hv 9885	23 410 ± 980	Stromatolite	Klein Awas Pan
Pta 4128	21 900 ± 260	Spring tufa	Aminuis Pan
Hv 9495	19 085 ± 1125	Mollusca	Molopo (Koopan Suid)
Hv 9501	> 17 540	Mollusca	Molopo River
Hv 9503	16 625 ± 195	Mollusca	Molopo River
Hv 9483	16 425 ± 300	Mollusca	Molopo River
Pta 2150	16 255 ± 700	Stromatolite	Urwi Pan
Pta 2146	16 000 ± 160	Stromatolite	Urwi Pan
Pta 2212	15 790 ± 110	Stromatolite	Urwi Pan

Appendix 1E: (*cont.*)

Lab. no.	^{14}C date	Material	Location
Pta 2213	15 610 ± 125	Stromatolite	Urwi Pan
Hv 8368	15 580 ± 350	Mollusca	Molopo (Koopan Noord)
Hv 9486	13 075 ± 1250	Mollusca	Molopo River
Hv 9485	12 765 ± 475	Mollusca	Molopo River
Hv 8372	12 480 ± 220	Mollusca	Molopo (Koopan Noord)
Pta 4144	12 050 ± 90	Spring tufa	Otjimaruru Pan
Hv 8385	8 705 ± 165	Calcrete	Twee Rivieren
Hv 9499	4 480 ± 290	Mollusca	Auob River
Hv 9500	3 440 ± 540	Mollusca	Auob River
Pta 3865	960 ± 140	Charcoal	Koopan Suid

Appendix 1F: Radiocarbon dates from the Southeastern Margin

The Gaap Escarpment (from Butzer *et al.*, 1978)

Lab. no.	^{14}C date	Material	Location
Si 3350	> 43 000	Tufa	Buxton-Norlim
Si 2040	38 600 ± 2350	Tufa	Mazelsfontein
Si 2034	36 800 ± 1800	Tufa	Gorrokop
Si 3379	32 700 ± 1190	Tufa	Buxton-Norlim
Si 1639	30 760 ± 1035	Tufa	Grootkloof
Si 1647	26 840 ± 520	Tufa	Mazelsfontein
Si 1301	26 130 ± 620	Tufa	Grootkloof
Si 2042	25 900 ± 400	Tufa	Mazelsfontein
Si 2033	24 400 ± 500	Tufa	Gorrokop
Si 2041	22 200 ± 300	Tufa	Mazelsfontein
Si 1640	20 825 ± 230	Tufa	Grootkloof
Si 2035	19 950 ± 300	Tufa	Gorrokop
Si 1642	15 980 ± 230	Tufa	Buxton-Norlim
Si 2043	15 150 ± 165	Tufa	Mazelsfontein
Si 1643	14 010 ± 170	Tufa	Buxton-Norlim
Si 1645	9 550 ± 115	Tufa	Gorrokop
Si 1641	9 310 ± 105	Tufa	Grootkloof
Si 1646	7 715 ± 90	Tufa	Mazelsfontein
Si 2036	3 155 ± 65	Tufa	Gorrokop
Si 1644	2 520 ± 70	Tufa	Gorrokop
Si 2038	1 375 ± 65	Tufa	Gorrokop
Si 2037	1 205 ± 60	Tufa	Gorrokop
Si 2044	1 070 ± 65	Tufa	Mazelsfontein
Si 2045	515 ± 80	Tufa	Mazelsfontein

Kathu Pan (from Beaumont *et al.*, 1984)

Lab. no.	^{14}C date	Material	Location
Pta 3591	32 100 ± 780	Calc. sand	Pit KP 5
Pta 3566	27 500 ± 530	Calc. sand	Pit KP 5
I 13040	26 930 ± 750	Calc. sand	Pit KP 5
Pta 3586	19 800 ± 280	Calc. sand	Pit KP 5
Pta 3073	7 350 ± 90	Peat	Pit KP 2
I 12093	5 980 ± 120	Calc. sand	Pit KP 5
Pta 2518	4 420 ± 60	Peat	Pit KP 2
I 13036	3 620 ± 80	Peat	Pit KP 2

Appendix 1F: (*cont.*)

Lab. no.	¹⁴C date	Material	Location
Pta 3510	2 980 ± 60	Peat	Pit KP 2
Pta 3582	2 690 ± 50	Peat	Pit KP 5
Pta 3504	1 830 ± 50	Peat	Pit KP 2

Equus Cave (from Beaumont *et al.*, 1984)

Lab. no.	¹⁴C date	Material	Location
Pta 2495	7 480 ± 80	Pollen	Strata 1a
Pta 2452	2 390 ± 55	Pollen	Strata 1a

Wonderwerk Cave (from Avery, 1981; Thackeray *et al.*, 1981; Beaumont *et al.*, 1984; Butzer, 1984*a*, *b*; Van Zindren Bakker, 1982*b*)

Lab. no.	¹⁴C date	Location
Pta 3441	12 400 ± 180	
Pta 2786	10 200 ± 90	Unit 4d
Pta 2790	10 000 ± 70	Unit 4d
Pta 2546	9 130 ± 90	Unit 4d
Pta 3366	8 000 ± 80	
Pta 2798	7 430 ± 60	Unit 4c
Pta 2545	5 970 ± 70	Unit 4c
Pta 2544	5 180 ± 70	Unit 4b
Pta 2797	4 890 ± 70	Unit 4a
Pta 2541	4 240 ± 60	Unit 4a
Pta 2785	3 990 ± 60	Unit 3a/3b
Pta 2543	2 910 ± 60	Unit 3
Pta 2542	1 890 ± 50	Unit 3
Pta 2779	1 210 ± 50	Unit 2b

Lobatse Cave II (from Shaw and Cooke, 1986)

Lab. no.	¹⁴C date	Material	Location
Grn 11832	44 700 ± 5100	Sinter	
Grn 11830	> 43 000	Sinter	
Wit 1331	26 060 ± 830	Sinter	Core 3A
Wit 1329	22 600 ± 550	Sinter	Core 1
Grn 11831	18 040 ± 230	Sinter	
Wit 1330	17 660 ± 300	Sinter	Core 2

Appendix 2 Radiocarbon dates from archaeological sites in the Kalahari without geomorphic contexts

(see also Appendices 1B and 1C)

Appendix 2A: The Middle Kalahari (after Denbow and Campbell, 1980, 1986; Denbow and Wilmsen, 1983, 1986; Helgren, 1984; Helgren and Brooks, 1983; Robbins, 1984, 1985; Robbins and Campbell, 1988)

Associations = C – Ceramics, D – Domestic animals, G – Grain cultivation, M – Metallurgy, T – External trade

Site	^{14}C date	Corrected date (AD)	Association
Depression Cave	18 910 ± 180		(LSA)
	13 060 ± 280		(LSA)
	10 900 ± 420		(LSA)
	7 100 ± 90		(LSA)
	3 540 ± 120		(LSA)
	1 860 ± 90	90	C
	470 ± 80	1480	CM
	370 ± 75	1580	CM
	305 ± 75	1645	CM
Drotsky's Cave	12 200 ± 125		(LSA)
Lotshitshi	1 660 ± 100	290	CD
Divuyu	1 400 ± 70	550	CDMT
	1 370 ± 60	580	CDMT
	1 330 ± 60	620	CDMT
	1 330 ± 60	620	CDMT
	1 220 ± 70	730	CDMT
	1 190 ± 70	760	CDMT
Matlapaneng	1 270 ± 80	680	CDGMT
	1 260 ± 60	690	CDGMT
	1 120 ± 110	830	CDGMT
	1 040 ± 50	910	CDGMT
	970 ± 50	980	CDGMT
Nqoma	1 290 ± 60	660	CDGMT
	1 220 ± 70	730	CDGMT
	1 100 ± 80	850	CDGMT
	1 000 ± 60	950	CDGM
	980 ± 60	970	CDGMT
	970 ± 70	980	CDGMT
	970 ± 50	980	CDGMT
	970 ± 50	980	CDGMT
	860 ± 60	1090	CDGM
Caecae	1 230 ± 50	720	CM
	1 150 ± 60	800	CDM
Serondela	1 220 ± 80	730	CD
	800 ± 80	1 150	CDMT
Chobe	1 190 ± 80	760	CD

Appendix 2A: (cont.)

Site	¹⁴C date	Corrected date (AD)	Association
Qogana	1 190 ± 80	760	C
Hippo tooth	1 120 ± 190	830	CD
Gci	810 ± 60	1140	
	495 ± 45	1455	
	110 ± 50	1840	
Qumqoisi	420 ± 50	1530	C
Toteng	400 ± 100	1550	
	< 185	1765	
Karo	360 ± 80	1590	CT
Gwi Pan	235 ± 370	1715	
Kgwebe Hills	195 ± 75	1755	CDGMT
	< 185	1765	CDGMT
Otjiserandu	140 ± 50	1810	CMT
Maun	120 ± 50	1830	CDMT

Appendix 2B: The Eastern Hardveld (after Lepionka, 1978; Denbow, 1980, 1981; Denbow and Campbell, 1980; Kiyaga-Mulindwa, 1983; Denbow and Wilmsen, 1986; Robbins, 1984, Campbell, 1988b, 1989)

Associations = C – Ceramics, D – Domestic animals, G – Grain cultivation, M – Metallurgy, T – External trade

Site	¹⁴C date	Corrected date (AD)	Association
Thamaga	4 510 ± 130	2560 (BC)	
Maunatlala	1 570 ± 140	380	C
Magagarape 1	1 510 ± 140	440	C
	1 340 ± 90	610	C
	1 220 ± 60	730	CG
	1 100 ± 60	850	C
Magagarape 3	1 350 ± 90	600	C
	870 ± 90	1080	C
Magagarape 2	180 ± 60	1770	C
	130 ± 60	1820	C
Baratani	1 430 ± 60	520	C
Letsibogo 15	1 420 ± 60	530	CDGM
Letsibogo 19	1 100 ± 50	800	CDGM
Bisoli	1 340 ± 50	610	CD
	1 240 ± 80	710	CDMT
Taukome	1 265 ± 80	685	CDM
	1 240 ± 80	710	CDMT
	995 ± 75	955	CDMT
Thamaga	1 190 ± 90	760	CD
Moeng	1 185 ± 120	765	CD
	1 007 ± 120	943	CD
	880 ± 80	1070	CD
	795 ± 75	1155	CD
	720 ± 125	1230	CD
Rraserura	1 130 ± 80	820	CDGMT
Thatswane	1 025 ± 80	925	CDGMT
	840 ± 75	1110	CDGMT
Toutswe	990 ± 75	960	CDMT
	860 ± 105	1090	CDMT
	755 ± 75	1195	CDMT
	750 ± .95	1200	CDMT
	645 ± 95	1305	CDMT
	450 ± 95	1500	CDGM

Appendix 2B: (*cont.*)

Site	^{14}C date	Corrected date (AD)	Association
Commando Kop	970 ± 40	980	CDM
	835 ± 55	1115	CDM
Maiphetwane	960 ± 50	990	CDMT
Kgaswe	960 ± 80	990	CDGMT
	940 ± 80	1010	CDGMT
	860 ± 80	1090	CDGMT
Moritsane	855 ± 75	1095	CDMT
Thatswane	840 ± 75	1110	CDGMT
Broadhurst	590 ± 50	1360	CDM
Shashe 3	570 ± 70	1380	CGMT
Shashe 6	200 ± 80	1750	CDGM
Domboshaba	490 ± 50	1460	CDMT
	450 ± 80	1500	CDMT

References

Acres, B., Blair Rains, A., King, R., Lawton, R., Mitchell, A. and Rackham, L. (1985). African dambos: their distribution, characteristics and use. *Zeitschrift für Geomorphologie*, Supplementband **52**: 63–86.

Aldiss, D. (1987a). *The Pre-Cainozoic Geology of the Okwa Valley near Tswaane Borehole*. Botswana Geological Survey Department, Bulletin **44**.

Aldiss, D. (1987b). A record of stone axes near Tswaane borehole, Ghanzi District. *Botswana Notes and Records* **19**: 41–3.

Alison, M.A. (1899). On the origin and formulation of pans. *Transactions of the Geological Society of South Africa* **4**: 159–61.

Anderson, G.D. and Walker, B.H. (1974). Vegetation composition and elephant damage in the Sengwa Wildlife Research Area of Rhodesia. *Journal of the Southern African Wildlife Management Association* **4**: 1–14.

Andersson, C.J. (1857). *Lake Ngami, or Explorations and Discoveries during Four Years' Wanderings in the Wilds of Southwest Africa*. Harper Brothers: New York. (Reprinted C. Struik, Cape Town, 1967.)

Arad, A. (1984). Relationship of salinity of groundwater to recharge in the southern Kalahari Desert. *Journal of Hydrology* **71**: 225–38.

Armstrong, A.L. and Jones, N. (1936). The antiquity of man in Rhodesia as demonstrated by stone implements of the ancient Zambezi gravels south of Victoria Falls. *Journal of the Royal Anthropological Institute* **66**: 331.

Arntzen, J.W. and Veenendaal, E.M. (1986). *A Profile of Environment and Development in Botswana*. IES, Free University: Amsterdam, and NIR: University of Botswana. 172 pp.

Arup-Atkins International (1988). Pandamatenga Development Study. Stage 1 Report. Ministry of Agriculture: Gaborone. Unpublished.

Ash, J.E. and Wasson, R.J. (1983). Vegetation and sand mobility in the Australian desert dunefield. *Zeitschrift für Geomorphologie*, Supplementband **45**: 7–25.

Astle, W.L. (1977). *Land Systems*. Investigation of the Okavango as a primary water resource for Botswana. UNDP/FAO Technical Note **32**.

Avery, D.M. (1981). Holocene micromammalian faunas from the northern Cape Province, South Africa. *South African Journal of Science* **77**: 265–73.

Baillieul, T.A. (1975). A reconnaissance survey of the cover sands in the Republic of Botswana. *Journal of Sedimentary Petrology* **45**: 494–503.

Baillieul, T.A. (1979). The Makgadikgadi pans complex of central Botswana. *Bulletin Geological Society of America* **90**: 133–6.

Baines, T. (1864). *Explorations in Southwest Africa*. Longmans, Green & Co.: London. (Reprinted Gregg Int. Pub., Farnborough, Hants, 1968.)

Baker, V.R. (1980). Some terrestrial analogs to dry valley systems on Mars. *NASA Technical Memorandum* **81776**: 286–8.

Baldock, J.W. (1977). *Resources inventory of Botswana: metallic minerals, mineral fuels and diamonds*. Botswana Geological Survey, Mineral Resources Report 4. 69 pp.

Baldock, J.W., Hepworth, J.V. and Marengwa, B.S. (1976). Gold, base metals and diamonds in Botswana. *Economic Geology* **71**: 139–56.

Balon, E.K. (1971). Replacement of *Alestes imberi* Peters, 1852 by *A. lateralis* Boulenger, 1900 in Lake Kariba, with ecological notes. *Fisheries Resources Bulletin of Zambia* **5**: 119–62.

Balon, E.K. (1974). Fish production of the drainage area and influence of ecosystem changes on fish distribution. In: E.K. Balon and A.G. Coche (eds), *Lake Kariba: a Manmade Tropical Ecosystem in Central Africa*, pp. 459–523. Junk: The Hague.

Barber, F.H. (1910). Is South Africa drying up? *Agricultural Journal of the Cape of Good Hope* **86**: 167–70.

Barker, J.F. (1983a). Towards a biogeography of the Kalahari, Part I. *Botswana Notes and Records* **15**: 85–92.

Barker J.F. (1983b). Towards a biogeography of the Kalahari, Part II. *Botswana Notes and Records* **15**: 93–8.

Beaumont, P.B. (1979). 'A first account of

recent excavations at Wonderwerk Cave.'
Paper presented to the South African Asso-
ciation of Archaeologists, University of Cape
Town, 1979.

Beaumont, P.B. (1986). Where did all the young
men go during 0–18 stage 2? *Palaeoecology of
Africa* 17: 79–93.

Beaumont, P.B., Van Zindren Bakker, E.M. and
Vogel, J.C. (1984). Environmental changes
since 32 000 BP at Kathu Pan, Northern Cape.
In: J.C. Vogel (ed.), *Late Cainozoic Palaeocli-
mates of the Southern Hemisphere*, pp. 329–38.
Balkema: Rotterdam.

Beetz, P.F.W. (1933). The geology of south west
Angola between Cunene and Lunda axis.
*Transactions of the Geological Society of South
Africa* 36: 136–76.

Bell-Cross, G. (1975). The fishes. In: D.W.
Phillipson (ed.), *Mosi-oa-Tunya: a Handbook
to the Victoria Falls*, pp. 185–7. Longmans:
London.

Bergström, R. and Skarpe, C. (1985). Characteristics of the Kalahari sand in western Botswana. *Meddelanden från Växtbiologiska Insti-
tutionen* 1985/1, 19 pp., Uppsala.

Besler, H. (1980). Die Dünen Namib: Entste-
hung und dynamik eines Ergs. *Stuttgarter
Geographisches Studien* 96. 208 pp.

Besler, H. (1983). The response diagram: distinction between aeolian mobility and stability of sands and aeolian residuals by grain
size parameters. *Zeitschrift für Geomorphologie*, Supplementband 45: 287–301.

Besler, H. and Marker, M. (1979). Namib
sandstone: a distinct lithological unit. *Transactions of the Geological Society of South Africa*
82: 155–60.

Beukes, N.J. (1970). Stratigraphy and sedimen-
tology of the Cape sandstone Stage Karoo
System. *Second Gondwanaland Symposium:
Proceedings and Papers*, pp. 321–41. CSIR:
Pretoria.

Bhalotra, Y.P.R. (1985a). *Rainfall Maps of
Botswana*. Department of Meteorological
Services: Gaborone.

Bhalotra, Y.P.R. (1985b). *The Drought of 1981–
85 in Botswana*. Department of Meteorological Services: Gaborone.

Biggs, R.C. (1976). The effects of the seasonal
flood regime on the ecology of Chiefs Island
and the adjacent floodplain system. In *Symposium on the Okavango Delta*, pp. 113–20.
Botswana Society: Gaborone.

Binda, P.L. and Hindred, P.R. (1973). Bimodal
grainsize distributions of some Kalahari type
sand from Zambia. *Sedimentary Geology* 10:
233–7.

Blair Rains, A. (1969). Unpublished draft
report on Central and Southern State Lands,
Botswana. Land Resources Division, Direc-
torate of Overseas Survey: London.

Blair Rains, A. and McKay, A.D. (1968). *The
Northern State Lands, Botswana*. Land
Resource Study 5. Ministry of Overseas
Development: London.

Blair Rains, A. and Yalala, A.M. (1972). *The
Central and Southern State Lands, Botswana*.
Land Resources Study 11. Land Resources
Division, Foreign and Commonwealth

Office, Overseas development Administration: London.

Bolus, H. (1905). Sketch of the floral regions of
South Africa. In: Flint, W. and Chilchrist, J.
(eds), *Science in South Africa*, pp. 198–240.
Maskew Miller: Cape Town.

Bond, G. (1946). The Pleistocene succession
around Bulawayo. *Occasional Papers of the
National Museum of Southern Rhodesia* 12:
104–15.

Bond, G. (1948). The direction of origin of the
Kalahari Sand of Southern Rhodesia. *Geological Magazine* 85: 305–13.

Bond, G. (1957). Quaternary Sands at the
Victoria Falls. In: J.D. Clark (ed.), *Proceedings
of the third Pan-African Congress on Prehistory,
Livingstone, 1955*, pp. 115–22. Chatto &
Windus: London.

Bond, G. (1963). Pleistocene environments in
southern Africa. In: F.C. Howell and F.
Bouliere (eds), *African Ecology and Human
Evolution*, pp. 308–34. Viking Fund Publications in Anthropology. Aldine: Chicago.

Bond, G. (1975). The geology and formation of
the Victoria Falls. In: D.W. Phillipson (ed.),
*Mosi-oa-Tunya – a Handbook of the Victoria
Falls Region*, pp. 19–48. Longman: London.

Bond, G. and Summers, R. (1954). A late
Stillbay hunting camp on the Nata River,
Bechuanaland Protectorate. *South African
Archaeological Bulletin* 9: 89–95.

Boocock, C. and Van Straten, O.J. (1961). *A
Note of the Development of Potable Water
Supplies at Depth in the Central Kalahari*.
Bechuanaland Protectorate Geological
Survey, Records 1957/58: pp. 11–14.

Boocock, C. and Van Straten, O.J. (1962).
Notes on the geology and hydrology of the
central Kalahari region, Bechuanaland Protectorate. *Transactions of the Geological Society
of South Africa* 5: 125–171.

Borchert, G. and Kempe, S. (1985). A Zambezi
Aqueduct. *Mitterlungen Geologische –
Palaeontologische Institute* SCOPE/UNEP sup-
plement 58: 443–57.

Bornhardt, W. (1900). *Zür Oberflächengestaltung
und Geologie Deutsch Ostafrikas*. Reimer:
Berlin.

Bothma, J. du P. and De Graff, G. (1973). A
habitat map of the Kalahari Gemsbok National Park. *Koedoe* 16: 181–8.

Botswana Society (1971). *Sustained Production
from Semi-arid Areas with Particular Reference
to Botswana*. Botswana Society: Gaborone.

Botswana Society (1984). *The Management of the
Botswana Environment*. Proceedings of a
Workshop, Gaborone, 1984.

Botswana Society (1986). *Developing our Environmental Strategy*. Proceedings of a Workshop, Gaborone, 1986.

Botswana Society (1988). *Botswana Society Social
Studies Atlas*. Botswana Society/Esselte Map
Service: Sweden. 49 pp.

Boughey, A.S. (1963). Interaction between animals, vegetation and fire in Southern Rhodesia. *Ohio Journal of Science* 43: 193–209.

Bourliere, F. and Hadley, M.J. (1970). The
ecology of tropical savannas. *Annual Review
of Ecology and Systematics* 1: 49–61.

Bowen, D.Q. (1978). *Quaternary Geology*. Pergamon Press: Oxford. 221 pp.

Bowler, J.M. (1973). Clay dunes: their occurrence, formation and environmental significance. *Earth Science Reviews* 9: 315–38.

Bowles, J. (1989). The use of odour baited traps and targets. *Kalahari Conservation Society Newsletter* 25: 10–11.

Bradshaw, B.F. (1881). Notes on the Chobe River, South Central Africa. *Proceedings of the Royal Geographical Society* 3: 208–13.

Breed C.S. and Grow, T. (1979). Morphology and distribution of dunes in sand seas observed by remote sensing. In: E.D. McKee (ed.), *A Study of Global Sand Seas*. US Geological Survey, Professional Paper 1052: 253–302.

Breed, C.S., Fryberger, S.G., Andrews, S., McCauley, C., Lennartz, F., Gebel, D. and Horstman, K. (1979). Regional studies of sand seas using Landsat (ERTS) imagery. In: E.D. McKee (ed.), *A Study of Global Sand Seas*. US Geological Survey, Professional Paper 1052: 305–97.

Breyer, J.I.E. (1982). Reconnaissance geomorphological terrain classification, lower Boteti region, northern Botswana. *ITC Journal* 192–3: 317–23.

Brind, W.G. (1955). 'The Okavango Delta. Report of the 1951–53 surveys.' Unpublished report to the Resident Commissioner, Mafeking.

Brooks, A.S. (1978). A note on the Late Stone Age features at #Gi: Analogies from historic San hunting practices. *Botswana Notes and Records* 10: 1–3.

Brooks, A.S. and Yellen, J.E. (1977). Archaeological excavations at #Gi: a preliminary report on the first two field seasons. *Botswana Notes and Records* 9: 21–30.

Bruno, S.A. (1985). Pan genesis in the southern Kalahari. In: D.G. Hutchins and A.P. Lynam (eds), *The Proceedings of a Seminar on the Mineral Exploration of the Kalahari*, Botswana Geological Survey, Bulletin 29: 261–77.

Buckley, D.K. (1984). *Groundwater Exploration at Lethlakeng*. Botswana Geological Survey Report GS/DBK/9/84. 17 pp.

Buckley, R. (1981). Parallel dunefield ecosystems – southern Kalahari and central Australia. *Journal of Arid Environments* 4: 287–98.

Buckley, R., Gubb, A. and Wasson, R. (1987a). Parallel dunefield ecosystems: predicted soil nitrogen gradient tested. *Journal of Arid Environments* 12: 105–10.

Buckley, R., Wasson, R. and Gubb, A. (1987b). Phosphorus and potassium status of arid dunefield soils in central Australia and southern Africa, and biogeographic implications. *Journal of Arid Environments* 13: 211–16.

Bull, P.A. (1978). A statistical approach to scanning electron microscope analysis of cave sediments. In: W.B. Whalley (ed.), *Scanning Electron Microscopy in the Study of Sediments*, pp. 201–26. Geobooks: Norwich.

Bultot, F. and Griffiths, J.F. (1972). The equatorial wet zone. In: J.F. Griffiths (ed.), *Climates of Africa. World survey of Climatology*, vol. 10, pp. 259–311. Elsevier: Amsterdam.

Burchell, W. (1822). *Travels in the Interior of Southern Africa*. 2 vols, London.

Burollet, P.F. (1984). Intracratonic and pericratonic basins in Africa. *Sedimentary Geology* 40: 1–11.

Butterworth, J.S. (1982). *The Chemistry of Mogatse Pan, Kgalakgadi District*. Botswana Geological Survey, Report JSB/14/82.

Butzer, K.W. (1974). Palaeoecology of South African Australopithecines: Taung revisited. *Current Anthropology* 15: 367–82.

Butzer, K.W. (1984a). Late Quaternary Palaeoenvironments in South Africa. In: R.G. Klein (ed.), *Southern African Palaeoenvironments and Prehistory*, pp. 1–64. Balkema: Rotterdam.

Butzer, K.W. (1984b). Late Quaternary environments in South Africa. In: J.C. Vogel (ed.), *Late Cainozoic Palaeoclimates of the Southern Hemisphere*, pp. 235–64. Balkema: Rotterdam.

Butzer, K.W., Issac, G.L., Richardson, J.L. and Washbourn-Kamau, C. (1972). Radiocarbon dating of East African lake levels. *Nature* 175: 1069–76.

Butzer, K.W., Fock, G.J., Stuckenrath, R. and Zilch, A. (1973). Palaeohydrology of Late Pleistocene Lake Alexandersfontein, South Africa. *Nature* 243: 328–30.

Butzer, K., Stuckenrath, R., Bruzewicz, A. and Helgren, D. (1978). Late Cenozoic palaeoclimates of the Gaap Escarpment, Kalahari Margin, South Africa. *Quaternary Research* 10: 310–39.

Cahen, L. and Lepersonne, J. (1952). Equivalence entre le système du Kalahari du Congo Belge et les Kalahari Beds d'Afrique australe. *Mémoires du Societé Belge Géologie, Paléontologie et Hydrologie* 8: 1–64.

Cahen, D. and Moeyersons, J. (1977). Subsurface movements of stone artefacts and their implications for the prehistory of Central Africa. *Nature* 266: 812–15.

Cairncross, B., Stanistreet, I.G., McCarthy, T.S., Ellery, W.N., Ellery, K. and Grobicki, T.S.A. (1988). Palaeochannels (stone-rolls) in coal seams: modern analogues from fluvial deposits of the Okavango Delta, Botswana, southern Africa. *Sedimentary Geology* 57: 107–18.

Calef, G. (1988a). Monitoring Botswana's collared elephants. *Kalahari Conservation Society Newsletter* 22: 6–7.

Calef, G. (1988b). *Aerial Census of Large Mammals in Northern Botswana 1988*. Department of Wildlife and National Parks, Botswana. Draft Report.

Campbell, A.C. (1970). Notes on some rock paintings at Savuti. *Botswana Notes and Records* 2: 15–23.

Campbell, A.C. (1980). *The Guide to Botswana* (2nd edn). Winchester Press: Johannesburg. 670 pp.

Campbell, A.C. (1981). A comment on Kalahari wildlife and the Khuke Fence. *Botswana Notes and Records* 13: 111–18.

Campbell, A.C. (1982). Notes on the prehistoric background to 1840. In: R.R. Hitchcock

and M. Smith (eds), *Settlement in Botswana –
The Historical Development of a Human Landscape*, pp. 13–20. Botswana Society/Heine-
mann: Gaborone.

Campbell, A.C. (1983). Traditional wildlife
populations and their utilization. In: *Which
Way Botswana's Wildlife?* pp. 13–19. Kalahari
Conservation Society: Gaborone.

Campbell, A.C. (1988a). 'Archaeological
Impact Assessment – Lower Shashe Dam.'
Unpublished report of the Snowy Mountains
Engineering Corporation/Department of
Water Affairs, Gaborone.

Campbell, A.C. (1988b). 'Archaeological
Review – Letsibogo Reservoir Area.' Unpub-
lished report of Sir M. MacDonald and
Partners/Department of Water Affairs,
Gaborone.

Campbell, A.C. (1988c). 'Archaeological
Impact Assessment – Maun reservoir.'
Unpublished report of the Snowy Mountains
Engineering Corporation/Department of
Water Affairs, Gaborone.

Campbell, A.C. (1989). 'Investigations of Early
Iron Age Settlement and Possible Tswana
Origins in South East Botswana'. Second
Interim report to NORAD, Gaborone.
(Unpublished.)

Campbell, A.C. and Child, G.F. (1971). The
impact of man on the environment of Botswana. *Botswana Notes and Records* **3**: 91–109.

Campbell, A.C. and Cooke, H.J. (eds) (1984).
The Management of Botswana's Environment.
Botswana Society: Gaborone. 49 pp.

Campbell, A.C., Hitchcock, R. and Bryan, M.
(1980). Rock art at Tsodilo, Botswana. *South
African Journal of Science* **76**: 476–8.

Carl Brothers International (1982). *An Evaluation of Livestock Management and Production
in Botswana with Special Reference to Communal Areas.* Final report (3 volumes). Ministry
of Agriculture.

Carter, J.M. (1983). The development of wildlife management areas. In: *Which Way Botswana's Wildlife?* pp. 63–73. Kalahari Conservation Society: Gaborone.

Central Statistics Office (1987). *Analytical
Report of the 1981 Botswana Census.* Government Printer: Gaborone.

Chapman, J. (1886). *Travels in the Interior of
South Africa* (2 vols). Bell & Daldy: London.
(Reprinted A.A. Balkema, Cape Town,
1971.)

Charlesworth, J.R. (1957). *The Quaternary Era*
(2 vols). Edward Arnold: London. 2291 pp.

Child, G.F. (1968). *An Ecological Survey of
North-eastern Botswana* UNDP/FAO: Rome.

Child, G.F. (1972). A wildebeest die-off at Lake
Xau. *Arnoldia Rhodesia* **5**: 1–13.

Claeys, E. (1947). Première étude de sables du
Kalahari du Congo Occidental. *Bulletin du
Société Belge Géologie* **56**: 372–8.

Clark, J.D. (1950). *The Stone Age Cultures of
Northern Rhodesia.* Southern African
Archaeological Society: Claremont.

Clark, J.D. (1970). *The Prehistory of Africa.*
Thames & Hudson: London.

Clark, J.D. (1975). Stone Age man at the
Victoria Falls. In: D.W. Phillipson (ed.),

*Mosi-oa-Tunya – a Handbook of the Victoria
Falls Region*, pp. 28–47. Longman: London.

Clark-Lowes, D.D. and Yeats, A.K. (1977). 'The
Hydrocarbon Prospects of Botswana.' Shell
Coal, Botswana. (Unpublished.)

Clement, A.J. (1967). *The Kalahari and its Lost
City.* Longman: Cape Town.

Clifford, B.E.H. (1928). *Report on a Journey by
Motor Transport from Mahalapye through the
Kalahari Desert, Ghanzi and Ngamiland to the
Victoria Falls.* Government Printer: Pretoria.

Clifford, B.E.H. (1931). Across the great Makar-
ikari Salt Lake. *Geographical Journal* **91**:
234–41.

Coates, J.N.M. (1980). 'The Karoo Sequence in
Botswana.' Geological Survey, Botswana.
(Unpublished.)

Coates, J., Davies. J., Gould, D., Hutchins, D.,
Jones, C., Key, R., Massey, N., Reeves, C.,
Stansfield, G. and Walker, I. (1979). *The
Kalatraverse One Report.* Botswana Geologi-
cal Survey Department, Bulletin 21. 321 pp.

Cochran, P. (1969). 'Rhodesian prospect – Dett
project – Wankie sections.' Dett Prospecting
Project. (Unpublished report.)

Cockcroft, M.J., Wilkinson, M.J. and Tyson,
P.D. (1987). The application of a present-day
climatic model to the Late Quaternary in
southern Africa. *Climatic Change* **10**: 161–91.

Cohen, G. (1974). Stone Age artefacts from
Orapa diamond mine, central Botswana.
Botswana Notes and Records **6**: 1–5.

COHMAP (1988). Climatic changes of the last
18 000 years: observations and model simulations. *Science* **241**: 1043–52.

Cooke, C.K. (1966). Re-appraisal of the industry hitherto known as Proto-Stillbay. *Arnoldia Rhodesia* **2** (22): 1–14.

Cooke, C.K. (1967). A preliminary report on
the Stone Age of the Nata River. *Arnoldia
Rhodesia* **2** (40): 1–10.

Cooke, C.K. (1979). The Stone Age in Botswana: a preliminary survey. *Arnoldia Rhodesia* **8** (27): 1–32.

Cooke, C.K. and Patterson, M.L. (1960a). A
Middle Stone Age open site: Ngamiland,
Bechuanaland Protectorate. *South African
Archaeological Bulletin* **15**: 36–9.

Cooke, C.K. and Patterson, M.L. (1960b). Stone
Age sites: Lake Dow area, Bechuanaland.
South African Archaeological Bulletin **15**:
119–22.

Cooke, H.B.S. (1957). The problem of Quaternary glacio-pluvial correlation in East and
southern Africa. In: J.D. Clark (ed.), *Proceedings of the third Pan-African Congress on
Prehistory, Livingstone, 1955*, pp. 51–5. Chatto
& Windus: London.

Cooke, H.B.S. (1958). Observations relating to
Quaternary environments in East and southern Africa. *Transactions of the Geological
Society of South Africa.* Annexure to vol. **61**.
73 pp.

Cooke, H.J. (1975a). The palaeoclimatic signifi-
cance of caves and adjacent landforms in the
Kalahari of western Ngamiland, Botswana.
Geographical Journal **141**: 430–44.

Cooke, H.J. (1975b). The Lobatse Caves. *Botswana Notes and Records* **7**: 29–33.

Cooke, H.J. (1976). The palaeogeography of the Middle Kalahari of northern Botswana. *Proceedings of the Symposium on the Okavango Delta and its Future Utilisation*, pp. 21–8. Botswana Society: Gaborone.

Cooke, H.J. (1979a). The origin of the Makgadikgadi Pans. *Botswana Notes and Records* 11: 37–42.

Cooke, H.J. (1979b). K. Heine: Radiocarbon chronology of Late Quaternary lakes in the Kalahari, Southern Africa: a discussion. *Catena* 6: 107.

Cooke, H.J. (1979c). Botswana's present climate and the evidence for past change. In: M.T. Hinchley (ed.), *Proceedings of the Symposium on Drought in Botswana*, pp. 53–8. Botswana Society: Gaborone.

Cooke, H.J. (1980). Landform evolution in the context of climatic change and neo-tectonism in the middle Kalahari of northern central Botswana. *Transactions of the Institute of British Geographers* NS 5: 80–99.

Cooke, H.J. (1984). The evidence from northern Botswana of climatic change. In: J. Vogel (ed.), *Late Cenozoic Palaeoclimates of the Southern Hemisphere*. pp. 265–78. Balkema: Rotterdam.

Cooke, H.J. (1985). The Kalahari today: a case of conflict over resource use. *Geographical Journal* 151: 75–85.

Cooke, H.J. and Baillieul, T. (1974). The caves of Ngamiland: an interim report on explorations and fieldwork 1972–74. *Botswana Notes and Records* 6: 147–56.

Cooke, H.J. and Verhagen, B.Th. (1977). The dating of cave development: an example from Botswana. *Proceedings of the Seventh International Speleological Congress* (Sheffield).

Cooke, H.J. and Verstappen, B. Th. (1984). The landforms of the western Makgadikgadi basin of northern Botswana, with a consideration of the chronology of Lake Palaeo-Makgadikgadi. *Zeitschrift für Geomorphologie* NF 28: 1–19.

Cooke, R.U. (1981). Salt weathering in deserts. *Proceedings of the Geological Association* 92: 1–16.

Cornwallis Harris, W. (1852). *Wild Sports of Southern Africa*. London.

Crowell, A.L. and Hitchcock, R. (1978). San ambush hunting in Botswana. *Botswana Notes and Records* 10: 37–51.

Cumming, D.H.M. (1981). The management of elephant and other large mammals in Zimbabwe. In: P.A. Jewell, S. Holt and D. Hart, (eds), *Problems in the Management of Locally Abundant Wild Mammals*, pp. 91–118. Academic Press: New York.

Cumming, D.H.M. (1982). The influence of large herbivores on savanna structure in Africa. In: B.J. Huntley and B.H. Walker (eds), *Ecology of Tropical Savannas*, pp. 217–45. Springer-Verlag: New York.

Cumming, D.H.M. (1983). The decision-making framework with regard to the culling of large mammals in Zimbabwe. *Proceedings, Management of Large Mammals in African Conservation Areas*, pp. 173–86. Haum Educational: Pretoria.

Cumming, R.G. (1850). *Five Years of a Hunter's Life in the Far Interior of South Africa*. John Murray: London.

Dart, R.A. (1926). *Australopithecus africanus*, the man-ape of South Africa. *Nature* 115: 195–9.

Davison, E., 1977. *Wankie. The Story of a Great Game Reserve*. Egal: Salisbury. 211 pp.

Deacon, H.J. (1975). Demography, subsistence and culture during the Acheulian in southern Africa. In: K.W. Butzer and G.L. Issac (eds), *After the Australopithecines*, pp. 543–69. Mouton: The Hague.

Deacon, H.J. (1988). The origins of anatomically modern people and the South African evidence. *Palaeoecology of Africa* 19: 193–200.

Deacon, J. (1984). Later Stone Age people and their descendants in southern Africa. In: R. Klein (ed.), *Southern African Prehistory and Paleoenvironments*, pp. 221–328. Balkema: Rotterdam.

Deacon, J. and Lancaster, N. (1988). *Late Quaternary Palaeoenvironments of Southern Africa*. Oxford Science Publications. 225 pp.

Debenham, F. (1948). Report on the resources of Bechuanaland, Nyasaland, Northern Rhodesia, Tanganyika, Uganda and Kenya. *Colonial Research Publication* 2: 31–9.

Debenham, F. (1952). The Kalahari today. *Geographical Journal* 118: 12–23.

De Dapper, M. (1979a). Le microrelief des surfaces de sommet des plateaux à couverture sableuse aux environs de Kolwezi (Shaba, Zaïre). *Bulletin de la Siciétié Belge de Géologie* 88: 97–104.

De Dapper, M. (1979b). The microrelief of the sand covered plateaux near Kolwezi (Shaba, Zaïre). I The microrelief of the all-over dilungu. *Geo-Eco-Trop* 3: 1–18.

De Dapper, M. (1981a). The microrelief of the sand-covered plateaux near Kolwezi (Shaba, Zaïre). II The microrelief of the crest dilungu. *Geo-Eco-Trop* 5: 1–12.

De Dapper, M. (1981b). Geomorfologishe studie van het plateaucomplex rond kolwezi (Shaba-Zaïre). *Klasse der Wetenschappen* 43, AWLSK Brussel. 203 pp.

De Dapper, M. (1985). Quaternary aridity in the tropics as evidenced from geomorphological research using conventional panchromatic aerial photographs (examples from Peninsular Malaysia and Zaïre). *Bulletin de la Siciétié Belge de Géologie* 94: 199–207.

De Dapper, M. (1988). Geomorphology of the sand-covered plateaux in southern Shaba, Zaïre. In: G.F. Dardis and B.P. Moon (eds), *Geomorphological Studies in Southern Africa*, pp. 115–35. Balkema: Rotterdam.

De Heinzelin, J. (1963). A tentative Paleogeographic map of Neogene Africa. In: F.C. Howell and F. Bouliere (eds), *African Ecology and Human Evolution*, pp. 648–54. Viking Fund Publications in Anthropology. Aldine: Chicago.

De Martonne, E. and Aufrère, L. (1928). Map of internal basin drainage. *Geographical Review* 17: 414.

Denbow, J.R. (1979). *Cenchus cillaris*: an ecological indicator of Iron Age middens using aerial photography in eastern Botswana. *South African Journal of Science* **75**: 405–8.

Denbow, J.R. (1980). Early Iron Age remains from the Tsodilo Hills, northwestern Botswana. *South African Journal of Science* **76**: 474–5.

Denbow, J.R. (1981). Broadhurst, a 14th century A.D. expression of the early Iron Age in southeastern Botswana. *South African Archaeological Bulletin* **36**: 66–74.

Denbow, J.R. (1982). The Toutswe tradition: a study in socio-economic change. In: R.R. Hitchcock and M. Smith (eds), *Settlement in Botswana – The Historical Development of a Human Landscape*, pp. 73–86. Botswana Society/Heinemann: Gaborone.

Denbow, J.R. (1983). 'Iron Age economics: herding, wealth and politics along the fringes of the Kalahari during the Early Iron Age.' Unpublished PhD thesis, University of Indiana, Bloomington.

Denbow, J.R. (1984). Prehistoric herders and foragers of the Kalahari: the evidence for 1500 years of interaction. In: C. Schrire (ed.), *Past and Present in Hunter–Gatherer Studies*, pp. 175–93. Academic Press: Orlando, Florida.

Denbow, J.R. and Campbell, A.C. (1980). National Museum of Botswana: archaeological research programme. *Nyame Akuma* **17**: 3–9.

Denbow, J.R. and Campbell, A.C. (1986). The early stages of food production in southern Africa and some potential linguistic correlations. *Sprache und Geschichte in Afrika* **7**: 83–103.

Denbow, J.R. and Wilmsen, E.N. (1983). Iron Age pastoralist settlements in Botswana. *South African Journal of Science* **79**: 405–7.

Denbow, J. and Wilmsen, E.N. (1986). Advent and course of pastoralism in the Kalahari. *Science* **234**: 1509–15.

De Queiroz, D.X. (1955). Valores médios e annais representativos de elementos climáticos em Angola. *Serviço Meteorologico, Luanda.* 8 pp.

De Queiroz, J.S. (1989). 'The soils of Botswana: a summary account of their distribution and characteristics'. USAID: Gaborone. (Unpublished.)

De Swardt, A.M.J. and Bennet, G. (1974). Structural and physiographic evolution of Natal since the Late Jurassic. *Transactions of the Geological Society of South Africa* **77**: 309–22.

De Vos, A. (1975). *Africa, the Devastated Continent?* Junk: The Hague.

De Vries, J.J. (1984). Holocene depletion and active recharge of the Kalahari groundwaters – a review and an indicative model. *Journal of Hydrology* **70**: 221–32.

De Vries, J.J. and Von Hoyer, M. (1988). Groundwater recharge studies in semi-arid Botswana – a review. In: I. Summers (ed.), *Estimates of Natural Groundwater Recharge*, pp. 339–47. NATO ASI Series C, vol. 222.

De Winter, B., De Winter, M. and Killick, D. (1966). *Sixty-six Transvaal Trees.* National tree list of South Africa.

DHV Consulting Engineers (1980). *Countrywide Animal and Range Assessment Project* (7 vols). European Development Fund and Ministry of Commerce and Industry, Gaborone.

Dincer, T., Hutton, L. and Khupe, B. (1978). Study, using stable isotopes, of flow distribution, surface–groundwater relations and evapotranspiration in the Okavango Swamp, Botswana. *Isotope Hydrology 1978.* International Atomic Energy Authority: Austria.

Dingle, R.V. (1982). Continental margin subsidence: a comparison between the east and west coasts of Africa. In: R.A. Scrutton (ed.), *Dynamics of Passive Margins*, pp. 59–71. Geodynamics series 6. American Geophysical Union: Washington, and Geological Society of America: Boulder.

Dingle, R.V., Siesser, W.G. and Newton, A.R. (1983). *Mesozoic and Tertiary Geology of Southern Africa.* Balkema: Rotterdam. 293 pp.

Dixey, F. (1938). Some observations on the physiolographical development of central and southern Africa. *Transactions of the Geological Society of South Africa* **41**: 113–71.

Dixey, F. (1943). The morphology of the Congo–Zambezi watershed. *South African Geographical Journal* **25**.

Dixey, F. (1945). The geomorphology of Northern Rhodesia. *Transactions of the Geological Society of South Africa* **48**: 9–45.

Dixey, F. (1950). Part 1. Geology. In: J.D. Clark (ed.), *The Stone Age Cultures of Northern Rhodesia*, pp. 9–29. South African Archeological Society.

Dixey, F. (1956a). Some aspects of the geomorphology of central and southern Africa. *Transactions of the Geological Society of South Africa*, annexure to vol. 58.

Dixey, F. (1956b). Water supply problems in arid regions in the British colonies and southern Africa. *Colonial Geology and Mineral Resources* **6**: 307–25.

Dregne, H.E. (1968). Appraisal of research on surface materials of desert environments. In: W.G. McGinnies, B.J. Goldman and P. Paylore (eds), *Deserts of the World. An Appraisal of Research into their Physical and Biological Environments*, pp. 287–377. University of Arizona Press: Tempe.

Du Toit, A.L. (1907). Geological survey of the eastern portion of Griqualand west. *Annual Report of the Geological Commission of the Cape of Good Hope.*

Du Toit, A.L. (1916). Notes on the Karoo system in the southern Kalahari. *Transactions of the Geological Society of South Africa* **19**: 1–13.

Du Toit, A.L. (1926). *Report of the Kalahari Reconnaissance of 1925.* Department of Irrigation: Pretoria.

Du Toit, A.L. (1927). The Kalahari and some of its problems. *South African Journal of Science* **24**: 88–101.

Du Toit, A.L. (1933). Crustal movements as a factor in the evolution of South Africa. *South African Geographical Journal* **16**: 3–20.

Du Toit, A.L. (1954). *The Geology of South Africa* (3rd edn). Oliver and Boyd: Edinburgh.

Ebert, J.I. (1977). *Mobility, Climate and Technological Specialisation in Middle and Late Stone Ages of Southern Africa.* Paper presented at the 42nd annual meeting of the Society of American Archaeology, Louisiana.

Ebert, J.I. (1978). Comparability between Hunter–Gatherer groups in the past and present: Modernisation versus Explanation. *Botswana Notes and Records* 10: 19–26.

Ebert, J.I., Ebert, M.C., Hitchcock, R.K. and Thoma, A. (1976). An Acheulian locality at Serowe, Botswana. *Botswana Notes and Records* 8: 29–37.

Ebert, J.I. and Hitchcock, R.K. (1978). Ancient lake Makgadikgadi, Botswana: mapping, measurement and palaeoclimatic significance. *Palaeoecology of Africa* 10/11: 47–56.

Edmonds, A.C.R. (1976). *Vegetation Map 1:500 000, the Republic of Zambia.* Government Printer: Lusaka.

Ehret, C. (1967). Cattle-keeping and milking in eastern and southern African history: the linguistic evidence. *Journal of African History* 8: 1–17.

Ellenburger, V. (1972). History and pre-history in Botswana. *Botswana Notes and Records* 4: 135–46.

Ellery, K. (1987). 'Wetland plant community composition and successional processes in the Maunachira River system of the Okavango Delta.' Unpublished MSc thesis, University of the Witwatersrand.

Ellis, C.J.R. (1978). 'Exploration of Block "W" Botswana.' Shell Coal Botswana Ltd. (Unpublished.)

Eloff, F.C. (1984). The Kalahari ecosystem. *Koedoe* 27: 11–20.

Elphick, R. (1977). *Kraal and Castle: Khoikhoi and the founding of White South Africa.* Yale University Press: New Haven.

Eriksson, P.G., Nixon, N., Snyman, C.P. and Bothma, J.du P. (1989). Ellipsoidal parabolic dune patches in the southern Kalahari Desert. *Journal of Arid Environments* 16: 111–24.

Eugster, H.P. and Kelts, K. (1983). Lacustrine chemical sediments. In: A.S. Goudie and K. Pye (eds), *Chemical Sediments and Geomorphology*, pp. 321–68. Academic Press: London.

Falconer, J. (1971). History of the Botswana Veterinary Services 1905–1966. *Botswana Notes and Records* 3: 74–8.

Farini, G.A. (1886). *Through the Kalahari Desert.* Sampson Low, Marston, Searie & Rivington: London.

Farr, J., Cheney, C., Baron, J. and Peart, R. (1981). *GS10 Project: Evaluation of Underground Water Resources*, final report. Botswana Geological Survey Department. 292 pp.

Farr, J., Peart, R., Nelisse, C. and Butterworth, J. (1982). *Two Kalahari Pans: a Study of their Morphometry and Evolution.* Botswana Geological Survey, Report GS10/10.

Field, D.I. (1977). 'Potential carrying capacity of rangeland in Botswana.' Unpublished report of the Land Utilisation Division, Ministry of Agriculture, Gaborone.

Field, D.I. (1978). 'A handbook of basic ecology for range management in Botswana.' Unpublished report of the Land Utilisation Division, Ministry of Agriculture, Gaborone.

Flint, R.F. (1959). Pleistocene climates in eastern and southern Africa. *Bulletin of the Geological Society of America* 70: 343–74.

Flint, R.F. and Bond, G. (1968). Pleistocene sand ridges and pans in western Rhodesia. *Bulletin of the Geological Society of America* 79: 299–314.

Folk, R.L. and Ward, W.C. (1957). Brazos River bar – a study in the significance of grainsize parameters. *Journal of Sedimentary Petrology* 27: 3–27.

Fosbrooke, H.A. (1971). Social implications of sustained livestock production in the Kalahari. In: Proceedings of the Conference on Sustained Production from Semi-arid Areas. *Botswana Notes and Records.* Special Edition 1: 181–92.

Foster, J.B. and Dagg, A.I. (1972). Notes on the biology of the giraffe. *East African Wildlife Journal* 10: 1–16.

Foster, S., Bath, A., Farr, J. and Lewis, W. (1982). The likelihood of active groundwater recharge in the Botswana Kalahari. *Journal of Hydrology* 55: 113–36.

Frakes, L.A. and Crowell, J.C. (1970). Late Palaeozoic Glaciation: III Africa exclusive of the Karoo Basin. *Bulletin of the Geological Society of America* 81: 2261–86.

Frommurze, H.F. (1953). Hydrological research in arid and semi-arid areas of the Union of South Africa and Angola. *Arid Zone Programme 41, Revue of Research in Arid Zone Hydrology* UNESCO, Paris: 58–77.

Fryberger, S.G. and Ahlbrandt, T.S. (1979). Mechanism for the formation of aeolian sand seas. *Zeitschrift für Geomorphologie* NF 23: 440–60.

Fryberger, S.G. and Goudie, A.S. (1981). Arid geomorphology. *Progress in Physical Geography* 5: 420–8.

Fryberger, S.G., Schenk, C.J. and Krystinik, L.F. (1988). Stokes surfaces and the effects of near-surface groundwater-table on aeolian deposition. *Sedimentology* 35: 21–41.

Gaitskell, A. (1954). *Report on Mission to Bechuanaland Protectorate to Investigate the Possibilities of Economic Development in the Western Kalahari, 1952.* HMSO: London.

Gardner, R.A.M. (1981). Reddening of dune sands – evidence from south east Asia. *Earth Surface Processes and Landforms* 6: 459–68.

Gibbons, A. St H. (1904). In remotest Barotseland. *Geographical Journal* 26: 444.

Gidskehaug, A., Creer, K.M. and Mitchell, J.G. (1975). Palaeomagnetism and K-Ar ages of the south-west African basalts and their bearing on the time of the initial rifting of the South Atlantic Ocean. *Geophysical Journal of the Royal Astronomical Society* 42: 1–20.

Goodwin, A.J.H. (1926). South African stone implement industries. *South African Journal of Science* 23: 784–8.

Goodwin, A.J.H. and Van Riet Lowe, C. (1929). The Stone Age cultures of South Africa. *Annals of the South African Museum* 27: 1–289.

Goudie, A.S. (1969). Statistical laws and dune ridges in southern Africa. *Geographical Jour-*

nal **135**: 404–6.

Goudie, A.S. (1970). Notes on some major dune types in southern Africa. *South African Geographical Journal* **52**: 93–101.

Goudie, A.S. (1971). 'Calcrete as a Component of Semi-arid Landscapes.' Unpublished PhD thesis. University of Cambridge.

Goudie, A.S. (1973). *Duricrusts in Tropical and Subtropical Landscapes.* Clarendon Press: Oxford. 174 pp.

Goudie, A.S. (1983). Calcrete. In: A.S. Goudie and K. Pye (eds), *Chemical Sediments and Geomorphology*, pp. 93–132. Academic Press: London.

Goudie, A.S. and Pye, K. (1983). *Chemical Sediments and Geomorphology.* Academic Press: London. 437 pp.

Goudie, A.S. and Thomas, D.S.G. (1985). Pans in southern Africa with particular reference to South Africa and Zimbabwe. *Zeitschrift für Geomorphologie* **29**: 1–19.

Goudie, A.S. and Thomas, D.S.G. (1986). Lunette dunes in southern Africa. *Journal of Arid Environments* **10**: 1–12.

Goudie, A.S. and Watson, A. (1984). Rock block monitoring of rapid salt weathering in southern Tunisia. *Earth Surface Processes and Landforms* **9**: 95–8.

Government of Botswana (1985a). *Botswana National Development Plan, 1985/91.* Government Printer: Gaborone.

Government of Botswana (1985b). *Department of Mines, Annual Reports.* Ministry of Mineral Resources and Water Affairs, Gaborone.

Government of Botswana (1989). *Annual Report of the Geological Survey Department for the Year 1988.* Government Printer: Gaborone.

Green, D. (1966). *The Karoo System in Bechuanaland.* Botswana Geological Survey Bulletin **2**. 27 pp.

Greenwood, P.G. and Carruthers, R.M. (1973). *Geophysical Surveys in the Okavango Delta, Botswana.* Geophysical Division, Institute of Geological Science, London, Report **15**. 23 pp.

Greeley, R. and Iversen, T.D. (1985). *Wind as a Geological Process on Earth, Mars, Venus and Titan.* Cambridge University Press. 333 pp.

Gregory, J.W. (1921). *The Rift Valleys and Geology of East Africa.* Seeley, Service & Co.: London.

Grey, D.R.C. (1976). 'The prospecting of the Mopipi area.' Unpublished report of the Anglo-American Corporation.

Grey, D.R.C. and Cooke, H.J. (1977). Some problems in the Quaternary evolution of the landforms of northern Botswana. *Catena* **4**: 123–33.

Grove, A.T. (1969). Landforms and climatic change in the Kalahari and Ngamiland. *Geographical Journal* **135**: 191–212.

Grove, A.T. (1977). The geography of semi-arid lands. *Philosophical Transactions of the Royal Society of London*, series B, **278**: 457–75.

Guy, P.R. (1976). The feeding behaviour of elephant (*Loxidonta africana*) in the Sengwa area, Rhodesia. *South African Journal of Wildlife Research* **6**: 55–63.

Gwosdz, W. and Modisi, M.P. (1983). *The Carbonate Resources of Botswana.* Botswana Geological Survey Mineral Resources Report **6**. 167 pp.

Hamilton, J. (1977). Sr isotope and trace element studies of the Great Dyke and Bushveld mafic phase and their relation to early Proterozoic magma genesis in southern Africa. *Journal of Petrology* **18**: 24–52.

Hancock, P.L. and Williams, G.D. (1986). Neotectonics. *Journal of the Geological Society, London* **143**: 325–6.

Hannah, L., Wetterberg, G. and Duvall, L., (1988). *Botswana Biological Diversity Assessment.* USAID Bureau for Africa.

Hare, F.K. (1977). *Climate and Desertification.* UN Conference on Desertification, A/Conference Proceedings, 74/5. 140 pp.

Harrison, S.P., Metcalfe, S.E., Street-Perrott, F.A., Pittock, A.B., Roberts, C.N. and Salinger, M.J. (1984). A climatic model of the last glacial/interglacial transition based on palaeotemperature and palaeohydrological evidence. In: J. Vogel (ed.), *Late Cainozoic Palaeoclimates of the Southern Hemisphere*, pp. 21–34. Balkema: Rotterdam.

Hastenrath, S. and Kutzbach, J.E. (1983). Palaeoclimatic estimates from water and energy budgets of East African Lakes. *Quaternary Research* **19**: 141–53.

Heath, D.C. (1972). Die geologie van die Sisteem Karoo in die gebied Mariantal-Asab, SuidwesAfrika. *Memoirs, Geological Survey of South Africa* 61. 36 pp.

Heathcote, R.L. (1983). *The Arid Lands: their Use and Abuse.* Longman: London. 323 pp.

Heaton, T.H.E., Talma, A.S. and Vogel, J.C. (1983). Origin and history of nitrate in confined groundwater in the western Kalahari. *Journal of Hydrology* **62**: 243–62.

Hecht, A.D., Barry, R., Fritts, M., Imbrie, J., Kutzbach, J., Mitchell, J. and Savin, S. (1979). Palaeoclimatic research: status and opportunities. *Quaternary Research* **12**: 6–17.

Heemstra, H.H. (1976). *The Vegetation of the Seasonal Swamps.* Investigation of the Okavango as a primary water source for Botswana. UNDP/Fao Technical Note **28**.

Heine, K. (1978a). Jungquartäre Pluviale und Interpluviale in der Kalahari (Sudliches Afrika). *Palaeoecology of Africa* 10: 31–9.

Heine, K. (1978b). Radiocarbon chronology of the Late Quaternary lakes in the Kalahari. *Catena* **5**: 145–9.

Heine, K. (1979). Reply to Cooke's discussion of 'K. Heine: Radiocarbon chronology of the Late Quaternary Lakes in the Kalahari'. *Catena* **6**: 259–66.

Heine, K. (1981). Aride und Pluviale Bedingungen während der letzen Kaltzeit in der Südwest-Kalahari (südliches Afrika). *Zeitschrift für Geomorphologie Supplement* **38**: 1–37.

Heine, K. (1982). The main stages of late Quaternary evolution of the Kalahari region, southern Africa. *Palaeoecology of Africa* **15**: 53–76.

Heine, K. (1987). Zum alter Jungquartärer seespiegelschwankungen in der Mittleren Kalahari, Südliches Afrika. *Palaeoecology of Africa* **18**: 73–101.

Heine, K. (1988a). Re-evaluation of Passarge's (1904) record of Late Quaternary environmental conditions in the Middle Kalahari. Paper presented to the VII SASQUA Symposium, Bloemfontein, April 1987.

Heine, K. (1988b). Southern African palaeoclimates 35–25 ka ago: A preliminary summary. *Palaeoecology of Africa* 19: 305–15.

Helgren, D.M. (1978). Environmental history of the Northwest Kalahari: Preliminary report. *Palaeoecology of Africa* 10: 65–6.

Helgren, D.M. (1979). *River of Diamonds: An Alluvial History of the Lower Vaal Basin, South Africa.* University of Chicago, Department of Geography Research Paper 185.

Helgren, D.M. (1984). Historical geomorphology and geoarchaeology in the southwestern Makgadikgadi Basin, Botswana. *Annals of the Association of American Geographers* 74: 298–307.

Helgren, D.M. and Brooks. A.S. (1983). Geoarchaeology at Gi, a Middle Stone Age and Later Stone Age Site in the Northwest Kalahari. *Journal of Archaeological Science* 10: 181–97.

Hellen, J.A. (1968). Rural economic development in Zambia, 1890–1964. *Afrika-Studien* Ifo-institut für Wirtschaftsforschung München. 297 pp.

Higgins, C.G. (1984). Piping and sapping: development of landforms by groundwater outflow. In: R.G. La Fleur (ed.), *Groundwater as a Geomorphic Agent*, pp. 18–58. Allan & Unwin: London.

Hills, E.S. (1940). The lunette: a new landform of aeolian origin. *Australian Geographer* 3: 1–7.

Hitchcock, R.K. (1977). 'Hunter–gatherers, boreholes and lands: a reconnaissance survey of northern Kweneng.' Unpublished report of the Ministry of Local Government and Lands, Gaborone.

Hitchcock, R.K. (1978). *Kalahari Cattle Posts: A regional study of Hunter–Gatherers, Pastoralists and Agriculturalists in the western Sandveld Region, Central District, Botswana.* Government Printer: Gaborone.

Hitchcock, R.K. (1982). Prehistoric hunter–gatherer adaptations. In: R.R. Hitchcock and M. Smith (eds), *Settlement in Botswana – The Historical Development of a Human Landscape*, pp. 47–65. Botswana Society/Heinemann: Gaborone.

Hitchcock, R.K. (1985). Water, land and livestock: the evolution of tenure and administrative patterns in the grazing areas of Botswana. In: L. Picard (ed.), *The Evolution of Modern Botswana*, pp. 84–121. Rex Collings/University of Nebraska Press.

Hodson, Lieut. A.W. (1912). *Trekking the Great Thirst. Travel and Sport in the Kalahari Desert.* Fisher Unwin: London. 359 pp.

Holmes, A. (1965). *Principles of Physical Geology* (2nd edn). Nelson: London.

Homewood, K. and Rodgers, W.A. (1987). Pastoralism, conservation and the overgrazing controversy. In: D. Anderson and R. Grove (eds), *Conservation in Africa: People, Policies and Practice*, pp. 111–28. Cambridge University Press.

Hubbard, M. (1981). Desperate games: Bongola Smith, the Imperial Cold Storage Company and Bechuanaland's beef, 1931. *Botswana Notes and Records* 13: 19–24.

Huntley, B.J. (1982). Southern African savannas. In: B.J. Huntley and B.H. Walker (eds), *Ecology of Tropical Savannas*, pp. 101–19. Springer-Verlag: New York.

Hutchins, D.E. (1888). *Cycles of Drought and Good Seasons in South Africa.* Times Office: Wynberg. 137 pp.

Hutchins, D.G., Hutton, S.M. and Jones, C.R. (1976a). The geology of the Okavango Delta. *Proceedings of the Symposium on the Okavango Delta and its Future Use*, pp. 15–19. Botswana Society: Gaborone.

Hutchins, D.G., Hutton, L.G., Hutton, S.M., Jones, C.R. and Loenhert, E.P. (1976b). *A Summary of the Geology, Seismicity, Geomorphology and Hydrogeology of the Okavango Delta.* Botswana Geological Survey Department Bulletin 7. 27 pp.

Hutton, L.G. and Dincer, T. (1979). Using satellite imagery to study the Okavango Swamp, Botswana. In: M. Deutsch, D. Wiesnet and A. Rango (eds), *Satellite Hydrology.* American Water Resource Association.

IUCN (1988). *The Nature of Zimbabwe, a Guide to Conservation and Development.* International Union for the Conservation of Nature and Natural Resources Field operations Division: Gland, Switzerland. 87 pp.

Jack, D.J. (1980). 'CEGB Botswana uranium in Calcrete Project.' Unpublished final report of the Union Carbide Exploration Corporation.

Jackson, P.B.N. (1961). *The Fishes of Northern Rhodesia. A Check List of Indigenous Species.* Government Printer: Lusaka. 140 pp.

Jarman, T.R.W. and Butler, K.E. (1971). Livestock management and production in the Kalahari. In: Proceedings of the Conference on Sustained Production from Semi-arid Areas. *Botswana Notes and Records.* Special Edition 1: 132–9.

Jeffares, J.L.S. (1938).'Ngamiland waterways surveys 1937.' Unpublished report to the Resident Commissioner, Mafeking.

Jennings, C.M.H. (1974). 'The hydrogeology of Botswana.' Unpublished PhD thesis, University of Natal, Pietermaritzburg.

Johns, C.C. (1985). Geological investigations in the Kalahari region of Zambia. In: D.G.Hutchins and A.P. Lynam, (eds), *The Proceedings of a Seminar on the Mineral Exploration of the Kalahari.* Botswana Geological Survey Department, Bulletin 29: 216–33.

Johnstone, P.A. (1973). Evaluation of a Rhodesian game ranch. *Journal of the South African Wildlife Management Association* 5: 43–51.

Jones, C.R. (1973). *Marico River area.* 1:125 000 QDS 2466B. Geological Survey of Botswana. Map.

Jones, C.R. (1982). The Kalahari of southern Africa. In: T.L. Smiley (ed.), *The Geological Story of the World's Deserts. Striae* 17: 20–34.

Jones, M.T. (1962). 'Report on a photogeological study of an area between the Okavango Swamps and the South West Africa border.'

Unpublished Botswana Geological Survey.

Jones, N. (1944). The climatic and cultural sequences at Sawmills, Southern Rhodesia. *Occasional Papers, Rhodesia Museum* **11**: 39–46.

Jones, N. (1946). The archaeology of Lochard. *Occasional Papers, Rhodesia Museum* **12**: 116–27.

Jubb, R.A. (1964). Some fishes of the Victoria Falls region. In: B.M. Fagan (ed.), *A Handbook to the Victoria Falls, the Batoka Gorge and part of the Upper Zambesi River*, pp. 129–40. Committee for the Preservation of Natural History and Monuments, Northern Rhodesia.

Junod, S.M. (1963). A note on pottery from Tsodilo, Bechuanaland. *South African Archaeological Bulletin* **18**: 20.

Kalahari Conservation Society (KCS) (1983). *Which Way Botswana's Wildlife?* Proceedings of a Symposium KCS: Gaborone. 108 pp.

Kalahari Conservation Society (KCS) (1984). KCS convenes meeting on CKGR problems. *Kalahari Conservation Society Newsletter* **5**: 1.

Kalahari Conservation Society (KCS) (1985). Central Kalahari Game Reserve. *Kalahari Conservation Society Newsletter* **9**: 1.

Kalahari Conservation Society (KCS) (1986a). Salvinia found in the Okavango. *Kalahari Conservation Society Newsletter* **13**: 8.

Kalahari Conservation Society (KCS) (1986b). Namibia to extract water from the Okavango. *Kalahari Conservation Society Newsletter* **13**: 8.

Kalahari Conservation Society (KCS) (1988a). *Management Plan for the Central Kalahari and Khutse Game Reserves.* 136 pp.

Kalahari Conservation Society (KCS) (1988b). *Sustainable Wildlife Utilisation: The Role of Wildlife Management Areas.* Proceedings of a workshop. KCS: Gaborone.

Kanthack, F.E. (1921). Notes on the Kunene River, southern Angola. *Geographical Journal* **57**: 5.

Kent, L.E. and Rogers, A.W., (1947). *Diatomaceous Deposits of the Union of South Africa.* Geological Survey of South Africa, Memorandum 42.

Kent, L.E. and Gribnitz, K.H. (1985). Freshwater shell deposits in the northwestern Cape Province: further evidence for a widespread wet phase during the late Pleistocene in Southern Africa. *South African Journal of Science* **81**: 361–70.

Key, R.M. and Hutton, S.M. (1976). The tectonic generation of the Limpopo Mobile Belt, and a definition of its western extremity. *Precambrian Research* **3**: 79–90.

Key, R.M. and Rundle, C. (1988). The regional significance of new isotopic ages from Precambrian windows through the Kalahari desert in northwestern Botswana. *Proceedings of the Geological Society of South Africa.*

King, L.C. (1942). *South African Scenery* (1st edn). Oliver & Boyd: London.

King, L.C. (1947). Landscape study in southern Africa. *Proceedings of the Geological Society of South Africa* **50**: 23–102.

King, L.C. (1951). The geology of Makapan and other caves. *Transactions of the Royal Society of South Africa* **33**: 121–50.

King, L.C. (1955). Pediplanation and isostasy: an example from South Africa. *Quarterly Journal of the Geological Society of London* **110**: 353–9.

King, L.C. (1962). *The Morphology of the Earth.* Oliver & Boyd: Edinburgh. 699 pp.

Kingston, J., Woodward, J.E., Malan, S.P. and Jennings, R.P. (1961). 'Geology and petroleum prospects of the Kalahari Basin.' Unpublished report of Mobil Oil, South Africa Pty Ltd.

Kiyaga-Mulindwa, D. (1983). Moeng I: a Tswana Early Iron Age site in the northern Kalahari Desert, Botswana. Paper presented at the 8th Pan-African Congress on Prehistory and Related Studies, Jos, Nigeria.

Klein, R.G. (1979). Stone Age exploitation of animals in southern Africa. *American Scientist* **67**: 151–60.

Klein, R.G. (1980). Environmental and ecological implications of large mammals from Upper Pleistocene and Holocene sites in southern Africa. *Annals of the South African Museum* **81**: 223–83.

Klein, R.G. (1981). Stone Age predation on small African bovids. *South African Archaeological Bulletin* **36**: 55–65.

Klein, R.G. (1984). The prehistory of Stone Age herders in South Africa. In: J.D. Clark and S.A. Brandt (eds), *From Hunters to Farmers*, pp. 281–9. University of California Press: Berkeley.

Knight. M.H., Knight-Eloff, A.K. and Bornman, J.J. (1988). The importance of borehole water lick sites to Kalahari ungulates. *Journal of Arid Environments* **15**: 269–81.

Kokot, D.F. (1948). *An Investigation into the Evidence bearing on Recent Climatic Change in Southern Africa.* Department of Irrigation. Government Printer: Pretoria. 160 pp.

Krinsley, D.H., and Doornkamp, J.C. (1973). *Atlas of Quartz Sand Surface Textures.* Cambridge University Press. 91 pp.

Kroner, A. (1977). The Precambrian geotectonic evolution of Africa: plate accretion versus plate destruction. *Precambrian Research* **4**: 163–214.

Kutzbach, J.E. (1983). Monsoon rains of the Late Pleistocene and Early Holocene: patterns, intensity and possible causes of changes. In: F.A. Street-Perrott, M. Beran and R. Ratcliffe (eds), *Variations in the Global Water Budget*, pp. 371–89. Reidel: Dordrecht.

Lamplugh, G.W. (1902). Calcrete. *Geological Magazine* **9**: 75.

Lamplugh, G.W. (1907). Geology of the Zambezi basin around the Batoka Gorge. *Quarterly Journal of the Geological Society* **63**: 162–216.

Lancaster, I.N. (1976). 'Pans of the southern Kalahari, Botswana.' Unpublished PhD thesis, University of Cambridge.

Lancaster, I.N. (1978a). The pans of the southern Kalahari, Botswana. *Geographical Journal* **144**: 80–98.

Lancaster, I.N. (1978b). Composition and formation of southern Kalahari pan margin dunes. *Zeitschrift für Geomorphologie* **22**: 148–69.

Lancaster, I.N. (1979). Evidence for a widespread late Pleistocene humid period in the Kalahari. *Nature* **279**: 145–6.

Lancaster, N. (1980). Dune systems and palaeoenvironments in southern Africa. *Palaeoentology Africana* **23**: 185–9.

Lancaster, N. (1981). Palaeoenvironmental implications of fixed dune systems in southern Africa. *Palaeogeography, Palaeoclimatology, Palaeoecology* **33**: 327–46.

Lancaster, N. (1986a). Grain-size characteristics of linear dunes in the south-western Kalahari. *Journal of Sedimentary Petrology* **56**: 395–400.

Lancaster, N. (1986b) Pans in the southwestern Kalahari: a preliminary report. *Palaeoecology of Africa* **17**: 59–67.

Lancaster, N. (1987). Grain-size characteristics of linear dunes in the southwestern Kalahari – a reply. *Journal of Sedimentary Petrology* **57**: 573–4.

Lancaster, N. (1988). Development of linear dunes in the southwestern Kalahari, southern Africa. *Journal of Arid Environments* **14**: 233–44.

Lancaster, N. (1989). Late Quaternary palaeoenvironments of the southwestern Kalahari. *Palaeogeography, Palaeoclimatology, Palaeoecology* **70**: 367–76.

Lancaster, N. and Ollier, C.D. (1983). Sources of sand for the Namib sand sea. *Zeitschrift für Geomorphologie Supplementband* **45**: 71–83.

Lawson, D. (1989). Wildlife utilisation in Botswana. *Kalahari Conservation Society Newsletter* **25**: 4–5.

Leakey, L.S.B. and Solomon, J.D. (1929). East African Archaeology. *Nature* **124**: 9.

Lee, R.B. (1972). The !Kung Bushmen of Botswana. In: M. Biccieri (ed.), *Hunters and Gatherers Today*, pp. 327–68. Holt, Rinehart & Winston: New York.

Lee, R.B. (1979). *Kalahari Hunter–gatherers: Men, Women and Work in a Foraging Society.* Harvard University Press: Cambridge.

Lee, R.B. and De Vore, I. (1976). *Kalahari Hunter–gatherers: Studies of the !Kung San and their Neighbours.* Harvard University Press: Cambridge.

Leistner, O.A. (1959). Notes on the vegetation of the Kalahari Gemsbok National Park with special reference to its influence on the distribution of antelopes. *Koedoe* **2**: 128–51.

Leistner, O.A. (1967). The plant ecology of the southern Kalahari. *Memoirs of the Botanical Survey of South Africa* **38**. 172 pp.

Leistner, O.A. and Werger, M.J.A. (1973). Southern Kalahari phytoecology. *Vegetatio* **28**: 353–99.

Lepionka, L. (1978). Excavations at Tautswemogala. *Botswana Notes and Records* **9**: 1–16.

Levin, M., Hambleton-Jones, B. and Smit, M. (1985). Uranium in the groundwater of the Kalahari regions south of the Molopo River. In: D. Hutchins and A. Lynham (eds), *Proceedings of a Seminar on the Mineral Exploration of the Kalahari.* Botswana Geological Survey Department, Bulletin **29**: 234–50.

Lewin, R. (1986). In ecology, change brings

stability. *Science* **234**: 1071–3.

Lewis, A.D. (1936). Sand dunes of the Kalahari within the borders of the Union. *South African Geographical Journal* **19**: 23–32.

Lister, L.A. (1979). The geomorphic evolution of Zimbabwe–Rhodesia. *Transactions of the Geological Society of South Africa* **82**: 363–70.

Lister, M.H. (1949). *The Journals of Andrew Geddes Bain.* The Van Riebeck Society: Cape Town.

Livingstone, D. (1858a). *Missionary Travels and Researches in South Africa.* Ward Lock: London. 617 pp.

Livingstone, D. (1858b). In: W. Monk (ed.), *Cambridge Lectures.* Deighton, Bell Co.: London.

Livingstone, I. (1986). Geomorphological significance of wind flow patterns over a Namib linear dune. In: W.G. Nickling (ed.), *Aeolian Geomorphology. The Binghampton Symposium in Geomorphology International Series* **17**, pp. 97–112. Allen & Unwin: Boston.

Lockett, N.H. (1979). The geology of the country around Dett. *Rhodesian Geological Survey Bulletin* **85**. 198 pp.

Longley, R.W. (1976). Weather and weather maps of South Africa. *Weather Bureau, Technical Bulletin* **3**. Pretoria.

Ludtke, G. (1986). *Geophysical, Geochemical and Geological investigations in the Ngami and Kheis Areas of Botswana 1980–1983* Final Report. Botswana Geological Survey Department, Bulletin **32**. 319 pp.

Lugard, E.J. (1909). The flora of Ngamiland. *Kew Bulletin* **3**.

Mabbutt, J.A. (1955). Erosion surfaces in Namaqualand and the ages of surface deposits in the south-western Kalahari. *Transactions of the Geological Society of South Africa* **58**: 13–29.

Mabbutt, J.A. (1957). Physiographic evidence for the age of the Kalahari Sands of the south west Kalahari. In: J.D. Clark (ed.), *The 3rd Pan-African Congress on Prehistory, Livingstone 1955.*

Mabbutt, J.A. and Wooding, R.A. (1983). Analysis of longitudinal dune patterns in the north western Simpson Desert, central Australia. *Zeitschrift für Geomorphologie, Supplementband* **45**: 51–70.

McCarthy, T.S. (1983). Evidence for the former existence of a major southerly-flowing river in Griqualand West. *Transactions of the Geological Society of South Africa* **86**: 37–49.

McCarthy, T.S., Ellery, W., Rogers, K., Cairncross, B. and Ellery, K., (1986a). The role of sedimentation and plant growth in changing flow patterns in the Okavango Delta. *South African Journal of Science* **82**: 579–84.

McCarthy, T.S., McIver, J. and Cairncross, B. (1986b). Carbonate accumulation on islands in the Okavango Delta. *South African Journal of Science* **82**: 588–91.

McCarthy, T.S., Ellery, W., Ellery, K.N. and Rogers, K.H. (1987). Observations on the abandoned Nqoga channel of the Okavango Delta. *Botswana Notes and Records* **19**: 83–90.

McCarthy, T.S., Stanistreet, I.G., Cairncross, B., Ellery, W.N., Ellery, K., Oelofse, R. and

Grobicki, T.S.A. (1988a). Incremental aggradation on the Okavango Delta-Fan, Botswana. *Geomorphology* 1: 267–78.

McCarthy, T.S., Rogers, K.H., Stanistreet, I.G., Ellery, W.N., Cairncross, B. and Ellery, K. (1988b). Features of channel margins in the Okavango Delta. *Palaeoecology of Africa* 19: 3–14.

McCarthy, T.S., Stanistreet, I. and Cairncross, B. (1990). The sedimentary dynamics of active fluvial channels on the Okavango (Delta) alluvial fan, Botswana. *Sedimentology*, in press.

McConnell, R.B. (1959). Notes on the geology and geomorphology of the Bechuanaland Protectorate. *XX Congress on Mineral Resources*, pp. 307–25.

McFarlane, M. (1976). *Laterite and Landscape.* Academic Press: London.

McFarlane, M.J. (1983). Laterite. In: A.S. Goudie and K. Pye (eds), *Chemical Sediments in Geomorphology*, pp. 7–58. Academic Press: London.

McGee, O.S. and Hastenrath, S.L. (1966). Harmonic analysis of the rainfall over South Africa. *Notos* 15: 79–90.

MacGregor, A.M. (1916). The Karroo rocks and later sediments north west of Bulawayo. *Transactions of the Geological Society of South Africa* 19: 14–32.

MacGregor, A.M. (1947). An outline of the geological history of southern Rhodesia. *Bulletin of the Southern Rhodesian Geological Survey* 38. 73 pp.

McKee, E.D. (1979). Introduction to a study of global sand seas. In: E.D. McKee (ed.), *A Study of Global Sand Seas. US Geological Survey professional paper* 1052: 1–19.

McLeod, G. (1986). Overgrazing or range management. Which way for Botswana? *Splash* 2: 3–5.

Mackel, R. (1974). Dambos: a study of morphodynamic activity on the plateau regions of Zambia. *Catena* 1: 327–65.

Mackenzie, I.A. (1946). *Report on the Kalahari Expedition, 1945.* Government Printer: Pretoria.

Maggs, T. (1984). The Iron Age south of the Zambezi. In: R. Klein (ed.), *Southern African Prehistory and Paleoenvironments*, pp. 329–60. Balkema: Rotterdam.

Main, M. (1987). *Kalahari. Life's Variety in Dune and Delta.* Southern Book Publishers: Johannesburg. 265pp.

Malaisse, F. (1975). Carte de la vegetation du bassin de la Luanza. In: J.J. Symoens (ed.), *Exploration Hydrobiologique du Bassin du Lac Bangweolo et du Luapula* 18: 1–41.

Malherbe, S.J. (1984). The Geology of the Kalahari Gemsbok National Park. *Koedoe* 27: 33–44.

Mallick, D.I.J., Habgood, F. and Skinner, A.C. (1981). A geological interpretation of Landsat imagery and air photography of Botswana. *Overseas Geology and Mineral Resources* 56. Institute of Geological Sciences: London. 36 pp.

Marloth, R. (1908). Das Kaapland, insonderheit das Reich der Kapflora, das Waldgebiet und die Karoo, pflanzengeografisch dargestellt. *Wiss. Ergebn. Deutsch. Tiefsee-Exped. 'Waldavia' 1898–1899*, Band 3. Fischer: Jena.

Martens, H.E. (1971). The effect of tribal grazing patterns on the habitat in the Kalahari. In: Proceedings of the Conference on Sustained Production from Semi-arid Areas. *Botswana Notes and Records* Special edition 1: 242–7.

Martin, H. (1968). Hydrology and water balance of some regions covered by Kalahari Sands in South West Africa. *Inter-African Conference on Hydrology, Nairobi.*

Martinelli, E. and associates (1980). 'Completion report on the trial borehole 'Peggy', Mongu, Western Province, Zambia.' Unpublished report for NORAD, Lusaka, Zambia.

Mason, R.J. (1967). Questions of terminology in regard to the study of Earlier Stone Age cultures in South Africa. In: W.W. Bishop and J.D. Clark (eds), *Background to Evolution in Africa*, pp. 765–70. University of Chicago Press.

Massey, N.W.D. (1974). *Contribution to the Kalatraverse I report.* Report of the Geological Survey of Botswana NWDM/13/74.

Maufe, H.B. (1920). The geological section between Bulawayo and Victoria Falls. *South African Journal of Science* 17: 113–15.

Maufe, H.B. (1930). Changes of climate in southern Rhodesia during late geological times. *South African Geographical Journal* 13: 12–16.

Maufe, H.B. (1935). Some factors in the geological evolution of Southern Rhodesia and neighbouring countries. *South African Geographical Journal* 18: 3–21.

Maufe, H.B. (1939). New sections in the Kalahari beds at the Victoria Falls, Rhodesia. *Transactions of the Geological Society of South Africa* 31: 211–24.

Mazonde, I.N. (1988). The inter-relationship between cattle and politics in Botswana's economy. In: G.J. Stone (ed.), *The Exploitation of Animals in Africa*, pp. 345–56. Aberdeen University African Studies Group: Aberdeen.

Mazor, E. (1982). Rain recharge in the Kalahari – a note of some approaches to the problem. *Journal of Hydrology* 55: 137–44.

Mazor, E., Verhagen, B., Sellschop, J., Robins, N. and Hutton, L. (1974). Kalahari groundwaters: their hydrogen, carbon and oxygen isotopes. In: *Isotope Techniques in Groundwater Hydrology*, pp. 203–25. IAEA: Vienna.

Mazor, E., Verhagen, B., Sellschop, J., Jones, M., Robins, N., Hutton, L. and Jennings, C. (1977). Northern Kalahari groundwaters: hydrologic, isotopic and chemical studies at Orapa, Botswana. *Journal of Hydrology* 34: 203–33.

Meigs, P. (1953). World distribution of homoclimates. In *Reviews of Research on Arid Zone Hydrology*, pp. 203–9. UNESCO: Paris.

Miller, O.B. (1939). The Mukusi Forests of Bechuanaland. *Empire Forestry Journal* 18: 193–201.

Mitchell, B.L. (1961). Some notes on the vegetation of a portion of the Wankie National Park. *Kirkia* 2: 200–9.

Mmusi, P.S. (1989). *Budget Speech delivered to the National Assembly on 15th February 1989.* Government Printer: Gaborone.

Moeyersons, J. (1978). The behaviour of stones and stone implements, buried in consolidating and creeping Kalahari Sands. *Earth Surface Processes* **3**: 115–28.

Molyneaux, T. and Reinecke, T. (1983). Ancient ruins and mines of the Tati region of north-east Botswana. *South African Archaeological Bulletin* **38**: 9.

Money, N.J. (1972). An outline of the geology of western Zambia. *Records of the Geological Survey, Republic of Zambia* **12**: 103–23.

Moore, A.E. (1988). Plant distribution and the evolution of the major river systems in southern Africa. *South African Geological Journal* **91**: 346–9.

Moss, C. (1989). *Elephant Memories. Thirteen Years in the Life of an Elephant Family.* Fontana: Glasgow. 336 pp.

Murray, M.L. (1978). 'Wildlife utilisation, investigation and planning in western Botswana.' Unpublished report of the Ministry of Commerce and Industry, Gaborone.

Murray, M.L. (1988). Carving up the Kalahari. *Kalahari Conservation Society Newsletter* **22**: 8.

Musonda, F.B. (1987). Surfaces textures of sand grains from the Victoria Falls region, Zambia: implications for depositional environments and local archaeological occurrences. *South African Archaeological Bulletin* **42**: 161–4.

Netterberg, F. (1969a). Ages of calcretes in southern Africa. *South African Archaeological Bulletin* **24**: 88–92.

Netterberg, F. (1969b). 'The geology and engineering properties of South African calcretes.' Unpublished PhD thesis, University of the Witwatersrand.

Netterberg, F. (1978). Dating and correlation of calcretes and other pedocretes. *Transactions of the Geological Society of South Africa.* **81**: 379–91.

Netterberg, F. (1980). Geology of Southern African calcretes: 1. Terminology, description, macrofeatures and classification. *Transactions of the Geological Society of South Africa* **83**: 255–83.

Newell, R.E., Gould-Stewart, S. and Chung, J.C. (1981). Possible interpretation of palaeoclimatic reconstructions for 18 000 BP in the region 60°N to 60°S, 60°W to 100°E. *Palaeoecology of Africa* **13**: 1–19.

Ngcongco, L. (1982). Precolonial migration in southeastern Botswana. In: R.R. Hitchcock and M. Smith (eds), *Settlement in Botswana – The Historical Development of a Human Landscape*, pp. 23–30. Botswana Society/Heinemann: Gaborone.

Nicholson, S.E. and Flohn, H. (1980). African environmental and climatic changes and the general atmospheric circulation in the late Pleistocene and Holocene. *Climatic Change* **2**: 313–48.

Norris, R.M. (1969). Dune reddening and time. *Journal of Sedimentary Petrology* **39**: 7–11.

Nugent, C. (1987). Can the Zambezi irrigate the Kalahari ? *Zimbabwe Science News* **21**: 68–9.

Obst, E. and Kayser, K. (1949). *Die grosse Randstufe auf der Ostseite Süsafrikas und ihr Vorland: ein beitrag zur Geschichte der Jungen aushebung des Subkontinents.* Geografishe Gesells: Hamburg. 342 pp.

Odell, M.L. (1980). 'Botswana's first livestock development project – an experiment in agricultural transformation.' Unpublished report of the Swedish International Development Agency, Gaborone.

Ollier, C.D. (1985). Morphotectonics of continental margins with great escarpments. In: M. Morisawa and J.T. Hack (eds), *Tectonic Geomorphology. The Binghamton Symposia in Geomorphology, International Series* 15, pp. 3–25. Allen & Unwin, Boston.

Oswell, W.E. (1900). *William Cotton Oswell, Hunter and Explorer. The Story of his Life.* Heinemann: London.

Owens, M. and Owens, D. (1980). The fences of death. *African Wildlife* **34**: 25–7.

Parson, J. (1981). Class, cattle and the state. *Botswana Journal of Southern African Studies* **7**: 242–63.

Parsons, Q.N. (1982). *A New History of Southern Africa.* Macmillan: London.

Partridge, T.C. and Maud, R.R. (1987). The geomorphic evolution of southern Africa since the Mesozoic. *South African Journal of Geology* **90**: 179–208.

Passarge, S. (1895). *Adamaua.* Reimer: Berlin.

Passarge, S. (1901). *Z. deutch Ges. Erdkunde Berlin* 36.

Passarge, S. (1904). *Die Kalahari.* Dietrich Riemer: Berlin. 823 pp.

Passarge, S. (1906). Waaserwirtscaftlichte probleme in der Kalahari. *Globus III: Länder und Völkerkunde* **190**: 299–302.

Passarge, S. (1911). Die pfannenformigen Hohlformen der Südafrikanischen Steppen. *Petermans Mitterlungen* **57**: 130–5.

Patterson, L. (1987). Cordon fences. *Kalahari Conservation Society Newsletter* **18**: 10–11.

Peabody, F.E. (1954). Travertines and cave deposits of the Kaap Escarpment of South Africa and the type locality of *Australopithecus africanus. Bulletin of the Geological Society of America* **65**: 671–706.

Pearce, E.A. and Smith, C.G. (1984). *The World Weather Guide.* Hutchinson: London. 480 pp.

Peel, R.F. (1941). Denudational landforms in the central Libyan Desert. *Journal of Geomorphology* **4**: 3–23.

Penck, A. and Bruckner, E. (1909). *Die Alpen in Eiszeitalten* (3 vols). Tauchnitz: Leipzig.

Petrov, M.P. (1976). *Deserts of the World.* John Wiley & Sons: New York. 447 pp.

Phillipson, D.W. (1977). *The Later Prehistory of Eastern and Southern Africa.* Heinemann: London. 323 pp.

Picard, L.A. (1980). Bureaucrats, cattle and public policy. *Comparative Political Studies* **13**: 350–1.

Pickford, M. and Mein, P. (1988). The discovery of fossiliferous Plio-Pleistocene cave fillings in Ngamiland, Botswana. *CR Acadamie Science Paris, series* 2, **307**: 1681–6.

Pike, J.G. (1971). The development of the water resources of the Okavango Delta. In: Proceedings of the Conference on Sustained

Production from Semi-arid Areas, *Botswana Notes and Records*, special edition **1**: 35–40.

Pim, A.W. (1933). *Financial and Economic Position of the Bechuanaland Protectorate; Report of the Commission Appointed by the Secretary of State for Dominion Affairs*. HMSO: London.

Place Names Commission (1984). *Third Report of the Place Names Commission*. Government Printer: Gaborone.

Poldervaart, A. (1957). Kalahari sands. In: J.D. Clark (ed.), *The 3rd Pan-African Congress on Prehistory, Livingstone 1955*, pp. 106–19. Chatto & Windus: London.

Pole Evans, I.B. (1948). An expedition to Ngamiland: June–July 1937. *Bechuanaland Survey, South African Memoirs* **21**: 75–203.

Pole Evans, I.B. (1950). The possibilities of beef production in southern Africa. *Imperial Journal of Experimental Agriculture (1950)*: 81–95.

Potten, D.H. (1976). Aspects of recent history in Ngamiland. *Botswana Notes and Records* **8**: 63–85.

Prins, G. (1980). *The Hidden Hippopotamus. Reappraisal in African History: the Early Colonial Experience in Western Zambia*. Cambridge University Press.

Procter, D.L.C. (1983). Biological control of the aquatic weed *Salvinia molesta* D.S. Mitchell in Botswana using the weevils *Cyrtobagous singularis* and *Cyrtobagous* sp. nov. *Botswana Notes and Records* **15**: 99–101.

Raffle, J.A. (1984). Mapping the Kalahari Desert. *Botswana Notes and Records* **16**: 107–16.

Range, P. (1912*a*). Geologie des Deutschen Namalandes. *Beitrage z. Geol Erforschungen d. Deutsch Schutzgebbiete*: 1–98.

Range, P. (1912*b*). Topography and geology of the German Kalahari. *Transactions of the Geological Society of South Africa* **15**: 63–73.

Rathbone, P.A. and Gould, D. (1982). *The Geology of the Molopo Farms Project Area*. Botswana Geological Survey, Report PAR/1/82.

Rayner, R.J. and McKay, I.J. (1986). The treasure chest of the Orapa Diamond Mine. *Botswana Notes and Records* **18**: 55–61.

Reeves, C.V. (1972). Rifting in the Kalahari? *Nature* **237**: 95–6.

Reeves, C.V. (1974*a*). *The Gravity Survey of Ngamiland 1970–71. Part 1: Field Observations and Data Reduction*. Botswana Geological Survey Department, Report CVR/13/73.

Reeves, C.V. (1974*b*). *The Gravity Survey of Ngamiland 1970–71. Part 2: Interpretation*. Botswana Geological Survey Department, Report CVR 20/74.

Reeves, C.V., (1978). The gravity survey of Ngamiland 1970–71. *Botswana Geological Survey Bulletin* **11**. 84 pp.

Reeves, C.V. and Hutchins, D.G. (1975). Crustal structures in central southern Africa. *Nature* **273**: 222.

Rey, C.F. (1988). In: N. Parsons and M. Crowder (eds), *Monarch of all I survey: Bechuanaland Diaries 1929–37*. Botswana Society: Gaborone.

Ringrose, S. and Mathieson, W. (1987*a*). Desertification in Botswana: progress towards a viable monitoring system. *Desertification Control Bulletin* **13**: 6–11.

Ringrose, S. and Mathieson, W. (1987*b*). Spectral assessment of indicators of range degradation in the Botswana Hardveld environment. *Remote Sensing of Environment* **23**: 379–96.

Robbins, L.H. (1984). Toteng, a Late Stone Age site along the Nchabe River, Ngamiland. *Botswana Notes and Records* **16**: 1–6.

Robbins, L.H. (1985). The Manyana Rock Painting Site. *Botswana Notes and Records* **17**: 1–14.

Robbins, L.H. and Campbell, A.C. (1988). The Depression Rock Shelter Site, Tsodilo Hills. *Botswana Notes and Records* **20**: 1–3.

Rogers, A.W. (1907). Geological survey of parts of Vryberg, Kuruman, Hay and Gordonia. *Annual Report of the Geological Commission of the Cape of Good Hope*, pp. 11–122.

Rogers, A.W. (1920). Geological survey and its aims; and a discussion on the origin of the Great Escarpment. *Transactions of the Geological Society of South Africa* **22**: 19–33.

Rogers, A.W. (1936). The surface geology of the Kalahari. *Transactions of the Geological Society of South Africa*. **24**: 57–80.

Rognon, P. (1980). Pluvial and arid phases in the Sahara: the role of non-climatic factors. *Palaeoecology of Africa* **12**: 45–62.

Rushworth, J.E. (1970). 'A preliminary checklist of the vascular plants of Wankie National Park and a summary of life-forms, families, genera and species.' Unpublished report. 26 pp.

Rust, I.C. (1975). Tectonic and sedimentary framework of Gondwana Basins in Southern Africa. In: K.S.W. Campbell (ed.), *Gondwana Geology*, pp. 537–64. ANU Press: Canberra.

Rust, U. (1984). Geomorphic evidence of Quaternary environmental changes in Etosha, South West Africa/Namibia. In: J. Vogel (ed.), *Late Cainozoic Palaeoclimates of the Southern Hemisphere*, pp. 279–86. Balkema: Rotterdam.

Rust, U. (1985). Die Enstehung der Etoschapfanne im Rahmen der Landschaftsentwicklung des Etoscha Nationalparks (nördliches Südwestafrika/Namibia). *Madoqua* **14**: 197–266.

Rust, U., Schmidt, H. and Dietz, K. (1984). Palaeoenvironments of the present day arid south western Africa 30 000–5000 BP: results and problems. *Palaeoecology of Africa* **16**: 109–48.

Rust, U. and Wienke, F. (1976). Geomorphologie der küstennahen Zentralen Namib (Südwestafrika). *Münchener Geographische Abhandlungen* **19**. 74 pp.

SACS (South African Committee for Stratigraphy) (1980). Stratigraphy of South Africa, Part 1. Lithostratigraphy of the Republic of South Africa, SWA/Namibia and the Republics of Boputhatswana, Transkei and Venda. *Handbook of the Geological Survey of South Africa* **8**.

Sampson, R. (1972). *The Man with the Toothbrush in his Hat*. Multimedia Publications:

Lusaka.

Sampson, C.G. (1974). *The Stone Age Archaeology of Southern Africa*. Academic Press: New York.

Savory, B.M. (1965). Sand of Kalahari type in Sesheke District, Northern Rhodesia. In: G.J. Snowball (ed.), *Science and Medicine in Central Africa*, pp. 189–200. Pergamon Press: London.

Schapera, I. (1930). *The Khoisan Peoples of South Africa*. Routledge & Kegan Paul: London.

Schapera, I. (1943). *Native Land Tenure in the Bechuanaland Protectorate*. Lovedale Press: Alice, South Africa.

Schmitz, G. (1968). *The Plateau Area of South Africa*. Schmitz: South Africa. 15 pp.

Scholz, C.H. (1975). *Seismicity, Tectonics and Seismic Hazard of the Okavango Delta*. UNDP/FAO Bot/71/506 Project Report: Gaborone.

Scholz, C.H., Koczynski, T.A. and Hutchins, D.G. (1976). Evidence for incipient rifting in southern Africa. *Geophysics Journal of the Royal Astronomical Society* **44**: 135–44.

Schönfelden, E. (1935). Südost-Angola und den Westliche Caprivi-Zipvel. *Petermanns Mitteilungen* **81**: 49.

Schulze, B.R. (1972). South Africa. In: J.F. Griffiths (ed.), *Climates of Africa. World Survey of Climatology*, vol. 10, pp. 501–86. Elsevier: Amsterdam.

Schulze, B.R. (1984). *Climate of South Africa: Part 8: General Survey* (5th edn). South African Weather Bureau: Pretoria. 330 pp.

Schulze, L. (1907). *Aus Namaland und Kalahari*. Jena.

Schwarz, E.H.L. (1919). The progressive desiccation of Africa: the cause and the remedy. *South African Journal of Science* **15**: 139–90.

Schwarz, E.H.L. (1920). *The Kalahari, or Thirstland Redemption*. Maskew Miller: Cape Town.

Seeley, H.G. (1892). Researches on the structure, organization and classification of the fossil reptilia. *Philosophical Transactions of the Royal Society* **182**: 311–70.

Seiner, V.F. (1909). Ergebnisse einer Bereisung des Gebiets zwischen Okawango und Sambesi (Caprivi-Zipvel) in der Jahren 1905 und 1906. *Mitteilungen aus den Deutschen Schutzgebieten* **22**: 48–9.

Selous, F.C. (1893). *Travel and Adventure in South East Africa*. Rowland Ward: London.

Senior, B.R. and Senior, D.A. (1972). Silcrete in southwest Queensland. *Bulletin BMR Geology and Geophysics Australia* **125**: 23–8.

Seitshiro, G. (1978). 'Gradient of five non-operational boreholes in the Kweneng district.' Unpublished report of the Range Ecology Unit, Ministry of Agriculture, Gaborone.

Shantz, H.L. (1956). History and problems of arid lands development. In: G.F. White (ed.), *The Future of Arid Lands*. American Association for the Advancement of Science: Washington DC. Publication **43**.

Shaw, P.A. (1983). Fluctuations in the level of Lake Ngami: the historical evidence. *Botswana Notes and Records* **15**: 79–84.

Shaw, P.A. (1984). A historical note on the outflows of the Okavango Delta System. *Botswana Notes and Records* **16**: 127–30.

Shaw, P.A. (1985a). Late Quaternary landforms and environmental change in northwest Botswana: the evidence of Lake Ngami and the Mababe Depression. *Transactions of the Institute of British Geographers*, NS **10**: 333–46.

Shaw, P.A. (1985b). The desiccation of Lake Ngami: an historical perspective. *Geographical Journal* **151**: 318–26.

Shaw, P.A. (1986). The palaeohydrology of the Okavango Delta: some preliminary results. *Palaeoecology of Africa* **17**: 51–8.

Shaw, P.A. (1988a). After the flood: the fluvio-lacustrine landforms of northern Botswana. *Earth Science Reviews* **25**: 449–56.

Shaw, P.A. (1988b). Lakes and Pans. In: B. Moon and G. Dardis (eds), *The Geomorphology of Southern Africa*, pp. 120–40. Southern Publishers: Johannesburg.

Shaw, P.A. (1989a). Fluvial systems of the Kalahari – a review. In: A Yair and S. Berkowicz (eds), *Arid and Semi-arid Environments – Geomorphological and Pedological Aspects*. Catena Supplement **14**: 119–26.

Shaw, P.A. (1989b). Sowa Pan: the soda ash project starts soon. *Kalahari Conservation Society Newsletter* **23**: 8.

Shaw, P.A. and Cooke, H.J. (1986). Geomorphic evidence for the late Quaternary palaeoclimates of the middle Kalahari of northern Botswana. *Catena* **13**: 349–59.

Shaw, P.A., Cooke, H.J. and Thomas, D.S.G. (1988). Recent advances in the study of Quaternary landforms in Botswana. *Palaeoecology of Africa* **19**: 15–26.

Shaw, P.A. and De Vries, J.J. (1988). Duricrust, groundwater and valley development in the Kalahari of southeast Botswana. *Journal of Arid Environments* **14**: 245–54.

Shaw, P.A. and Thomas, D.S.G. (1988). Lake Caprivi: a late Quaternary link between the Zambezi and middle Kalahari drainage systems. *Zeitschrift für Geomorphologie* **32**: 329–37.

Shaw, P.A. and Thomas, D.S.G. (1989). Pans playas and salt lakes. In: D.S.G. Thomas (ed.), *Arid Zone Geomorphology*, pp. 184–205. Belhaven Press: London.

Siderius, W. (1972). *Soils of Eastern and Northern Botswana*. FAO/UNDP technical document **3**, Gaborone.

Siesser, W.G. and Dingle, R.V. (1981). Tertiary sea level movements around South Africa. *Journal of Geology* **89**: 83–96.

Silberbauer, G.B. (1965). *Bushman Survey Report*. Bechuanaland Government: Gaborone.

Silberbauer, G.B. (1972). The G/wi Bushmen. In: M. Biccieri (ed.), *Hunters and Gatherers Today*, pp. 271–335. Holt, Rinehart & Winston: New York.

Silberbauer, G.B. (1981). *Hunter and Habitat in the Central Kalahari Desert*. Cambridge University Press: New York.

Sims, D. (1981). 'Agroclimatological information, crop requirements and agricultural zones for Botswana. Unpublished report of the Land Utilization Division, Ministry of

Agriculture, Botswana.

Skarpe, C. (1980). Observations on two bush-fires in the western Kalahari, Botswana. *Acta Phytogeographica Suedica* **68**: 131–40.

Skarpe, C. (1981). *A Report on the Range Ecology Project, Western Kalahari.* Ministry of Agriculture, Gaborone.

Skarpe, C. (1983). Cattle grazing and the ecology of the western Kalahari. In: *Which Way Botswana's Wildlife?* pp. 21–4. Proceedings of a symposium. Kalahari Conservation Society: Gaborone.

Skarpe, C. and Bergström, R. (1986). Nutrient content and digestability of forage plants in relation to plant phenology and rainfall in the Kalahari, Botswana. *Journal of Arid Environments* **11**: 147–64.

Sleep, N.H. (1971). Thermal effects of the formation of Atlantic continental margins by continental break up. *Geophysical Journal of the Royal Astronomical Society* **24**: 325–50.

Smale, D. (1968). The occurrence of clinoptilo-lite in pan sediments in the Nata area, northern Botswana. *Transactions of the Geological Society of South Africa* **71**: 147–53.

Smale, D. (1973). Silcretes and associated silica diagenesis in southern Africa and Australia. *Journal of Sedimentary Petrology* **43**: 1077–89.

Smit, P.J. (1977). 'Die Geohidrologie in die opvanggebied van die Molopo rivier in die noordlike Kalahari.' Unpublished PhD thesis, University of the Orange Free State, Bloemfontein.

Smith, P.A. (1985). *Salvinia molesta*: an alien water weed in Botswana. *Kalahari Conservation Society Newsletter* **7**: 9–11.

Smith, R.A. (1984). *The Lithostratigraphy of the Karoo Supergroup in Botswana.* Botswana Geological Survey, Bulletin **26**. 239 pp.

Snowy Mountains Engineering Corporation (SMEC) (1986). *Southern Okavango Integrated Water Development* Phase 1. Inception Report. Department of Water Affairs/ SMEC, Gaborone.

Snowy Mountains Engineering Corporation (1987). *Southern Okavango Integrated Water Development* Phase 1. Technical Study, Final Report. 5 vols. Department Water Affairs/ SMEC, Gaborone.

Snowy Mountains Engineering Corporation (1989). *Ecological Zoning – Okavango Delta.* 2 vols. Kalahari Conservation Society/Ministry of Local Government and Lands, Gaborone.

Sohnge, P.G., Visser, D.J.L. and Van Riet Lowe, C. (1937). *The Geology and Archaeology of the Vaal River Basin.* Union of South Africa Geological Survey Memoir **35**, pp. 5–59. Government Printer: Pretoria.

Stansfield, G. (1973). *The Geology of the Area around Dukwe and Tlalamabele, Central District, Botswana.* Geological Survey of Botswana, District Memoir **1**.

Steel, E.A. (1917). The Zambezi–Congo watershed. *Geographical Journal* **50**, 157 and 180.

Stigand, A.G. (1912). Notes on Ngamiland. *Geographical Journal* **39**: 376–9.

Stigand, A.G. (1923). Ngamiland. *Geographical Journal* **62**: 401–19.

Story, R. (1964). Plant lore of the Bushmen. In: D.H.S. Davis (ed.), *Ecological Studies of Southern Africa.* Junk: The Hague.

Streckler, M.S. and Watts, A.B. (1982). Subsidence history and tectonic evolution of atlantic-type continental margins. In: R.A. Scrutton (ed.), *Dynamics of Passive Margins*, pp. 184–96. Geodynamics Series 6. American Geophysical Union: Washington, and Geological Society of America: Boulder.

Street, F.A. (1980). The relative importance of climate and local hydrogeological factors in influencing lake-level fluctuations. *Palaeoecology of Africa* **12**: 137–58.

Street, F.A. and Grove, A.T. (1976). Environmental and climatic implications of late Quaternary lake-level fluctuations in Africa. *Nature* **261**: 385–90.

Street-Perrott, F.A. and Roberts, N. (1983). Fluctuations in closed-basin lakes as an indicator of past atmospheric circulation patterns. In: F.A. Street-Perrott, M. Beran and R. Ratcliffe (eds), *Variations in the Global Water Budget*, pp. 331–45. Reidel: Dordrecht.

Street-Perrott, F.A., Roberts, N. and Metcalfe, S. (1985). Geomorphic implications of late Quaternary hydrological and climatic changes in the Northern Hemisphere tropics. In: I. Douglas and T. Spencer (eds), *Environmental Change and Tropical Geomorphology*, pp. 165–83. Allen & Unwin: London.

Street-Perrott, F.A. and Harrison, F.P. (1985). Lake levels and climate reconstruction. In: A.D. Hecht (ed.), *Palaeoclimate Analysis and Modelling*, pp. 291–340. John Wiley & Sons: New York.

Streten, N.A. (1980). Some synoptic indices of the Southern Hemisphere mean sea level circulation 1972–77. *Monthly Weather Review* **108**: 18–36.

Summerfield, M.A. (1978). 'The nature and origin of silcrete with particular reference to Southern Africa' Unpublished DPhil thesis, University of Oxford.

Summerfield, M.A. (1982). Distribution, nature and probable genesis of silcrete in arid and semi-arid southern Africa. In: D.H. Yaalon (ed.), *Aridic Soils and Geomorphic Processes. Catena* Supplement **1**: 37–65.

Summerfield, M.A. (1983a). Silcrete as a palaeoclimatic indicator: evidence from southern Africa. *Palaeogeography, Palaeoclimatology, Palaeoecology* **41**: 65–79.

Summerfield, M.A. (1983b). Silcrete. In: A.S. Goudie and K. Pye (eds), *Chemical Sediments and Geomorphology*, pp. 59–92. Academic Press: London.

Summerfield, M.A. (1985a). Plate tectonics and landscape development on the African continent. In: M. Morisawa and J.T. Hack (eds), *Tectonic Geomorphology. The Binghamton Symposia in Geomorphology, International Series 15*, pp. 27–51. Allen & Unwin: Boston.

Summerfield, M.A. (1985b). Tectonic background to long-term landform development in tropical Africa. In: I. Douglas and T. Spencer (eds), *Environmental Change and Tropical Geomorphology*, pp. 281–94. Allen &

Unwin: London.

Summerfield M.A. (1987). Neotectonics and landform genesis. *Progress in Physical Geography* **11**: 384–97.

Summerfield, M.A. (1988). Global tectonics and landform development. *Progress in Physical Geography* **12**: 389–404.

Sutton, E.R. (1979). The geology of the Mafungabusi area. *Bulletin of the Rhodesian Geological Survey* **81**. 318pp.

Sweet, C.P. (1971). 'Report on the soils of the NW section of Wankie National Park.' Unpublished report CS/3/2/32, Branch of Chemistry and Soil Science, Department of Research and Specialist Services, Salisbury. 16 pp.

Sweet, J. (1986). Boreholes in the Kalahari: the pros and cons. *Kalahari Conservation Society Newsletter* **11**: 10–11.

Tabler, E.C. (1960). The narrative of Frederick Hugh Barber and the journal of Richard Frewen. *Zambezia and Matabeleland in the Seventies*. Chatto & Windus: London.

Tabler, E.C. (1963). *Trade and Travel in Early Barotseland*. Robins Series 2. Chatto & Windus: London.

Tabler, E.C. (1973). *Pioneers of South West Africa and Ngamiland 1738–1880*. Balkema: Cape Town.

Taljaard, J.J. (1981). Upper-air circulation, temperature and humidity over southern Africa. *South African Weather Bureau Technical Paper* **10**. 94 pp.

Tanaka, J. (1976). Subsistence ecology of Central Kalahari San. In: R.B. Lee and I. DeVore (eds), *Kalahari Hunter–Gatherers: Studies of the !Kung San and their Neighbours*, pp. 98–119. Harvard University Press: Cambridge (Mass.).

Tankard, A.J., Jackson, M.P.A., Eriksson, K.A. Hobday, D.K., Hunter, D.R. and Minter W.E.L. (1982). *Crustal Evolution of Southern Africa. 3.8 Billion Years of Earth History*. Springer Verlag: New York. 522 pp.

Taylor, R.D. and Martin, R.B. (1987). Effects of veterinary fences on wildlife conservation in Zimbabwe. *Environmental Management* **11**: 327–34.

Taylor, R.D. and Walker, B.H. (1978). Comparisons of vegetation use and herbivore biomass on a Rhodesian game and cattle ranch. *Applied Ecology* **15**: 565–81.

Thackeray, A., Thackeray, J.F., Beaumont, P. and Vogel, J. (1981). Dated rock engravings from Wonderwerk Cave, South Africa. *Science* **214**: 64–7.

Thomas, C.M. (1973). *South Ngamiland*. 1:125 000 QDS 2022D. Geological Survey of Botswana Map.

Thomas, D.S.G. (1983/4). Geomorphic evolution and river channel orientation in north west Zimbabwe. *Proceedings of the Geographical Association of Zimbabwe* **15**: 12–22.

Thomas, D.S.G. (1984a). Ancient ergs of the former arid zones of Zimbabwe, Zambia and Angola. *Transactions of the Institute of British Geographers* NS **9**: 75–88.

Thomas, D.S.G. (1984b). 'Late Quaternary environmental change in central Southern Africa with particular reference to extensions of the arid zone.' Unpublished DPhil thesis, University of Oxford.

Thomas, D.S.G. (1985). Evidence of aeolian processes in the Zimbabwean landscape. *Transactions, Zimbabwe Scientific Association* **62**: 45–55.

Thomas, D.S.G. (1986a). Dune pattern statistics applied to the Kalahari Dune Desert, Southern Africa. *Zeitschrift für Geomorphologie* **30**: 231–42.

Thomas, D.S.G. (1986b). Evidence of Quaternary palaeoclimates in western Zimbabwe – a preliminary assessment. In: G.J. Williams and A.P. Wood (eds), *Geographical Perspectives on Development in Southern Africa*, pp. 9–22. Commonwealth Geographical Bureau, James Cook University, Queensland.

Thomas, D.S.G. (1986c). Ancient deserts revealed. *Geographical Magazine* **58**: 11–15.

Thomas, D.S.G. (1987a). Discrimination of depositional environments, using sedimentary characteristics, in the Mega Kalahari, central southern Africa. In: L.E. Frostick and I. Reid (eds), *Desert Sediments, Ancient and Modern*, pp. 293–306. Geological Society of London Special Publication **35**. Blackwell Scientific Publications: Oxford.

Thomas, D.S.G. (1987b). Research strategies and methods for Quaternary Science: the case of southern Africa. *School of Geography Research Paper* **39**. University of Oxford.

Thomas, D.S.G. (1988a). The nature and depositional setting of arid to semi-arid Kalahari sediments, southern Africa. *Journal of Arid Environments* **14**: 17–26.

Thomas, D.S.G. (1988b). Analysis of linear dune sediment-form relationships in the Kalahari Dune Desert. *Earth Surface Processes & Landforms* **13**: 545–53.

Thomas, D.S.G. (1988c). The geomorphological role of vegetation in the dune systems of the Kalahari. In: G.F. Dardis and B.P. Moon (eds), *Geomorphological Studies in Southern Africa*, pp. 145–58. Balkema: Rotterdam.

Thomas, D.S.G. (1988d). The biogeomorphology of arid and semi-arid environments. In: H.A. Viles (ed.), *Biogeomorphology*, pp. 193–221. Blackwell Scientific Publications: London.

Thomas, D.S.G. (1988e). Environmental management and environmental change: the impact of animal exploitation on marginal lands in central southern Africa. In: G.J. Stone (ed.), *The Exploitation of Animals in Africa*, pp. 5–22. Aberdeen University African Studies Group: Aberdeen.

Thomas, D.S.G. (1989). Aeolian sand deposits. In: D.S.G. Thomas (ed.), *Arid Zone Geomorphology*, pp. 232–61. Belhaven Press: London.

Thomas, D.S.G. and Goudie, A.S. (1984). Ancient ergs of the Southern Hemisphere. In: J. Vogel (ed.), *Late Cainozoic Palaeoclimates of the Southern Hemisphere*, pp. 407–18. Balkema: Rotterdam.

Thomas, D.S.G. and Martin, H.E. (1987). Grain-size characteristics of linear dunes in the southwestern Kalahari – discussion. *Journal of Sedimentary Petrology* **57**: 231–42.

Thomas, D.S.G. and Shaw, P.A. (1988). Late Cainozoic drainage evolution in the Zambezi basin: geomorphological evidence from the Kalahari rim. *Journal of African Earth Sciences* **7**: 611–18.

Thomas, D.S.G. and Shaw, P.A. (1990). The deposition and development of the Kalahari Group sediments, central southern Africa. *Journal of African Earth Sciences* **10**, 187–97.

Thomas, D.S.G. and Tsoar, H. (1990). The geomorphological role of vegetation in desert dune systems. In: J. Thornes (ed.), *Vegetation and Erosion*, pp. 471–89 John Wiley: Chichester.

Thornthwaite, C.W. (1948). An approach towards a rational classification of climate. *Geographical Review* **38**: 55–94.

Timberlake, J.R. (1980). 'Vegetation map of South-eastern Botswana.' Division of Land Utilisation, Ministry of Agriculture, Gaborone. Unpublished.

Tlou, T. (1972). The taming of the Okavango Swamps. *Botswana Notes and Records* **4**: 147–59.

Tlou, T. (1985). *A History of Ngamiland 1750–1906: the History of an African State*. Macmillan: Botswana.

Tlou, T. and Campbell, A. (1984). *History of Botswana*. Macmillan: Botswana.

Tobias, P.V. (1964). Bushman hunter–gatherers: a study in human ecology. In: D.H.S. Davis (ed.), *Ecological Studies in Southern Africa*, pp. 69–89. Junk: The Hague.

Tobias, P.V. (1967). 'Foreword' in A.J. Clement: *The Kalahari and its Lost City*. Longman: Johannesburg.

Tolsma, D.J., Ernst, W.H.O. and Verwey, R.A. (1987). Nutrients in soil and vegetation around two artificial waterholes in eastern Botswana. *Journal of Applied Ecology* **24**: 991–1000.

Torrance, J.D. (1972). Malawi, Rhodesia and Zambia. In: J.F. Griffiths (ed.), *Climates of Africa. World Survey of Climatology*, vol. 10, pp. 409–60. Elsevier: Amsterdam.

Trapnell, C.G. and Clothier, J.N. (1957). *The Soils, Vegetation and Agricultural Systems of North-west Rhodesia* (2nd edn). Government Printer: Lusaka. 69 pp.

Tree, I. (1989). How to hittapottamus. *Geographical Magazine* **61/4**: 4–5.

Trewatha, G.T. (1962). *The Earth's Problem Climates*. University of Wisconsin Press: Madison. 334 pp.

Tripp, W.B. (1888). Rainfall of South Africa, 1842–1886. *Quarterly Journal of the Royal Meteorological Society* **14**: 108–23.

Tsoar, H. (1978). *The Dynamics of Longitudinal Dunes*. Final technical report, European research office. US Army: London. 171 pp.

Tsoar, H. and Møller, J.T. (1986). The role of vegetation in the formation of linear sand dunes. In: W.G. Nickling (ed.), *Aeolian Geomorphology. The Binghampton Symposium in Geomorphology, International Series 17*, pp. 75–95. Allen & Unwin: Boston.

Tumkaya, N. (1987). Botswana's population trends: past and future. *Botswana Notes and Records* **19**: 113–28.

Turner, G. (1987a). Early Iron Age herders in northwestern Botswana: the faunal evidence. *Botswana Notes and Records* **19**: 7–23.

Turner, G. (1987b). Hunters and herders of the Okavango Delta, Northern Botswana. *Botswana Notes and Records* **19**: 25–40.

Twidale, C.R. (1981). Granite inselbergs: domed, block-strewn and castellated. *Geographical Journal* **147**: 54–71.

Twidale, C.R. (1988), Granite landscapes. In: B.P. Moon and G. Dardis, (eds), *The Geomorphology of Southern Africa*, pp. 198–230. Southern Publishers: Johannesburg.

Tyson, P.D. (1979). Southern African rainfall: past, present and future. In: M.T. Hinchley (ed.), *Proceedings of the Symposium on Drought in Botswana*, pp. 45–52. Botswana Society: Gaborone.

Tyson, P.D. (1986). *Climatic Change and variability in Southern Africa*. Oxford University Press. 220 pp.

Tyson, P.D. and Dyer, T.G.J. (1975). Mean annual fluctuations of precipitation in the summer rainfall region of south Africa. *South African Geographical Journal* **57**: 104–10.

Tyson, P.D., Dyer, T.G.J. and Mametse, M.N. (1975). Secular changes in South African rainfall: 1880 to 1972. *Quarterly Journal of the Royal Meteorological Society* **101**: 817–33.

UNEP, (1984). 'Report of the UNEP Clearing House Technical Mission to Botswana, 3 November – 11 December 1983.' Unpublished UNEP report, Nairobi.

UNDP/FAO (1977). *Investigation of the Okavango Delta as a Primary Water Resource for Botswana*. DP/Bot/71/506 Technical report. 3 vols.

Van der Meulen, F. and Van Gils, H.A.M.J. (1983). Savannas of southern Africa: attributes and use of some types along the tropic of Capricorn. *Bothalia* **14**: 675–81.

Van Der Post, L. (1958). *The Lost World of the Kalahari*. Hogarth Press: London.

Van Eeden, O.R. (1973). The correlation of the sub-divisions of the Karoo system. *Transactions of the Geological Society of South Africa* **76**: 201–6.

Van Horn, L. (1977). The agricultural history of Barotseland 1840–1964. In: R. Palmer and N. Parsons (eds), *The Roots of Rural Poverty in Central and Southern Africa*. Heinnemann: London.

Van Rensburg, H.J. (1971). Range ecology in Botswana. *Technical Document No 2, Surveys and Training for the Development of Water Resources and Agricultural Production*. FAO/UNDP: Gaborone.

Van Riet Low, C. (1930). Further notes on the archaeology of Sheppard Island. *South African Journal of Science* **26**: 665–83.

Van Riet Lowe, C. (1935). Remains of the Stone Age. *Annals of the Transvaal Museum* **16**: 495–6.

Van Rooyen, T.H. and Verster, E. (1983). Granulometric properties of the roaring sands in the south-eastern Kalahari. *Journal of Arid Environments* **6**: 215–22.

Van Rooyen, T.H., Van Rensberg, D.J., Theron, G.K. and Bothma, J. du P. (1984). A preliminary report on the dynamics of the vegetation of the Kalahari Gemsbok National Park. *Koedoe* **27** (supplement): 83–102.

Van Straten, O.J. (1955). *The Geology and Groundwater of the Ghanzi Cattle Route.* Bechuanaland Protectorate Geological Survey, Bulletin **5**. 132 pp.

Van Waarden, C. (1988). 'Makalamabedi Drift 03-B4-1, Mitigation report' Unpublished report by Joint Venture Consultants/Ove Arup & Partners/Stewart Scott & Partners, Gaborone.

Van Zindren Bakker, E.M. (1967). Upper Pleistocene and Holocene stratigraphy and ecology on the basis of vegetation changes in sub-Saharan Africa. In: W.W. Bishop and J.D. Clark (eds), *Background to Evolution in Africa*, pp. 125–47. University of Chicago Press.

Van Zindren Bakker, E.M. (1976). The evolution of Late Quaternary palaeoclimates of southern Africa. *Palaeoecology of Africa* **9**: 160–202.

Van Zindren Bakker, E.M. (1980). Comparison of Late Quaternary climatic evolutions in the Sahara and Namib-Kalahari region. *Paleoecology of Africa* **12**: 381–94.

Van Zindren Bakker, E.M. (1982a). African Palaeoenvironments 18 000 yrs B.P. *Palaeoecology of Africa* **15**: 77–99.

Van Zindren Bakker, E.M. (1982b). Pollen analytical studies of the Wonderwerk Cave, South Africa. *Pollen et Spores* **24**: 235–50.

Van Zindren Bakker, E.M. and Clark, J.D. (1962). Pleistocene climates and culture in northeastern Angola. *Nature* **196**: 639–42.

Van Zuidam, R.A. (1975). Calcrete. A review of concepts and an attempt to a new genetic classification. In: T. Vogt (ed.), *Colloque 'Types de crouts calcaires et leur répartition régional'*, pp. 92–8. Université Louis Pasteur: Strasbourg.

Veatch, A.C. (1935). Evolution of the Congo basin. *Memoirs of the Geological Society of America* **3**. 183 pp.

Verbeck, W.A. (1968). Problems and advances under arid conditions. *Proceedings of the Second World Conference on Animal Production*. University of Maryland.

Verboom, W.C. (1974). The Barotse loose sands of Western Province, Zambia. *Zambian Geographical Magazine* **27**: 13–17.

Verboom, W.C. and Brunt, M.A. (1970). An ecological survey of Western Province, Zambia. *Zambian Geographical Association Magazine* **27**: 13–17.

Verhagen, B., Mazor, E. and Sellschop, J. (1974). Radiocarbon and tritium evidence for direct rain recharge to groundwaters in the northern Kalahari. *Nature* **249**: 643–4.

Verhagen, B., Smith, P., McGeorge, I. and Dziembowski, Z. (1978). Tritium profiles in Kalahari sands as a measure of rain-water recharge. *Proceedings of the International Atomic Energy Agency (IAEA) Symposium*, Neuherberg, pp. 733–51.

Verhagen, B. Th. (1983). Environmental isotope study of a ground water supply project in the Kalahari of Gordonia. *Proceedings of the International Atomic Energy (IAEA) Symposium, Vienna*, pp. 1–15.

Vernay, A.S. (1931). The Great Kalahari Sandveldt: experiences of the Vernay–Lang zoological expedition in the vast arid plains of southern Africa known as the Kalahari Desert. *Natural History* **31**: 169–82, 262–74.

Visser, J.N.J. (1983). An analysis of the permo-Carboniferous glaciation in the marine Kalahari basin, southern Africa. *Palaeogeography, Palaeoclimatology, Palaeoecology* **44**: 295–315.

Voigt, E. (1983). *Mapungubwe: An Archaeozoological Interpretation of an Iron Age Community.* Transvaal Museum, Pretoria.

Volman, T.P. (1984). Early prehistory of southern Africa. In: R. Klein (ed.), *Southern African Prehistory and Paleoenvironments*, pp. 169–220. Balkema: Rotterdam.

Von Hoyer, M., Keller, S. and Rehder, S. (1985). *Core Borehole Lethlakeng 1.* Botswana Geological Survey Department Report MVH/4/85. 7 pp.

Vossen, P. (1986). 'Growth season lengths in Botswana – a comparison of some methods of approach.' Unpublished report of the Land Utilisation Division, Department of Meteorological Services, Gaborone.

Vossen, R. (1984). Studying the linguistic and ethno-history of the Khoe-speaking (central Khoisan) peoples of Botswana: research in progress. *Botswana Notes and Records* **16**: 19–36.

Wagner, P.A. (1916). The geology and mineral industry of South-West Africa. *Memoirs of the Geological Survey of South Africa* **7**.

Walker, N. (1983). The significance of an early date for pottery and sheep in Zimbabwe. *South African Journal of Science* **38**: 88–92.

Walker, T.R. (1979). Red color in dune sand. In: E.D. McKee (ed.), *A Study of Global Sand Seas*, pp. 61–81. US Geological Survey Professional Paper **1052**.

Wallis, A.H. (1935). Is our rainfall getting less? *The 1820 Magazine* **6**: 357.

Wasson, R.J. (1984). Late Quaternary palaeoenvironments in the desert dunefields of Australia. In: J. Vogel (ed.), *Late Cainozoic Palaeoclimates of the Southern Hemisphere*, pp. 419–32. Balkema: Rotterdam.

Wasson, R.J. and Hyde, R. (1983). Factors determining desert dune type. *Nature* **304**: 337–9.

Wasson, R.J. and Nanninga, P.M. (1986). Estimated wind transport of sand on vegetated surfaces. *Earth Surface Processes and Landforms* **11**: 505–16.

Watson, R.L.A. (1960). The geology and coal resources of the country around Wankie, Southern Rhodesia. *Southern Rhodesian Geological Survey Bulletin* **48**.

Watts, N.L. (1977). Pseudo-anticlines and other structures in some calcretes of Botswana and South Africa. *Earth Surface Processes* **2**: 63–74.

Watts, N.L. (1980). Quaternary pedogenic calcretes from the Kalahari (southern Africa): mineralogy, genesis and diagenesis. *Sedimentology* **27**: 661–86.

Wayland, E.J. (1934). Rifts, rivers, rains and early Man in Uganda. *Journal of the Royal Anthropological Institute* **64**: 333–52.

Wayland, E.J. (1944). Drodsky's Caves. *Geographical Journal* 103: 230–3.

Wayland, E.J. (1950a). Two boulder-on-boulder flaking techniques. *South African Archaeological Bulletin* 5: 99–100.

Wayland, E.J. (1950b). From an archaeological notebook. *South African Archaeological Bulletin* 5: 4–14.

Wayland, E.J. (1953). More about the Kalahari. *Geographical Journal* 119: 49–56.

Wayland, E.J. (1954). Outlines of prehistory and Stone Age climatology in the Bechuanaland Protectorate. *Memoires d'Académie Royale des Sciences Coloniales: Sciences Naturale et Medicales* 25: 1–47.

Wayland, E.J. (undated). 'Stone Age research in the Bechuanaland Protectorate.' Unpublished papers, Botswana National Museum.

Weare, P.R. (1971). Vegetation of the Kalahari in Botswana. In: Proceedings of the Conference on Sustained Production from Semi-arid Areas. *Botswana Notes and Records*, special edition 1: 88–95.

Weare, P.R. and Yalala, A. (1971). Provisional vegetation map. *Botswana Notes and Records* 3: 131–52.

Weir, J.S. (1966). A possible course of evolution of animal drinking holes (pans) and reflected changes in their biology. *Proceedings of the 1st Federal Science Congress, Salisbury, 18–22 May 1966*, pp. 301–5.

Weir, J.S. (1969). Chemical properties and occurrences on Kalahari Sands of salt licks created by elephants. *Journal of Zoology* 138: 292–310.

Welbourne, R. (1975). Tautswe Iron Age site: its yield of bones. *Botswana Notes and Records* 7: 1–16.

Wellington, J.H. (1929). The Vaal–Limpopo watershed. *South African Geographical Journal* 12: 36–45.

Wellington, J.H. (1938). The Kunene river and the Etosha Plain. *South African Geographical Journal* 20: 21–32.

Wellington, J.H. (1939). The Greater Etosha Basin. *South African Geographical Journal* 21: 47–8.

Wellington, J.H. (1955). *Southern Africa, a Geographical Study, Volume 1: Physical Geography*. Cambridge University Press. 528 pp.

Werger, M.J.A. (1978). Biogeographical division of southern Africa. In: W. Werger (ed.), *Biogeography and Ecology of Southern Africa*, pp. 145–70. Junk: The Hague.

Westerhof, A.B. (1976). 'Occasional report on the potential for salt deposits in the Western Province, Zambia'. MINDECO Ltd, MINDEX Department, Lusaka. Unpublished.

White, F. (1965). The savanna-woodlands of the Zambezian and Sudanian Domains. *Webbia* 19: 651–81.

Wilczewski, N. and Martin, H. (1972). Algenstromatolithen aus der EtoschPfanne Sudwestafrikas. *Neues Jahrbuch Geologie Paläontologie Monaschefte* 12: 720–6.

Wild, H. and Fernandez, A. (eds) (1967). *Vegetation Map of Flora Zambesica Area*. Collins: Salisbury.

Williams, G.J. (1975). Geomorphology of the Southern Province of Zambia. *Zambian Geographical Association Handbook Series* 4: 19–40.

Williams, G.J. (1986). A preliminary Landsat interpretation of the relict landforms of western Zambia. In: G.J. Williams and A.P. Wood (eds), *Geographical Perspectives on Development in Southern Africa*, pp. 23–33. Commonwealth Geographical Bureau, James Cook University, Queensland.

Williams, M.A.J. (1985). Pleistocene aridity in tropical Africa, Australia and Asia. In: I. Douglas and T. Spencer (eds), *Environmental Change and Tropical Geomorphology*, pp. 219–33. Allen & Unwin: London.

Williamson, D.T. and Williamson, J.E. (1981). An assessment of the impact of fences on large herbivore mass in the Kalahari. *Botswana Notes and Records* 13: 107–10.

Williamson, D.T. and Williamson, J.E. (1983). Cattle versus wildlife: optimal use of marginal areas. In: *Which Way Botswana's Wildlife?* pp. 91–4. Kalahari Conservation Society: Gaborone.

Williamson, D.T. and Williamsom, J.E. (1985). Botswana's fences and the depletion of Kalahari wildlife. *Parks* 10: 5–7.

Wilmsen, E.N. (1978). Prehistoric and historic antecedents of a contemporary Ngamiland community. *Botswana Notes and Records* 10: 5–18.

Wilson, B.H. (1973). Some natural and man-made changes in the channels of the Okavango Delta. *Botswana Notes and Records* 5: 132–53.

Wilson, B.H. and Dincer, T. (1977). An introduction to the hydrology and hydrography of the Okavango Delta. *Symposium on the Okavango Delta*, pp. 33–46. Botswana Society: Gaborone.

Wilson, J.F. (1865). Water supply in the basin of the River Orange or 'Gariep South Africa'. *Journal of the Royal Geographical Society* 35: 106–29.

Wood, A.P. (1983). 'Report on a socio-economic study of the feasibility of moving the Western Province cattle cordon line to the international boundary.' Unpublished report of the Rural Development Studies Bureau, University of Zambia.

Wood, A.P. (1988). Cattle and development in western Zambia. In: G.J. Stone (ed.), *The Exploitation of Animals in Africa*, pp. 317–43. Aberdeen University African Studies Group: Aberdeen.

Woolard, J. (1986). The active chemical components of the Basarwa arrow poison. *Botswana Notes and Records* 18: 139–40.

Wright, E.B. (1958). *Geology of the Area South of Lake Ngami*. Records of the Geological Survey of the Bechuanaland Protectorate for 1956, pp. 29–35.

Wright, E.B. (1978). Geological studies in the northern Kalahari. *Geographical Journal* 144: 235–50.

Yeager, R. (1989). Demographic pluralism and ecological crisis in Botswana. *Journal of Developing Areas* 23: 385–404.

Yellen, J.E. (1971). Archaeological excavations in western Ngamiland. *Botswana Notes and Records* 3: 276.

Yellen, J.E. (1977). *Archaeological Approaches to the Present: Models for Reconstructing the Past.* Academic Press: New York.

Yellen, J.E., Brooks, A.S., Stuckenrath, R. and Welbourne, R. (1987). A terminal Pleistocene assemblage from Drotsky's Cave, Western Ngamiland, Botswana. *Botswana Notes and Records* 19: 1–6.

Yellen, J.E. and Harpending, H. (1972). Hunter–gatherer populations and archaeological inference. *World Archaeology* 4: 244–53.

Yellen, J.E. and Lee, R.B. (1976). The Dob-/du/da environment: considerations for a hunting and gathering way of life. In: R.B. Lee and I. DeVore (eds), *Kalahari Hunter–Gatherers: Studies of the !Kung San and their Neighbours*, pp. 27–46. Harvard University Press: Cambridge (Mass.).

Young, R.B. (1926). The calcareous tufa deposits of the Campbell Rand, from Boetsap to Taungs Native Reserve. *Transactions of the Geological Society of South Africa* 28: 55–67.

Index